Signals and Communication Technology

More information about this series at http://www.springer.com/series/4748

Amit Kumar Mishra · David Lloyd Johnson
Editors

White Space Communication

Advances, Developments and Engineering Challenges

Springer

Editors
Amit Kumar Mishra
Department of Electrical Engineering
University of Cape Town
Rondebosch
South Africa

David Lloyd Johnson
Meraka, CSIR
Rosebank
South Africa

ISSN 1860-4862 ISSN 1860-4870 (electronic)
ISBN 978-3-319-08746-7 ISBN 978-3-319-08747-4 (eBook)
DOI 10.1007/978-3-319-08747-4

Library of Congress Control Number: 2014949605

Springer Cham Heidelberg New York Dordrecht London

Printed on acid-free paper

Springer is part of Springer Science+Business Media (www.springer.com)

Foreword

Affordable and ubiquitous access to communication for all is a challenge that has consumed me for many years. In my journey over the past decade from being a research funder for the International Development Research Centre supporting connectivity projects in Africa to founding a social enterprise that builds low-cost WiFi mesh VoIP technologies, I remain convinced that lack of access to wireless spectrum is a critical roadblock to the development of affordable rural access and to entrepreneurship in general.

So why is wireless access so essential? In many emerging markets, wired technologies, especially in rural areas, are simply not practical. Traditional copper-based last mile solutions present particular challenges in poor countries. The roll-out cost of copper infrastructure is comparatively high, and often not practical in lower-income countries, even less so in sparsely populated rural areas. Worse, the steadily increasing value of copper as a commodity makes it an appealing target for thieves. Fibre optic infrastructure is another important connectivity technology especially for national and international communication backbones but it is currently only viable as a consumer solution in wealthy communities. Thus when talking about affordable, ubiquitous access to communication in developing countries, wireless technologies offer the most hope for effectively bridging the digital divide. And indeed, the mobile phone has become more or less synonymous with access in the South.

But much has changed since the first large-scale roll-outs of mobile networks began in the mid-nineties in Africa. When the first mobile operators were granted licenses to use wireless spectrum to build their networks, they were typically given the spectrum at no charge. There was plenty of available spectrum to go around. Today popular spectrum bands are auctioned for large amounts of money, often running into the billions of dollars. These are costs that are inevitably passed on to the consumer. Because so much money is now at stake concerning spectrum, the process of making new spectrum available has become complex. It is increasingly hard to ensure that spectrum is made available in a timely manner and to the entities that are most likely to serve a country's strategic interest.

As communication technologies and networks become increasingly valuable to government, industries, and citizens alike, demand for spectrum has steadily increased. Yet the *process* of making spectrum available and managing it as a resource often still fails to feature significantly in strategic broadband discussions.

In 2005, I met David Lloyd Johnson when he was a masters' student working on a wireless mesh connectivity project in Mpumulanga, South Africa that the International Development Research Centre was supporting. We shared a common vision that low-cost wireless technologies, such as WiFi-based mesh networks, could offer an alternative to traditional mobile network solutions. Through its unlicensed status, WiFi represented a kind of loophole to traditional spectrum licensing methods. That loophole has spawned a multi-billion dollar industry that has come to represent essential complementary network infrastructure for mobile data networks.

Now, nearly 10 years later, David and I continue to be involved in solving the challenge of affordable wireless access but the game has moved on to include new options—many of which have been uncovered through David's interactions with Professor Amit Mishra at UCT. The evolution of dynamic spectrum approaches to spectrum management is creating new possibilities for access. In particular dynamic spectrum management applied to the VHF and UHF television spectrum bands which has come to be know as "Television White Space" or often just "White Space" is an innovation of particular relevance to low-income countries where there is a lack of affordable rural access technologies and business and where television spectrum is comparatively lightly occupied.

Yet the early days of any new wireless technology bring many uncertainties and White Space is no exception. Its success depends on a complex dance among manufacturers, communication regulators, and standards organisations. Understanding the potential of White Space technology is something that requires knowledge of physics, engineering, economics, and politics. As a technology that is still emerging, there remains a dearth of resources to inform researchers, regulators and policy-makers. This work assembled by Amit and David provides insight into some of the most important aspects of White Space technology including emerging standards, lessons learned from early pilots, and recent research into spectrum occupancy. It also examines methods for estimating spectrum availability which is integral to the use of geo-location databases that underpin White Space regulation. Finally, it explores the realm of Software Defined Radio and emergent hardware platforms that may well be a game-changer in how we think about wireless hardware. Read on, and explore the future.

July 2014 Steve Song

Preface

White spaces refers to licensed frequencies that are unoccupied in a specific local area. The bands with the most significant white space are the VHF and UHF bands historically assigned to analog television broadcasting services and now also being used for Digital TV broadcasting. The migration to Digital TV (DTV) frees up a significant amount spectrum due to a much higher level of spectral efficiency in DTV transmissions. This freed up spectrum is known as the digital dividend and blocks in the 700 and 800 MHz bands have been or will be auctioned off for new 4G licensed services.

White spaces, on the other hand, opportunistically makes use of unused licensed spectrum in specific geographic locations. Legislation will either mandate licensed or unlicensed operation of white space devices operating as secondary users in unused frequency bands licensed to primary broadcast users. The current key application for white spaces is broadband and M2M communication. The FCC in the USA approved the unlicensed use of white space in 2008 and since this ruling a number of countries including Canada, UK, Malawi, Kenya, India and South Africa are running trial networks to decide if the technology provides a significant advantage for new markets such as rural broadband and wide-area M2M communication.

Many wireless standards are emerging for White space communications, these include IEEE 802.22, 802.11af, ECMA-392, and Weightless. Each have their own niche application areas. IEEE 802.22 is focused on providing long range connectivity in rural areas with a range of 30 km (even 100 km has been promised). 80211af and ECMA-3392 are focused on the wireless LAN market, providing improved range and penetration in urban environments. However, long range outdoor applications may also be possible with high gain antennas and some small adjustments to the MAC protocol to deal with longer round-trip times.

The white space standards stand apart from other wireless standards in that they require cognitive radios built on software defined radio platforms and spectrum databases to avoid interference with primary users. Since the introduction of white space standards, there are a number of key research areas being studied. These range from debates around the best propagation models for geo-location spectrum databases (US and Europe) to co-existence strategies for secondary users and channel allocation algorithms. This monograph will detail some of the major developments leading towards the implementation of white space radios and cover some of the key research areas in the field. In editing this monograph, the main focus has been on adding contributions which address implementation issues. We expect this book to act as a critical source of information for researchers and engineers involved in designing and installing white space-based systems. This book will also serve as a good introduction to researchers who are starting work in areas related to white space communication. And lastly, we believe this book will also help policy makers to grasp some of the key issues in this field. A field that has the potential to extend the reach of wireless communication to populations without Internet access—especially those in rural areas.

The layout of this monograph is as follows:

We start with chapters that review the current White space engineering developments and platforms that are available for building White space solutions.

Chapter 1, titled Engineering Review of IEEE 802.22 Standards on Cognitive Radio by Hiremath et al., reviews IEEE 802.22, the white space communication standard developed for long-range rural communication. This chapter expounds the standard and summarises the major technical specifications as proposed by the standard.

Chapter 2, titled A Review of Hardware Platforms for White-space Communication by Hazarika et al., describes the availble set of hardware platforms, which can be used for white-space communication research and prototyping.

In the second set of chapters we present field measurements done in white space bands in Southern Africa and India. These chapters will provide insights into the amount of white space available in rural and urban areas of these two countries.

Chapter 3, titled A Study on White and Gray Spaces in India, by Naidu et al., analyses spectrum measurements run in few a states in India in the bands corresponding to TV white spaces and GSM gray spaces.

Chapter 4, titled The Television White Space Opportunity in Southern Africa: From Field Measurements to Quantifying White Spaces by Lysko et al., presents spectrum scans in the TV white space bands in rural and urban areas of South African using a mobile setup and long term measurements done during the TV white space trial in Cape Town.

In the next set of chapters, we present two works in the broad area of spectrum sensing and spectrum database management related to white space bands. There are two methods to avoid interference to primary users. The first is to use dynamic spectrum sensing and this is described in Chap. 5. The second is through the use of a geo-location database and this is described in Chap. 6.

Chapter 5, titled Spectrum Sensing for Cognitive Radio Networks by Srinu et al., describes the spectrum sensing algorithms which have been prescribed for white space detection for cognitive networks in general and for white space communication in particular.

Chapter 6, titled Geo-location White Space Spectrum Databases: Review of Models and Design of a Dynamic Spectrum Access Coexistence Planner and

Manager by Mfupe et al., describes models and methods used to design a geo-location white space spectrum database (GL-WSDB) in the US and Europe. The chapter also describes a GL-WSDB developed by the CSIR in South Africa and compares its performance to field measurements and other results from other commercial GL-WSDBs.

Creating white space device solutions requires software defined radio hardware that is capable of carrying out computationally complex operations for spectrum sensing and OFDM communication used by white space protocols. In the next set of chapters, solutions for some of the implementation issues in white-space communication have been expounded. These discuss a hardware design capable of running a white space protocol as well as some channel characterization and equalization in white spaces.

Chapter 7, titled White RHINO—A New SDR Platform for Whitespace Communication by Hazarika et al., describes the development of a new software defined radio board fine-tuned to cater for the strict requirements set by the white-space communication standards.

Chapter 8, titled Challenges and Solution for TV White Space Channel Estimation and Equalisation by Bhatia, lists some of the established algorithms for channel equalisation for white space communication systems. Channel equalisation is a major challenge in all communication systems. OFDM mitigates the issue of fading in the channel to some extent but it is not a full proof solution.

In the last set of chapters, we focus on the use of white spaces for rural communication. Due to the long-range nature of rural connectivity, we focus on channel characterization and allocation for multiple long-range links. We also examine the interesting question of the value of white space spectrum for rural operators.

Chapter 9, titled VillageLink: A Channel Allocation Technique for Wide-Area White Space Networks by Pejovic et al., describes the challenges in selecting optimal channels in white space networks. The chapter then discusses VillageLink, an extension to 802.22 to allocate channels to base stations in a distributed manner using a Gibbs sampling-based method.

The last chapter, titled A Non-Cooperative TV White Space Broadband Market Model for Rural Entrepreneurs by Nleya et al., discusses the challenges of producing an optimal pricing scheme for leasing spectrum from primary users. It proposes a game-theory based solution to the problem of several primary users competing with each other to sell spectrum to secondary users that use QoS measures of delay and throughout to regulate price.

In conclusion, we wish to thank all the contributors for their thorough and practical set of chapters—crucial to the current worldwide debate on the use of white spaces. We also wish to thank Springer for their timely and professional service in editing the book.

Amit Kumar Mishra
David Lloyd Johnson

Contents

Chapter 1
Engineering Review of the IEEE 802.22 Standard on Cognitive Radio

Shrishail Hiremath, Amit Kumar Mishra and Sarat Kumar Patra

Abstract A shift of most emerging electronic devices to wireless access has created extensive demand for radio spectrum. Regulatory committees in different parts of the world, after rigorous studies on wireless spectrum, have found that most licensed spectrum is underutilized in time, frequency and space. This has caused regulatory bodies to permit access of unused spectrum on a license-free basis so long as no interference is caused to primary users of the spectrum. Cognitive Radio technology has emerged as a key technology in solving the spectrum scarcity problem. Recent studies have shown that the majority of the TV band is vacant in remote, rural areas and the characteristics of this band are ideal for providing broadband in these areas to bridge the digital divide. Regulatory committees and the IEEE society have started establishing standards for providing wireless broadband connectivity in rural and remote areas in licensed TV spectrum using cognitive radios. The IEEE 802.22 Wireless Regional Area Network (WRAN) standard for rural broadband connectivity has been published and provides details of the Physical and MAC layer specification and deployment process. This article discuss the overall IEEE 802.22 specification and the capabilities it offers.

1.1 Introduction

1.1.1 IEEE 802.22 Overview and Architecture

Most of electronic devices are shifting towards wireless for flexibility and portability to overcome garbage created by wires as well as limitations in mobility. This

S. Hiremath (✉) · S.K. Patra
National Institute of Technology Rourkela, Rourkela, India
e-mail: hiremaths@nitrkl.ac.in

A.K. Mishra
University of Cape Town, Cape Town, South Africa

A.K. Mishra and D.L. Johnson (eds.), *White Space Communication*,
Signals and Communication Technology, DOI 10.1007/978-3-319-08747-4_1

phenomena is mostly seen in unlicensed spectrum due to advances in wireless broadband communication, wireless LAN/PAN/MAN, health care, public safety and soon. This explosion has created crunch of radio resource. It is well known that radio spectrum is controlled and allocated by regulatory bodies, leading to being economically precious. This has instigated regulatory bodies like Federal communications and commissions (FCC), through it task force to reanalyze the licensed spectrum and recommend modification of Licensed spectrum use in 2004 [1, 14]. With special Task Force of FCC and some research universities in USA, reported that spectrum utilization in some band, like TV band it is very low where as medium utilization in Mobile band and is very heavily utilized in unlicensed band like in some time and place. It was also found that shift from Analog TV to Digital TV led to lower utilization of licensed band. This significant result made FCC to revisit traditional spectrum management issues which is creating hindrance in spectrum management for advances in wireless technologies in semi-urban and rural wireless broadband, health care and public safety [16]. FCC created golden opportunity, in its Notice of Proposed Rule Making (NPRM), released in May 2004 and next version of 2008, proposing to allow to that licensed TV channels 513 in the VHF band and 1451 in the UHF band to be used for fixed broadband access systems with cognitive radio technologies as mechanism for co-existence [13, 33]. This provided a large frequency band from 53 to 837 MHz as additional resource in High VHF and low UHF TV band. TV band has inherent characteristics, which suits the broadband network connectivity with added advantages. TV band characteristics are relatively low levels of industrial noise and ionosphseric reflections, reasonable antenna sizes, and good non-line-of-sight (NLOS) propagation and its large distance coverage can be suitable for wireless data connectivity in rural and remote areas [3].

It is found that most of rural and remote place in world are facing digital divide compared to their urban counterparts. This mainly attributed to infrastructure cost in form of installation Optical cable and Digital subscriber Line (DSL) to low population density areas like rural and remote place. So this has instigated researcher to use of TV spectrum for rural broadband connectivity in an unobstructed way. This is likely to be made possible by using cognitive concept which was pioneered by Joseph Mitola-III of Software Define Group [26]. Cognitive Radio is radio technology in which the radio hardware observes the environment scenario and adapts to the channel characteristics on the fly. It uses main concept like radio scene analysis, channel state estimation, predictive modeling, transmit power control and spectrum management commands to provide good QoS service to unlicensed spectrum without obstructing the licensed user [7, 33, 35]. IEEE 802.22–Wireless Regional Area Network (WRAN) working group was formed in 2004 for framing the Physical (PHY) and Medium Access Layer (MAC), air interface protocol for rural wireless broadband connectivity without affecting the incumbent users [1] in TV white space. Incumbent users are the Licensed TV operators and low power auxiliary device specified in Title 47, Part 74 of the Code of Federal Regulations (CFR) of FCC, USA like unlicensed Microphone devices operating in VHF and UHF band from 57 to 867 MHz. This standard ensure interoperbilty of multi-vendor products developed for wireless broadband accessibility to diverse geographic areas, including sparsely populated rural areas [3]. Draft for PHY and MAC procedure for IEEE 802.22 has been completed in July 2011. Other two sub drafts include IEEE 802.22.1 that explains the low power auxiliary protection procedure using beacon completed in 2010 and IEEE 802.22.2 that provides information about recommend practice and deployment of IEEE 802.22 system completed in 2013. Two more sub drafts of IEEE 802.22.2a and IEEE 802.22.2b are expected to framed later to modify the management and control plane and extending WRAN applications to real-time and near real-time monitoring, emergency broadband services, remote medical services etc., requiring higher data rates and greater number of devices [1]. Recently many articles published are reporting IEEE 802.22 standards application, MAC and PHY configuration before draft publication [16, 32, 33, 35]. In this paper we discuss out important key aspects of PHY, MAC and Cognitive plane embedded in WRAN system. Paper highlights PHY and MAC specification that are required to develop TV white space radio. We try to put most of the concept as mentioned in the standards [2, 3].

Following this introduction, Sect. 1.2 discuss overview of IEEE 802.22 standard and its architecture. Section 1.3 discusses the PHY layer in WRAN system including RF analysis. Section 1.4 discusses the MAC layer in WRAN system with important aspects involved like incumbent protection, self-coexistence, channel and quite period management. Section 1.5 discuss the cognitive capability in WRAN system. Sections 1.6 and 1.7 discuss the security aspects and IEEE 802.22.1 beaconing used to protect low power auxiliary devices. Last section covers the Discussion and conclusion.

1.2 IEEE 802.22 Overview and Architecture

IEEE 802.22 is the first Wireless regional area network (WRAN) standard. Novelty about this standard is, coexistence mechanism between licensed TV band,unlicensed broadband network, cognitive radio concept, super frame structure specially designed for incumbent protection and two layer security concept [2]. The standard serves wireless broadband connectivity to the low density populated rural areas up to 17–30 Km coverage radius and for better case it can go up to 100 Km based on EIRP and Antenna height. It supports point to multi point connection between Base station (BS) and Customer Premises Equipment (CPE) upto 512 fixed and portable with different mode QoS of data streaming. It incorporates the OFDMA as Physical layer access method and intern helps in accommodate multipath propagation delay upto 30 Km and for 100 Km range special scheduling techniques are used in MAC. It uses adaptive modulation based on channel characteristics as shown in Fig. 1.1 [5]. It support peak data rate up to 18 Mbps in 6 MHz bandwidth of TV band allocated by FCC. Minimum supported data rate is 1.5 Mbps downstream and 384 Kbps upstream when CPE is located at edge of cell.

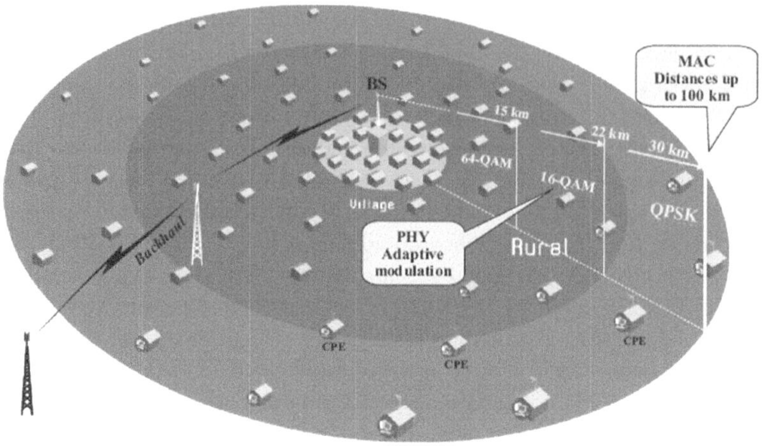

Fig. 1.1 WRAN cell deployment scenario [3]

1.2.1 System Architecture

IEEE 802.22 standard's unique characteristic is cognitive components. TV band is fragmented between licensed TV user, low power microphone user and WRAN users, so availability of channel is continuously varying. In such case, cognitive component of system architecture must efficiently keep record of the available multiple channels, segregate channels occupied by incumbent users and available channels for WRAN use and dynamically select frequency on real time basis to protect the incumbent users. Figure 1.2 shows the protocol reference module for BS and CPE. It contains three planes of operation namely data plane, Cognitive plane and management/control plane. Most of the functionality of Data and Management/control plane is retained as of WAN/MAN standards and cognitive plane newly added [9]. From Fig. 1.2 data plane extends in the Physical Layer (PHY), the Medium Access Control (MAC) layer and the Convergence sublayer (CS), Service Access Points (SAP) are added between these layers so that further new features can be added by vendors [3, 33]. The Data and Control/Management plane of the MAC is classified into three sublayers: service-specific CS, the MAC Common Part sublayer (CPS), and the security sublayer 1. The service-specific CS shall provide transformation or mapping of external network data that is received through the CS SAP, into MAC Service Data Units (SDUs) and data that is received by the MAC CPS through the MAC SAP. The Packet Convergence sublayer (CS) which resides on MAC Common Part Sublayer process the higher layer Service Data Unit (SDU) so that they can be sent and received by the IEEE802.22 BS and CPE.

MAC CPS carries all the core functions of MAC such as system access, connection establishment, and connection maintenance. The MAC security sublayer 1 takes care of authentication, secure key exchange, encryption, etc. The MAC CPS and the PHY exchange data, PHY control information and monitoring statics through

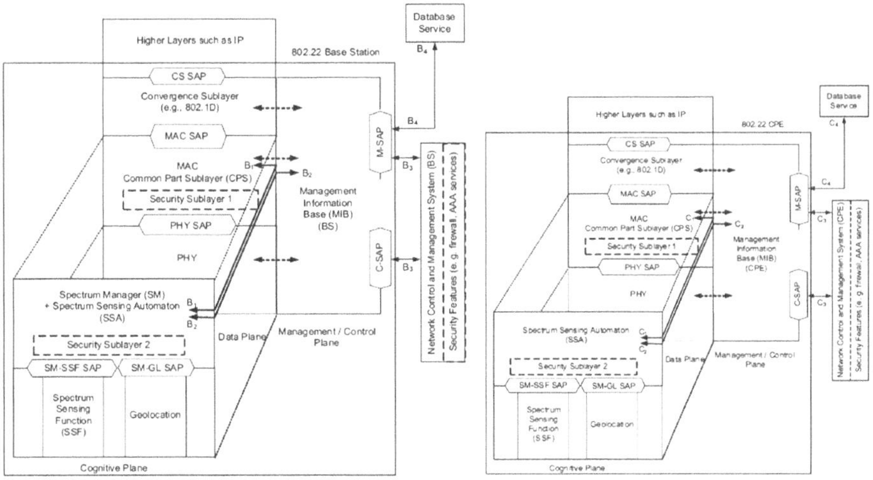

Fig. 1.2 Protocol reference model for BS and CPE [3]

PHY SAP. Management/control plane consists of the Management Information Base (MIB), which connects and Controls CPE,BS,router and switch through Simple Network Management Protocol (SNMP). MIB primitives functionality includes system configuration, monitoring statistics, notifications, triggers, CPE and session management, Radio Resources Management (RRM), communication with the database service, spectrum sensing and geolocation reporting, etc. [8]. Cognitive plane consists of Spectrum Sensing Function (SSF), the Geolocation (GL)function, the Spectrum Manager/Spectrum Sensing Automation (SM/SSA) and a dedicated security sublayer 2. Later all these parts are explained in detail. As IEEE 802.22 WRAN is a multiple cell network so cell operation, interfacing for management and control is done through standard Network Control and Management System (NCMS). NCMS allows the PHY/MAC layers specified in this standard to be independent of the network architecture, the transport network, and the protocols used at the back-end and therefore allows greater flexibility. More details about this is found in clause 5 of WRAN standard [3].

1.3 Physical Layer (PHY) for WRAN

In PHY layer information to down stream from BS or upstream from multiple CPEs is transmitted using orthogonal frequency division multiple access (OFDMA). Physical parameters are used in WRAN system are tabulated in Table 1.1. As the information about OFDMA is found in most of literature, so we give relevant information regarding symbols and their timing information [3, 25]. Operation frequency range is from 56 to 867 MHz with single air interface of 2048 sub carriers. One symbol

Table 1.1 System parameters for WRAN

Parameters	Specification	Remarks
Frequency range	54 862 MHz	
Channel bandwidth	6, 7, or 8 MHz	According regulatory domain
Data rate	4.54 to 22.6 Mbit/s	
Spectral efficiency	0.76 to 3.78 bit/(sHz)	
Payload modulation	QPSK, 16-QAM, 64-QAM	BPSK used for preamble pilots and CDMA codes
Transmit EIRP	4 W maximum for CPEs 4 W maximum for BSs in the USA regulatory domain	Maximum EIRP for BSs may vary in other regulatory domains
Multiple access	OFDMA	
FFT size (NFFT)	2048	
Cyclic prefix modes	1/4, 1/8, 1/16, 1/32	
Duplex	TDD	

of OFDM is combination of IFFT output with cyclic prefix. One Symbol duration of OFDM is $T_{IFFT} + T_{CP}$. Cyclic prefix mode vary from $T_{IFFT}/4$ to $T_{IFFT}/32$, so based on that symbol duration also changes. This conveyed to all CPE by BS in terms of T_{CP} to T_{IFFT} through super frame control header. In frequency domain an OFDMA symbols is defined in terms of its sub carrier. Based on functionality sub carries are grouped as (1) data sub carriers, (2) pilot sub carriers, (3) guard and null (including DC) sub-carriers [3, 21]. A sub-channel is the basic unit used for sub carrier allocation in both downstream and upstream. Among the 2048 sub carriers in each OFDM symbol, 384 sub carriers are null sub-carriers (left guard band, right guard band and DC subcarriers) with 0 amplitude and 0 phase. The remaining 1680 sub carriers are partitioned into 60 sub channels. Each sub channel is of 28 sub carrier with 24 for data and 4 for pilot [3, 21, 33].

Binary data to be transmitted is passed from MAC layer to PHY layer. Self-coexistence and normal mode of operation of WRAN system explained later, first frame of super frame will contain super frame preamble as first symbol,followed by a frame preamble symbol, than third symbol will be super frame control header (SCH) and at last fourth symbol contain the frame control header (FCH) and, when needed, DS-MAP, US-MAP, DCD and UCD, and data bursts if there is some room left [3, 6, 33]. Later in Fig. 1.7 shows that normally first frame is short of two symbol due to presence of SCH so that frame period of 10 ms. SCH takes care of quite period scheduling for incumbent protection and mapping the frames of WRAN cell so that self-coexistence is possible. Than remaining 15 frames normally carries frame preamble, FCH, DS-MAP, US-MAP, DCD and UCD if required, and data bursts. In Fig. 1.5 Transmit-receive Turnaround Gap (TTG) is inserted in each frame between the downstream and upstream bursts to allow the CPE to switch between the receive

Table 1.2 PHY Modes and their related modulations, coding rates, data rates

PHY mode	Modulation	Coding rate	Data rate (Mb/s)	Spectral (for 6 MHz bandwidth)
1	BPSK	Uncoded	6	6
2	QPSK	1/2 Repetition: 4	6	6
3	QPSK	1/2 Repetition: 3	6	6
4	QPSK	1/2 Repetition: 2	6	6
5	QPSK	1/2	4.54	0.76
6	QPSK	2/3	6.05	1.01
7	QPSK	3/4	6.81	1.13
8	QPSK	5/6	7.56	1.26
9	16-QAM	1/2	9.08	1.51
10	16-QAM	2/3	12.10	2.02
11	16-QAM	3/4	13.61	2.27
12	16-QAM	5/6	15.13	2.52
13	64-QAM	1/2	13.61	2.27
14	64-QAM	2/3	18.15	3.03
15	64-QAM	3/4	20.42	3.40
16	64-QAM	5/6	22.69	3.78

mode and transmit mode and to absorb the signal propagation time for a distance of up to 30 Km and a CPE turnaround time of 10 s. As previously mentioned beyond this distance proper scheduling of the downstream bursts and upstream grant is done at BS to absorb the time. At the end of each frame Receive-transmit turnaround gap (RTG) is inserted so that BS switches between receive mode to transmit mode. In USA pattern of 6 MHz bandwidth, various type of PHY modulation, encoding rate and resultant data rate are tabulated in Table 1.2 [3, 35]

1.3.1 Preamble, Control Header, MAP Definitions

1.3.1.1 Preamble Definition

There are three types of preambles, super frame preamble, frame preamble and Co-existence Beacon Protocol (CBP) preamble. These preambles are formed by two types of burst sequence defined in PHY layer as short time sequence (STS) and long term sequence (LTS). These frequency domain sequence facilitate burst detection, synchronization and channel estimation at a WRAN receiver. STS is formed using 512 bits of non zero binary sequence inserted at every 4th sub-carrier of an OFMD symbol and in time domain it is 4 repetition of 512 samples and LTS is formed using 1024 bits of non zero binary sequence inserted at every 2nd subcarrier of an OFDM

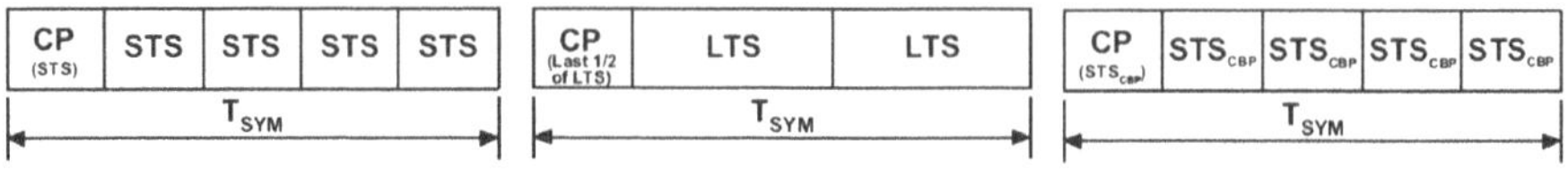

Fig. 1.3 Protocol reference model for BS and CPE [3]

symbol, so time domain it is two repetitions of 1024 samples per OFDM symbol. Super frame preamble symbol consist of four STS sequence and following one CP. Super-frame preamble used for downstream frequency and time synchronization. Frame preamble is one OFDM symbol made of last half part LTS as CP with two LTS sequence. The frame preamble is used for synchronization, channel estimation, frequency offset estimation, and received power estimation. CBP preamble is used for CBP detection, synchronization, frequency offset estimation, and CBP channel estimation [3, 33, 35]. CBP preamble is one OFDM symbol of five STS. All three types of Preamble are shown in Fig. 1.3. These sequence generated using algorithm ensuring that low peak-to-average-power-ratio (PAPR) in OFDMA. CBP preamble generated to have low cross-correlation with super frame preamble so that Super frame and CBP preamble can differentiated easily.

1.3.1.2 Control Header Definition

There are two control headers one SCH and other FCH. SCH provides information about WRAN cell is operating in normal mode or self-coexistence mode. Decoding of SCH critical,as it provides information about WRAN cell so that incumbent protection, self-coexistence mechanism and supporting inter-frame and intra-frame mechanism to quite period sensing is done efficiently. SCH is constitutes one OFDM symbol and it is transmitted using PHY mode 2 given in Table 1.2 and $T_{CP} = 1/4T_{IFFT}$ used. SCH is 45 bytes in length and is encoded by 1/2 rate convolution coder. FCH which is 3 byte in length, mainly gives the information about the frame length in OFDM symbols from start of frame to all preambles and length of MAP information sent after the FCH. The frame control header is transmitted as part of downstream protocol data unit in the downstream sub frame. FCH is transmitted using PHY mode 5 listed in Table 1.2. The FCH is sent in the first two sub channels of the symbol immediately following the preamble symbol in the frame.

1.3.1.3 MAP Definition

The MAP (Medium Access Parameters) are classified as DS-MAP and US-MAP. These are the message that defines the media access parameters like burst start time, burst length and sub channel usage for the OFDMA upstream sub frame for the CPEs scheduled to transmit toward the BS and BS towards CPEs. AS these are transmitted as part of PDU using the PHY mode 5 in the logical sub channel immediately

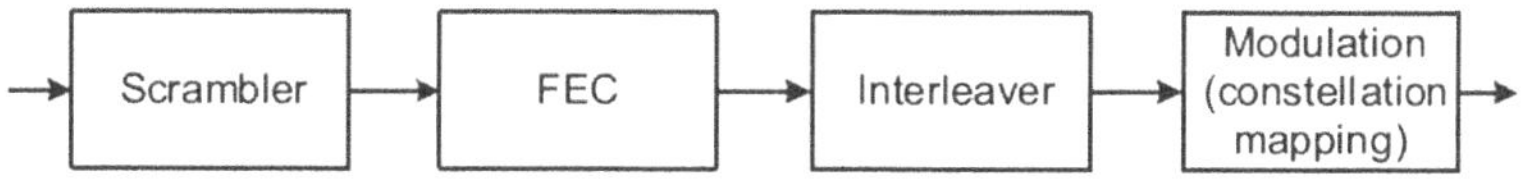

Fig. 1.4 Channel coding process [3]

following the FCH. As previously mentioned length of DS-MAP PDU contained in FCH, if it is not present than it is length of US-MAP PDU. Presence of Down link Channel Descriptor (DCD) and Up link Channel Descriptor (UCD) messages sent through DS-MAP. Length of the messages in OFDM slot decides the number of sub channels. PHY mod 5 in Table 1.2 used for transmission. If the number of subchannels required exceeds, than they are sent in next symbol. And unused sub channels in frame header used for data transmission [3, 35].

1.3.2 Modulation, Coding

Channel coding process involves four steps which is shown in the Fig. 1.4. That is scrambling, Forward error coding (FEC) in which binary convolution coding or advanced coding like low density parity check (LDPC) codes, and shortened block turbo codes (SBTCs) are included than puncturing , bit interleaving and constellation mapping based on different modulation used. Channel coder process only Physical Layer service data unit (PSDU) portion of PHY burst and control header not the preamble. PSDU data in frame is scrambled using pseudo random binary sequence (PRBS) polynomial of $1 + X^{14} + X^{15}$ in which control header and preamble of PHY burst left intact. PHY data burst is encoded using a rate: 1/2 binary convolutional encoder, which is mandatory coding defined in IEEE 802.22. Output of convolutional coder is punctured to obtain the different coding rate [3, 7, 15]. Advanced codes enhance the correction capability but at the cost of the increased pay load and receiver complexity, So CTC, LDPC and STBC are optional. Bit interleaving is used to obtain better frequency diversity, which is done using Turbo structured based repetitive interleaving. Table 1.2 shows various modulation techniques used. Interleaved data is constellation mapped based on modualtion type QPSK, 16 QAM or 64 QAM. Constellation mapping is done using Gray encoding. Usually pilot sub carrier are modulated using BPSK. Mode 1 of Table 1.2 is used for transmission of code-division multiple access (CDMA) ranging/bandwidth (BW) request message/urgent coexistence situation (UCS) notification. Mode 2 used for CBP and remaining mode 3–16 used for data transmission.

1.3.3 RF Analysis [3]

In USA, fixed regulatory class maximum effective isotropic radiated power (EIRP) is fixed as 4 W, with maximum antenna height above ground level is 30 m for both BS and CPE. For case of portable mode EIRP is 100 mW. IEEE 802.22 standard supports the Transmit Power Control (TPC) mechanism so that minimum EIRP can be used to maintain the link, and hence interference to incumbent protection is avoided and at the same time fast reliable connection between CPE is maintained. Coarse ranging is employed to minimize the dynamic range between the near and far receiver at BS by controlling CPE transmit EIRP. It is expected that IEEE 802.22 transmitter supports monotonic power level control over a range of at least 60 dB, with a resolution of 0.5 dB. Power control is done using initial ranging process and feedback mechanism. TPC algorithm also supports the EIRP adjustment at the rate 6dB/s with power fluctuations and propagation loss. CPE transmission and receiving is done through separate antennas with tracking accuracy of 0.5 dB of each other in all azimuths of the main lobe. Gain of sensing antenna in horizontal azimuth will be equivalent to an omni directional antenna gain. Where as BS uses the same antenna for reception and transmission. Standard specify that WRAN device must comply with RF mask specified in regulation. Average power detector is used to measure the power spectral density (PSD) over the measurement bandwidth of 100 KHz and video bandwidth of 100 KHz. Receiver sensitivity using robust modualtion of QPSK rate : 1/2 at 2×10^{-4} BER in 6 MHz TV channel bandwidth for BS and CPE −94.5 dBm and −91.3 dBm respectively. And receiver selectivity defined for same modualtion is 50.7 dB.

1.4 Medium Access Control (MAC) Layer in WRAN

MAC in WRAN process the centralized point to multi-point access of BS and CPE through connection oriented service. It derives the most of the characteristics from WMAN 802.16,with added special functionality of cognitive plane like incumbent protection, coexistence and QoS [33]. The MAC regulates downstream from BS to CPE through Time Division Multiplexing (TDM) and upstream where many CPEs are shared based on demand, uses Dynamically Assigned Multiple Access (DAMA)/OFDMA [17]. Based on Class of Service (CoS) used, CPE may be asked continue or discontinue or dynamically assigned to other CPE based on request. WRAN draft specifies three types access originated from MAC they are Unicast, Multicast and Broadcast. MAC uses four different types upstream scheduling schemes to solve the competition of CPEs within cell and overlapping cell which are operating on same channel with considering delay absorption and bandwidth requirement based on application scenario. Scheduling schemes are implemented using unsolicited bandwidth grants, polling, and contention procedures based on MAC header and CDMA. Authentication between BS and CPE for initial verification is done through 48 bit universal MAC address at time of network association. Connection is established between BS and CPE using connection ID (CID) of 12 bits. CID contains two information that is station ID and flow ID, which are part of DS/US MAP and MAC PDU respectively. This mechanism of connection supports overall 512 station, with each station supporting maximum eight upstream and downstream connections [29].

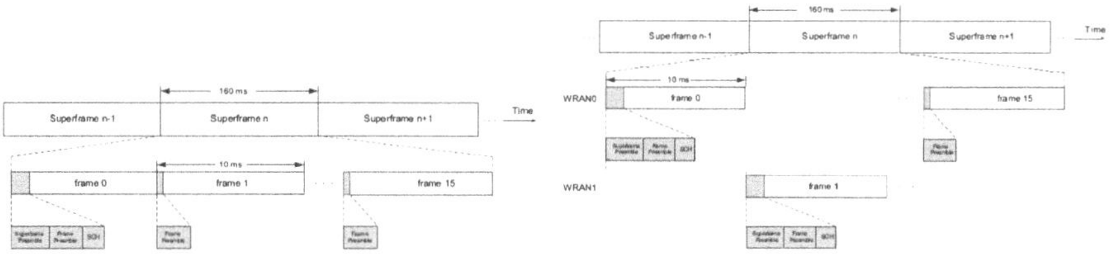

Fig. 1.5 Super frame structure [3]

1.4.1 Super-Frame and Frame Structure

WRAN draft specifies two mode of operation normal and self-coexistence mode. In normal mode one WRAN cell that is one BS and many CPE occupies one channel and operate on all frames of super-frame. Where as self-coexistence mode multiple WRAN cells occupies same channel and each coexisting cell operate on one or several frames exclusively. Figure 1.5 depicts the super frame structure in normal and self-coexistence mode of operation. Each super-frame consists of 16 frames of each 10 ms. In normal mode, first frame consists of a PHY super frame preamble, followed by frame preamble, super frame control Header (SCH) and rest of the frames include frame header and data payload. Self-coexistence is the case where frequency reuse factor is not maintained because of the overlapping nature of multiple BSs. So to avoid the interference, channel is assigned based on the frame basis to each BS.

In self-coexistence mode first frame of super frame allocated to each BS carries the super frame preamble, frame preamble and SCH as the Fig. 1.5 depicts. Frame structure is classified as downstream and upstream sub frame based on Time Division Duplex (TDD) so that dynamic channel availability is efficiently used. US and DS sub-frame are separated by TTG and RTG, which are discussed previously. Figure 1.6 shows the frame structure with relevant information sent. End part of upstream is further allotted for Self-Coexistence Window (SCW), that carries opportunistic coexistence beacon protocol (CBP) bursts scheduled by BS on demand. If necessary time buffers are also incorporated in SCW to accommodate delay between far and close

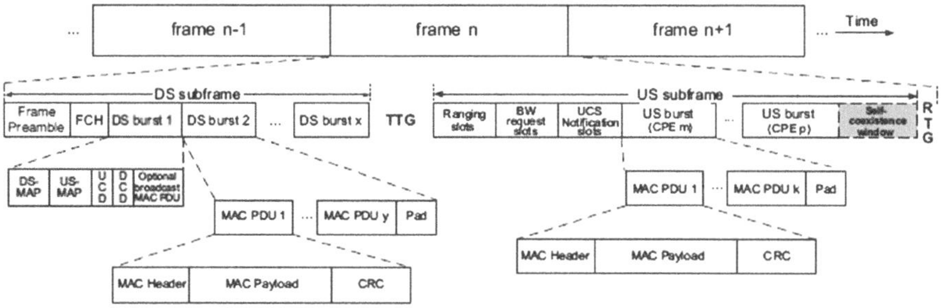

Fig. 1.6 MAC frame structure [3]

base stations and CPE operating on same channel. DS and US sub frames boundaries are adjustable based on relative capacity. US sub frame includes PHY-protocol data unit (PDU) for traffic and contention interval schedules from different CPEs if needed. Contention interval schedules includes : (a) CPE association (initial ranging) (b) CPE link synchronization, power control and geolocation (periodic ranging) (c) Bandwidth request (d) Urgent coexistence situation (UCS) notification (e) Quiet period resource adjustment [3, 35].

Two dimensional view of frame structure in terms of time and frequency is shown in Fig. 1.7, in which horizontal axis represent symbol and vertical axis represents sub channels. One frame of 10 ms can accommodate 26 OFDM symbols in 6 MHz TV channel bandwidth with 1/4 CP in time domain and 60 sub channels in frequency domain. MAC packets are linearly structured in time division multiplexing (TDM) way and PHY packets mapped to two dimensional time-frequency OFDMA. From BS downstream burst after FCH, frame carries DS/US-MAP, downstream and upstream channel descriptor (DCD and UCD) and MAC-PDU. As previously discussed in Sect. 1.2 MAP messages provides information about channel parameters

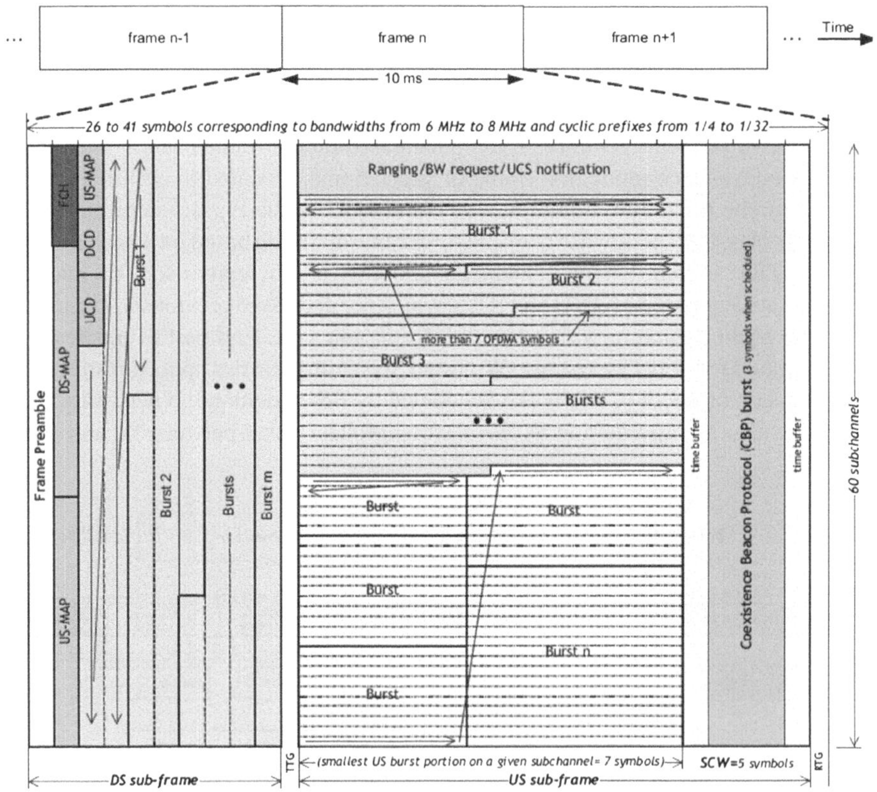

Fig. 1.7 Two dimensional MAC Frame structure [3]

and resource allocation in DS/US frames. DCD and UCD sent periodically, gives the information about physical channel characteristic in upstream and downstream. MAC PDU contains MAC header, payload and CRC, which takes care of connection flow between intra-cell CPEs. Figure 1.7 shows that, DS MAC burst are laid first vertically along sub channels and than advanced horizontally in time direction. This phenomena of mapping give added advantage of reducing delay by early scheduling of DS burst accessed by far CPEs and avoid estimated interference due to CPE in overlapping cell of different US/DS capacity.

In upstream, contention scheduling request are sent first and data burst later. Upstream mapping of MAC data elements are done by two process. MAC data are laid horizontally on sub channels OFDM symbol by symbol, if all sub-channels than it is next list of sub-channels. If MAC data elements are sufficient to fill the sub-channels, than padded with zero. This longer upstream burst occupying one sub channel will help in reducing EIRP required for upstream transmission. In other way of mapping is horizontally with at least 7 symbols per sub-channel until all sub channels are occupied and usually last vertical busts will be between 7 and 13 symbols depending on number of symbols. This alternative process small burst reduces the latency in upstream burst at cost of higher EIRP required and efficiency, which is shown in Fig. 1.7. As previously, said end part of frame is SCW window which carries CBP packets from selected CPEs or BS. CBP packets provide self-coexistence signaling information to adjacent and overlapping WRAN cells, device identification for resolving interference with incumbents, on local regulation request and to carry out the terrestrial geo-location between the CPE and BS of same cell.

1.4.2 Scheduling and Bandwidth Management [3, 35]

Scheduling is process in which MAC scheduler, directs transport of data connection based on set of QoS parameters that quantify data behavior. Scheduling is classified as data transmission scheduling, which looks after data transmission in particular frame/bandwidth and taken care by BS for downstream and CPE by upstream. And other is upstream request/grant scheduling which is performed by BS, which looks after providing bandwidth to associated CPE for upstream transmission or opportunities to request bandwidth. Based on scheduling service and its associated QoS, scheduler signals the appropriate grant or polling command. There are four types of services present in WRAN system:

- Unsolicited Grant Service (UGS): supports real time fixed size data packets sent at periodic intervals like T1/E1 or voice over-IP without silence suppression. Its related QoS parameters required for scheduling include Maximum Sustained Traffic Rate, Maximum Latency, Tolerated Jitter, and Request/Transmission Policy.
- Real-time Polling Service (rtPS): It is associated with real-time variable packet size data sent periodically for example MPEG video. Manadatory QoS parame-

ters include Minimum Reserved Traffic Rate, Maximum Sustained Traffic Rate, Maximum Latency, and Request/Transmission Policy.
- Non-real-time Polling Service (nrtPS): Its design support delay tolerant data like FTP. And its mandatory QoS parameters include Minimum Reserved Traffic Rate, Maximum Sustained Traffic Rate, Traffic Priority, and Request/Transmission Policy.
- Best effort (BE): It does not have any minimum service level and transmitted based on space availability. Its QoS parameters for scheduling are Maximum Sustained Traffic Rate, Traffic Priority, and Request/Transmission Policy.

Bandwidth management done using control message sent through three FID during network entry and initialization, where bandwidth request are sent by CPE to BS. There are two types bandwidth requests available in MAC layer: Contention-based request which is sent during any upstream allocation, except during any initial ranging interval, UCS notification interval, and SCW and bandwidth request may be incremental or aggregate. And PHY also supports CDMA-request for upstream bandwidth allocation using ranging codes available in PHY. Through Polling BS allocates bandwidth to CPEs bandwidth for making bandwidth requests.

1.4.3 Incumbent Protection and Self-coexistence

As previously mentioned incumbent protection and detection is done through justified spectrum management, quiet period management, distributed spectrum sensing, detection algorithms, and measurements. MAC incorporates huge set of spectrum management and measurement commands that allows BS to make flexibility in controlling CPEs to have reliable spectrum map in cell and change their parameters [3, 23]. Measurements are classified as, in-band and out-band. In-band means measurements made on operating channel and its first adajcent channels and out-band is other than above three channels. MAC measurements are manged by hierarchical structure of management messages used between SM at the BS and SSA at the CPE and BS for incumbent detection and self-coexistence listed in clause 7.20 of [3]. Reports received by SM throng spectrum sensing automation (SSA) are two types regular and urgent. Regular reports are explicitly requested by BS to SSA periodically and also autonomous reporting of completed data collection by SSA. Urgent reports refer to sudden detection incumbent in operating or adjacent channel by SSA during in-band sensing, which activate UCS message. Incumbent detection is done in MAC through periodic sensing through quite period management, explained later [3, 11].

In IEEE 802.22 network scenario, many BSs and CPEs operate in the vicinity as shown in Fig. 1.8, self interference is measure drawback which may make network dysfunctional. It requires efficient air interference between BS and CPE. One solution may be using directional antennas by CPE, but they have relative capability to discriminate only TV broadcast bands not many BSs antennas directing towards them. And also ability of IEEE 802.22 BS coverage potentially reaching

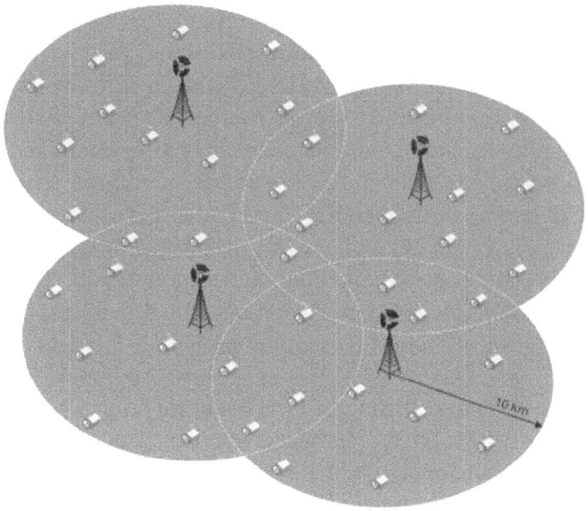

Fig. 1.8 self-coxistence scenario in IEEE 802.22 [3]

to 100 Km, further deteriorates phenomena of self-coexistence. Thus co-existence is critical for WRAN air-interface along with incumbent protection and detection and self-coexistence measure also have equal importance. Both of these mechanisms in MAC, make cognitive effect of flexibility, adaptability and reaction to environment changes stiff. Self-coexistence mechanism efficiently carried out using CBP protocol through exchanging coexistence beacons between overlapping WRAN cells. MAC layer uses mandatory mechanisms that includes spectrum etiquette and on-demand frame contention elements to address self-coexistence shown in Fig. 1.9.

As previously stated, Coexistence beacon protocol (CBP) is transport mechanism through which coexistence possible in IEEE 802.22 and CBP packets transmitted through air as discussed in PHY layer or back haul by encapsulating them into IP packets. Initially WRAN cell opearte in normal mode till it can detect and decode SCH or CBP bursts from neighbor cell and than shift to self-coexistence mode. CBP MAC PDU that contains SCH, by which transmitting BS or CPE, provide information about neighbor network discovery and coordination of quite periods (QP) and SCWs

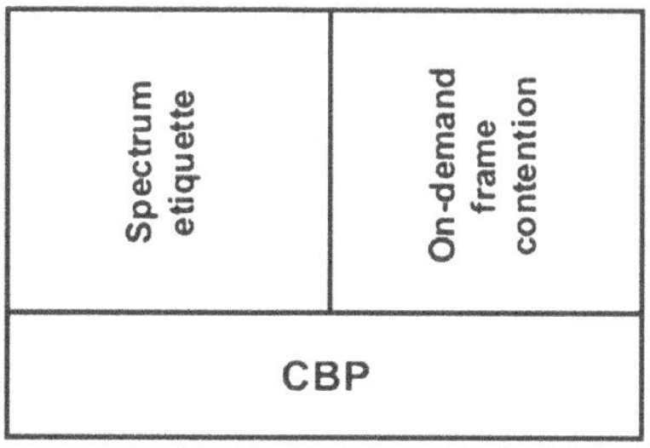

Fig. 1.9 CBP transport mechanisam for inter WRAN communication and self-coexistence [3]

that is how schedule of QP and SCWs advertised to CPEs of neighbor cells. Much about process of CBP communication, network discovery through CBP and SCW and inter-BS self-coexistence much details is found in self-coexistence clause of [3, 18]. Spectrum etiquette is resource sharing mechanism in case of overlapping BSs, in which it ensure how back and candidate channels described later are prioritized and it ensure that operating channel does interfere with other existing WRAN. On Demand Frame Contention Protocol (ODFC) used to resolve the contentions of frame with neighboring cell. When Bs switches from normal mode to self-coexistence mode it will schedule at least one contention SCW for potential frame request from overlapping cell. Than those overlapping cells can than schedule more SCW as their needs and contend for more frames by sending proper CBP bursts in SCW and it is not required every frame need to SCW. Fair spectrum resources is possible by using distributed, cooperative, and real-time spectrum sharing protocol called On-Demand Spectrum Contention (ODSC)that employs interactive MAC messaging on an inter-network communication channel to provide efficient, scalable, and fair inter-network spectrum sharing among the coexisting 802.22 cells [20].

1.4.4 Quite Period and Sensing

The Quite periods (QP) are network-wide time periods during, which incumbent detection on operating channel possible through in-band sensing by suspending all network traffic between BS and CPE [22]. So BS must schedule these QPs such way that effective amount of time is spent in incumbent protection and detection with maintaining reliable QoS as traffic is suspended during QPs [3]. So IEEE 802.22 BS and CPE employs two stage intra frame and inter frame sensing based quite period management mechanism that dynamically adjust the repetition rate and duration of quiet periods. Repetition rate and duration of QP depend on type of sensing algorithm used and signal type to be sensed. Based on above information quite period scheduled by implicitly through sensing related field of SCH or explicitly through channel request (CHQ-REQ) MAC message. Where as out of band sensing is usually carried out by CPE when there is no communication between BS and CPE in normal operation, so no network-wide scheduling of QPs required but depends on quite period of channel for sensing [30]. And also out of band sensing may be requested by BS during UCS of Fig. 1.10.

Two way QPs management depends two stage sensing mechanism employed:

1. Intra-frame sensing: Sensing stage which require one sensing period frame and it employs the sensing algorithmic that have quite periods less than one frame. Based on its measurement analysis on intra frame sensing over many frame next stage or longer period required sensing stage is excited
2. Inter-frame sensing. Sensing stage that incorporates sensing algorithms of longer quite periods than frame length. As said previously longer QPs effect QoS of network traffic, so BS has ability to dynamically control the allocation and dura-

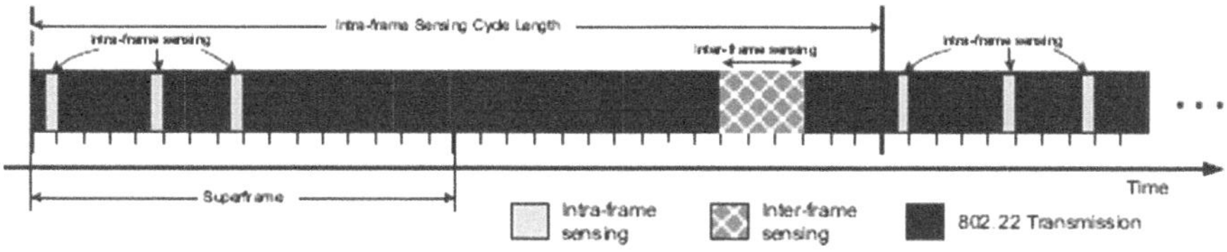

Fig. 1.10 Example of intra and inter frame sensing stage

tion inter frame sensing and also it takes feedback from intra frame sensing to decide whether go for inter frame sensing stage or not, so that QOS between CPE and WRAN BS maintained with protecting the incumbent user. Coarse sensing is done using intra-frame sensing and if fine sensing like finding signature incumbents required go for the inter frame sensing.

There are various sensing methods are proposed for the both case in the standard appendix part C and following literature [3, 6, 32]

1.4.5 Channel Management

Radio resource in IEEE 802.22 changes frequently so, MAC incorporates dynamic and flexible channel management system so that efficiently radio resource is utilized. BS and CPE MAC is facilitated with two modes of channel management that is embedded and explicit and decision regarding which and when to be used decided by BS. Embedded mode is much effective in spectrum utilization, as it is addressed to all CPE in cell and no individual command required to be sent. It in turn helps for all CPEs in cell to take corrective measurements when incumbent is detected in operating channel. Other hand explicit mode has flexibility based on different granular based channel management and relatively independent of MAC protocol [3, 23, 27].

1.5 Cognitive Capability

Cognitive capability is an essential in protection incumbents and efficient working of IEEE 802.22 WRAN system. Cognitive capability in WRAN system implemented using BS's spectrum manager (SM), Spectrum Sensing Automation (SSA), Access to the database service, Channel set management, CPE registration and tracking, policies, spectrum sensing services, and Geo location services. Cognitive capability assist WRAN devices to make decision about their radio operating behavior through communicating with database services,polices, geolocation or through direct sensing incumbents [4, 23].

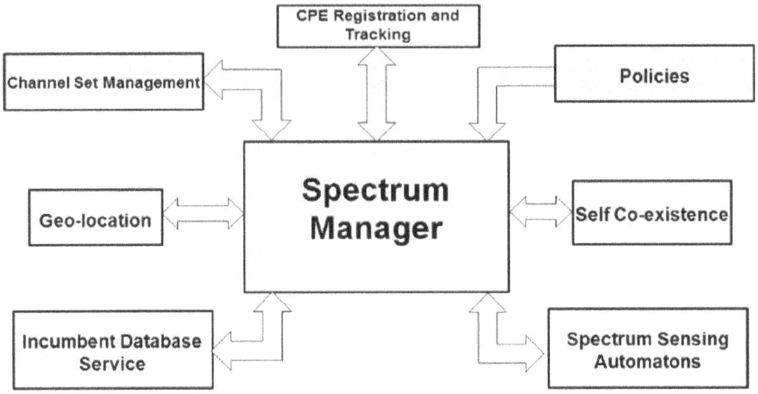

Fig. 1.11 SM and logical interfaces [3]

1.5.1 Spectrum Manager

As shown in Fig. 1.11, which is like brain of WRAN BS, present in cognitive plane IEEE 802.22 BS reference model. SM centralizes all decisions within WRAN cell regarding spectrum availability and utilization to ensure that protection of incumbents and efficient spectrum utilization while complying with regulatory policies is possible. BS and CPE both contain entity called SSA which intereface with spectrum sensing function (SSF) and that carries sensing process based SM command. The key functionality of SM are listed below:

- Maintain spectrum availability information: SM maintains the status of channel availability for WRAN operation at its location according to policies and rules established by local regulatory organization. Based on channel occupancy by incumbents or other WRANs in the area are used as input by SM for channel selection, channel management and self-coexistence. Information about the channel status are obtained by any of three means, database service, geolocation and sensing.
- Channel classification and selection: SM assigns operating channels to MAC/PHY modules and also define backup channels and their corresponding priority. Available channels are classified by following type

  1. **Disallowed**: Channels that are forbid from use by the operator due to operational or local regulatory constraints.
  2. **Operating**: This is the current channel used by CPE and BS for communication and it is regularly sensed after every 2 s for signal types and 802.22.1 beacon.
  3. **Backup**: Channel that can be immediately converted to operational as soon as WRAN plans to switch to new channels. This channels is scanned for at least once every 6 s and remains in backup list until incumbent is not found.

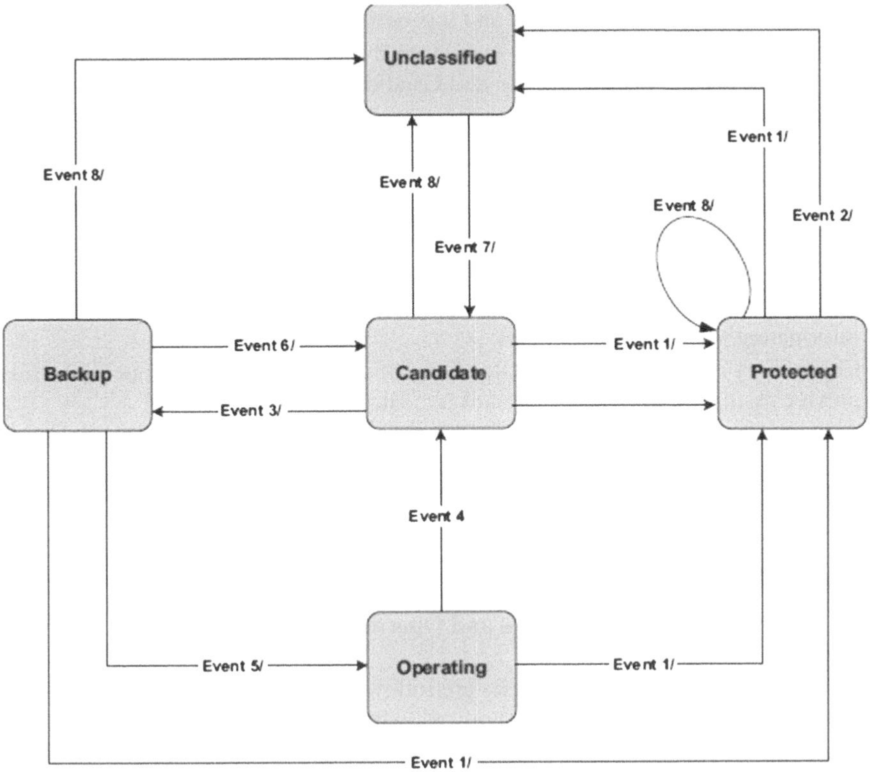

Fig. 1.12 Channel set station diagram [3]

4. **Candidate**: Channels possible to be listed in backup. These are channels, BS may request CPE for sensing to upgrading to Backup. These are sensed infrequently for at least every 6 s but not more than 30 s for incumbent presence.
5. **Protected**: Channel in which incumbent or WRAN operation detected through sensing. As soon as incumbents or WRAN vacated can be moved to candidate or backup based sensing at least every 6 sec but less than 30 sec.
6. **Unclassified**: Channels which are sensed but may be classified as candidate or protected based on sensing results.

Channels can not belong to more than one set any time, but operating channel at one WRAN system can also be operating channel for other because of self-coexistence mechanism in IEEE 802.22. As per the standard, specific algorithm for selecting of operational and prioritizing of backup and candidates is considered out of scope until sensing requirement are met. However algorithms are implemented based input from channel status, traffic requirement, location information and self-coexistence with neighboring WRAN system.

Fig. 1.12 shows, how SM moves channels to various group as they are returned from database service based incumbent users and channel quality at the end of quite periods. There are 5 states and 9 events that are based on sensing outputs during network initialization and operation. Events definition can be found in sub clause 10.2.3.1 of [3] and Backup and Candidate channel prioritization are done based on spectrum etiquette. The spectrum etiquette rules are chosen such that channel does not interfere, or interferes with a minimum number of channels to be used by its neighbor cells. Spectrum etiquette details procedure is found in sub clause 10.2.3.2 of [3].

- Association control is process by which CPE register with BS. SM can grant or denay association rights to CPE on its request based on database accessing for available channel list, maximum EIRP limits at CPE's location and received information to protect incumbents.
- Channel set management is how to share efficiently available channel, which is previously discussed in channel management part of MAC.
- Accessing the database service.
- Scheduling quiet periods for spectrum sensing.
- Enforcing IEEE 802.22 and regulatory domain policies.
- Making channel move decisions for one or more CPEs or the entire cell.

1.5.1.1 Spectrum Manger Policies and Operation

SM ensures that IEEE 802.22 policies are followed within the cell in order to guarantee the required protection of incumbents while supporting QoS for the WRAN users. There three types of policies which are put in enumeration 1, 2 and 3.

- Enmueration 1 policies discusses the events initiated from the database service but not from a locally generated operator database.
- Enumeration 2 policies are related to an event that a TV signal is detected.
- Enumeration 3 policies are related to an event that a wireless microphone signal or the IEEE 802.22.1 beacon signal is detected.

Based on above policies SM takes following action switch the entire cell to a new operating channel, direct a single CPE or a group of CPEs to a different operating channel when possible and terminate operation in a given channel for a single CPE, a group of CPEs or the entire. SM has two state of operation that is network initialization and network operation. During initialization SM primary duty is find the operating channel. In network operation mode all previous function discussed are carried out [3, 23].

1.5.2 Spectrum Sensing Automation–SSA

SSA is the entity present in BS and CPE of IEEE 802.22 system and it is interfaced with spectrum sensing function (SSF) and operates based on SM command to enable spectrum sensing. SSA has two states of operation SSA in band sensing and out of band sensing. SSA is normal by controlled by BS but under following conditions SSA controls local sensing

- At the initial turn-on of the BS before it starts to transmit any signal.
- At initial turn-on of the CPE before association is established with the base station.
- During the quiet periods defined by the SM and signaled by the BS through the SCH for in-band sensing.
- During out-of-band sensing at the BS when the base station is not transmitting.
- During idle time at the CPE when the base station has not attributed any specific task to the CPE sensing signal path through the BLM-REQ message or when the CPE does not transmit.
- When the CPE loses contact with its base station.

1.5.2.1 Spectrum Sensing

Spectrum sensing is the process of scanning TV RF channel to detect the occupancy of incumbents or other WRANs. This process is implemented through spectrum sensing Function (SSF) entity present in PHY part of CPE and BS that is driven by SSA in MAC of WRAN devices. There are MAC management frames that allow BS to control SSF within each of CPEs. SSF takes input from SM via SSA and output fed back to SM via SSA that is shown in Fig. 1.13. Standard doesn't specify any specific sensing algorithms as long as algorithms meets inputs, outputs and behavior of SSF. Some algorithms are specified in standards annexure part C but they are optional [3, 32, 37].

RF input is connected from RF stage to WRAN antenna. Country code specific to regulatory domain operation that to particular country. Channel number is the relative channel number that the SSF is to sense. Signal Type Array (STA) indicates types of signals to be sensed by SSF. STA is one dimensional binary array of bits from 0 to STALength-1.Table 1.3 specifies types of signal with STA number.

Sensing Window Specification Array (SWSA) provides information about number of sensing periods, sensing period duration in terms of symbols and adjacent sensing period separation in terms of symbols for each type signal in STA. Sensing window occupy portion of quite period or entire or multiple quite period. As per standard QP aggregate duration should not exceed 200 ms for spectrum sensing techniques that meets detection threshold and processing latency must not above 2 s. There are three sensing mode that specifies which outputs of the SSF are valid and sometimes its behavior. Sensing Mode 0 SSF generates binary decision whether signal is present or not in TV channel. Sensing Mode 1 with addition sensing mode 0 it provides confidence metric for binary decision and in case Sensing Mode 2 SSF generates estimate of mean and standard deviations for upto 255 instantaneous RSSI measurements on specified channel. Special consideration is given to IEEE 802.22.1 beacon detection in SSF, as beacons are designed to be detected within reasonable time window with keeping effect on QoS minimum. Some acceptable methods that comply detection of beacons include energy detection, baud rate detection, correlation of spreading

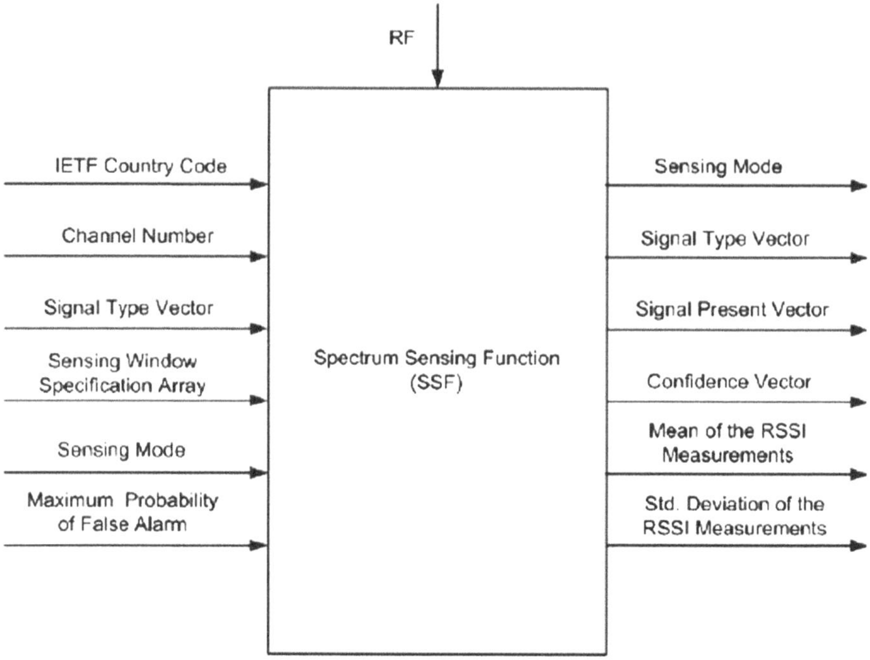

Fig. 1.13 Channel set station diagram [3]

sequence or synchronization and determination of the frame index. Also beacons provides additional information like location by (MSF1), signature (MSF2) and its authentication with larger sensing periods (MSF3).

1.5.2.2 Spectrum Sensing Requirements as per Regulation

Article doesn't discuss about spectrum sensing techniques as many articles are found addressing this topic and there are no mandatory techniques specified by standard and according to standard any SS techniques can be used until it meets regulatory requirements. So only regulatory required sensing requirements are specified for sensing. Table 1.3 lists the Channel sensing requirements for different signal type in USA according to FCCs report and order. IEEE 802.22 prefers distributed sensing with "OR" fusion rule and omnidirectional antenna with 0 dBi [3, 31].

According regulatory domain specification monitoring to acquire channel must be completed within 30 s and in service monitoring must be within 60 s [35].

Table 1.3 Signal type array indices

STA index	Signl type
0	Undetermined
1	IEEE 802.22 WRAN
2	ATSC
3	DVB-T
4	ISDB-T
5	NTSC
6	PAL
7	SECAM
8	Wireless microphone
9	IEEE 802.22.1 Sync burst
10	IEEE 802.22.1 PPDU MFS1
11	IEEE 802.22.1 PPDU MSF2
12	IEEE 802.22.1 PPDU MSF3
13 32	Reserve

1.5.3 Geolocation

Geolocation technology is essential in finding position of device in network and their relative moment greater than specified by regulation [3, 33, 34]. Two modes of geolocation are used in which satellite-based geolocation is mandatory. And other is terrestrial based location that is assisted by CDMA ranging, super-frame preamble, frame preamble and the coexistence beacon protocol (CBP). BS and CPE uses Satellite based geo-location antenna to determine the longitude and latitude of transmitting antenna within 50 m radius. Terrestrial geolocation involves two step process that is first range between BS and associated CPE is obtained with high accuracy and than next step is to use triangulation method establish geolocation of CPE. Coherent multi carrier OFDM modulation is efficiently used in precisely geo-locating and mapping.

1.6 Security Aspects in WRAN System [10, 28]

In order to provide protection of IEEE 802.22 users, service providers and incumbents, standard defines two security sub layers one for non cognitive and other cognitive functionality and their interaction between the two illustrated in reference module. Security sub layer 1 that is present in data plane provides subscribers with authentication, or confidentiality for user data and MAC management messages transmitted across the broadband wireless network by applying cryptographic transforms and also provide operators with strong protection from theft of service. It employs client/server key management protocol authentication process in which

BS centrally controls keying aspects. Security sub layer 2 present in cognitive plane enhance the security for cognitive functionality. It employs the distributed sensing, signal type determination, correlation with geolocation information from various sensors (CPEs) located in the network and policy enforcement to validate availability of spectrum for primary and secondary user. It avoids proxy incumbents by authentication of incumbents sensing information so that denial of service (DoS)attack is overcome [3, 12]. And it also looks after authentication of IEEE 802.22.1 beacon frame utilizing the security features that are already embedded in it, authentication of the geo location and co-existence information etc. Cognitive plane capability like SM, SSA, SSF and geolocation have security features embedded in their process. So IEEE 802.22 security layers ensures that security of following aspects are met availability, authentication, authorization, identification, integrity, confidentiality, and privacy.

Distributed sensing is process where more than one Radio frequency sensor locally sense the spectrum environment and send local information to BS for validating information about spectrum opportunity. It also uses knowledge of geo location of sensing devices to identify incumbent based on coverage. Standard specify the three types of fusion rule like simple OR and AND, also Voting based rule ensure that more security and flexibility in protection. Distributed sensing ensures that low miss detection and overcome DoS attack due to proxy incumbents.

1.7 802.22.1 Beacons [2]

To provide enhanced security to protected low power auxiliary device that are not broadcasting TV signals, such as microphones (auxiliary devices details found in Title 47, Part 74 of the Code of Federal Regulations (CFR) of FCC, USA), operating in TV channel, WRAN standard committee has released separate IEEE std 802.22.1 in 2010 that provide data and protocol format for communication of beacon networks [2, 36]. Beacons are installed by licensed operator at appropriate place so that they can send identifiable sync bursts, location information and operational parameters of low power auxiliary devices for protection. Beacon networks classified as Primary protecting Device (PPD), Secondary Protecting Device (SPD), Next Line Protecting Device (NLP). PPD is transmit beacon data every other frames and it may or may not include the data collected from inter-device communication. SPD is PD chosen on behalf of other PD in which security information exchanged between PPD through inter-device communication and it broadcast information as part of regular PPD's transmission. PPD ceases to operate NLP is chosen as SPD. Selection of these mode and operation on TV channel is taken care by next higher levels of protocol of OSI.

Beacon device also employs super frame structure as in MAC of WRAN and PHY layer contains radio frequency (RF) transceiver along with its low-level control mechanism. Wireless microphone occupies 200 KHz bandwidth and based on 6 or 7 or 8 MHz TV channel bandwidth divided into 30–40 sub channels. So beacons are also expected operate on sub channel other than wireless microphone channel to avoid interference and their bandwidth is also restricted to less than 200KHz. PHY employs the simple transceiver system based on direct sequence spread spectrum (DSSS) of with differential quadrature phase-shift keying (DQPSK). Because of DQPSK, channel estimator is not used that simplifies demodulation. Beacon device operate with following parameters: center frequency is same as Advanced Television Systems Committee-Digital Television (ATSC-DTV) pilot signal of the channel that is offset TV lower channel edge 309.4 KHZ, occupies bandwidth of 77 kHz at −3 dB and 110 at −20 dB, chip rate of: 76.873 kchips/s and Symbol rate: 9.6091 kBaud. Beacon transmit maximum output power of 250 mW for UHF bands and 50 mW for VHF bands, with high sensitivity antenna compared to low power auxiliary devices.

Super frame structure consists of 31 slots with each slot made of 32 DQPSK symbols. It consists of synchronization channel that hold succession of synchronization burst and beacon channel packed with PHY Packet Data Unit (PPDU), which contains the MAC beacon frame are transmitted continuously in parallel as I and Q channel. MAC becon frame consists of information relevant to the device or devices protected by the protecting device, including the physical location of the beaconing device and the estimated duration of TV channel occupancy. It is further subdivided into three MAC sub frames (MSF) that is MSF1 contains the source address field, location field and three MAC parameter fields, MSF2 contains the channel/subchannel map and signature fields and MSF3 contains the certificate field [24]. It is notifiable that even after acquiring super frame time information of beacon by WRAN device still they are not in sync. Synchronize without changing the transmission time of WRAN, beacon is designed that required quite period timing will be less than one super frame of WRAN, so it waits until despaired part of beacon frame doesn't interrupt. Based on super frame structure three level of authentication possible by WRAN and quite periods also varies. First level of authentication that is based on I channel 15 bit PN sequence sync burst, WRAN device can detect presence of beacon and QP required is 5.1ms [3]. First level of detection doesn't affect the communication of WRAN device. Second level authentication that requires identify MSF1 contents and requires QP of 30 ms so it is more reliable at cost longer QP. Third level of authentication that requires decoding of MSF2 content and require 72.43 ms which requires longer intra and inter frame sensing schedules [2, 19]. MSF3 mainly decoded when it is not possible to decode public key through back haul network that requires long QP of 100 ms,but is not accounted as level.

1.8 Discussion and Engineering Aspects of Cognitive Standard

IEEE 802.22 is the first standard that is drafted for use of TV white space for rural broadband communications. Regulatory domains classified into four geographical area USA, Canad, United Kingdom and Asia, Africa and Pacific. Approval authority in USA is FCC, Canada is IC and UK is OFCOM. TV white space frequency range with maximum bandwidth used in different regulatory region are mentioned in the Table. 1.4. Important Physical system parameters used in WRAN systems for design of White space radio are tabulated in Table. 1.1. There are two type of regulatory

Table 1.4 Regulatory regions and corresponding Frequency Operations

Regulatory region	Bandwidth in MHz	Frequency range in MHz
USA	6	57–695
Western Europe	7	50.5–226.5
Africa, Asia Western Europe Africa, Asia	8	474–858

classes one is stationary fixed and personal portable. Personal portable are of Mode I and II as defined in FCC Report and Order. IEEE std 802.22 operates in nomadic portable mode II [3].

Transmit power limits and EIRP requirements for WRAN system in USA regulatory domain are specified in Table 1.4. Multi-input Multi output (MIMO) is not possible in IEEE 802.22 system because of antennas size in TV spectrum is larger. According to Std geo location query has to be done based on certain change in distance through database query as in Table 1.6. There is minimum distance of 1Km is fixed as microphone protection radius (MPR) for WRAN device to operate from wireless microphone. Device identification requirement as for USA regulation domain using CBP burst is as mentioned in Table 1.6. Minimum time is time between transmission of a CBP packet carrying a device ID and serial Number for identification by spectrum sensing device to acknowledge the potential interference situation [3]-appendix A. Spectrum sensing regulatory requirement are explained in cognitive capability part. Any Spectrum sensing algorithms can be used as long regulatory constraints are met according to Tables 1.5 and 1.7.

MAC novel super frame structure looks after proper synchronization, self-coexistence mechanism and protection of incumbents and other WRANs. MAC special protocol like CBP and SCW ensures that efficient incumbent detection by raising contention alarms. Though hand off mechanism is not employed in WRAN system, but still MAC spectrum manger polices ensures that good QoS. Most of the cognitive capability are centrally controlled by spectrum manager in MAC of BS. Working of slave master module ensures in protection to incumbents and the WRAN system as discussed in security aspects. In our article special emphasis has been given to frame structures and essential components within MAC, important MAC concepts like self-coexistence, incumbent detection, quite period and bandwidth scheduling are explained thoroughly. Three engineering challenges that are main to cognitive radio are efficient distributed sensing and decision making so that regulatory specified requirement of channel sensing and detection probability are met as in Table 1.8. Distributed sensing is the compulsion in IEEE stn 802.22 with fusion option kept open. It is to open to manufacture to improve the reliability by using complex algorithm as long sensing timing and detection probability are met. If manufacture considers geo location information of sensing device, correlation of sensing devices which have sent UCS message to base station and their respective location along with advanced artificial intelligence can enhance the decision making

Table 1.5 Tansmit power limits and EIRP requirements

Regulatory class	Maximum BS EIR	Maximum antenna Height BS	Maximum CPE EIR	Maximum antenna Height CPE
Stationary fixed	4W	30m Above ground level (AGL) 76 m Ground height above average terrain (GHAAT)	4W	30m AGL 76 m GHAAT
Personal Portable	100 mW	NA	100 mW	NA
Stationary fixed (CAN)	500 W	$\le$ 60 m Antenna height above average terrain (AHAAT)	4 W	10 m AGL
	250 W	$\le$ 90 m AHAAT		
	125 W	$\le$ 120 m AHAAT		
	66 W	$\le$ 180 m AHAAT		
	33 W	$\le$ 240 m AHAAT		
	4W	$\le$500 m AHAAT		

Table 1.6 WRAN device location and accuracy and distance threshold

Location accuracy	Confidence level	Distance threshold for portable devices
100 m	67	50 m
300 m	95	50 m

Table 1.7 Device identification requirement

Minimum period	Maximum period	Signal processing	Timer
8 s	15 min	CBP burst	T_{CBP}

Table 1.8 Sensing requirements

Type of signal	Sensing detection threshold in dBm
ATSC	−114 (Averaged over 6 MHz)
NTSC	−114 (Averaged over 100 KHz)
Wireless microphone	−107 (Averaged over 200 KHz)

and reliability. Next challenge is quite period scheduling for sensing. Quite period and idle times are used for sensing. So quite period must be scheduled efficiently so that QoS of WRAN system is not deteriorated. As mentioned previously IEEE std 802.22 provides two mode of sensing that is intra and inter frame sensing. Presently WRAN is in testing mode, but when real time infrastructure is established large number of WRAN users demand has to be taken care. So this requires adaptive quite period scheduling based on previous channel usage pattern and as per the demand of users pattern. Third challenge is self-coexistence between WRAN users. Though standard provides spectrum etiquette and manger polices but can these help in efficient channel management,when real time infrastructure is established need to be checked. Above three engineering challenges can be answered effectively once the WRAN infrastructure with installation for real time operation is done.

1.9 Conclusion

IEEE std 802.22 first of the standard which helps to overcome the digital divide between rural and urban. Presently it is viable solution for rural and remote areas for broadband connectivity. WRAN cognitive capability can be extended to upcoming mobile networks like LTE advanced and femto cell for both data and voice services. There are similar related standards like DYSPAN, ECMA-392 and M2M that are working concept of white space radio. For all upcoming standards on TV white space this standard will be guiding protocol. our review on IEEE 802.22 std covers PHY, MAC and cognitive capability. It highlights the necessary engineering and regulatory requirements for WRAN device design in research forum.

References

1. http://www.ieee802.org/22/
2. IEEE standard for local and metropolitan area networks - specific requirements part 22.1: Standard to enhance harmful interference protection for low-power licensed devices operating in tv broadcast bands. IEEE Std 802.22-2010, pp. 1–300, (1 2010)
3. IEEE standard for information technology-telecommunications and information exchange between systems wireless regional area networks(wran)-specific requirements part 22: Cognitive wireless ran medium access control (mac) and physical layer (phy) specifications: Policies and procedures for operation in the tv bands. IEEE Std 802.22-2011, pp. 1–680 (1 2011)
4. Akyildiz, I.F., Lee, W.Y., Vuran, M.C., Mohanty, S.: A survey on spectrum management in cognitive radio networks. IEEE Commun. Mag. **46**(4), 40–48 (2008)
5. Apurva Mody, G.C.: Overview of the ieee 802.22 standard on wireless regional area networks (wran) and core technologies http://www.ieee802.org/22/Technology/22-10-0073-03-0000-802-22-overview-and-core-technologies.pdf
6. Arslan, H. (ed.): Cognitive Radio, Software Defined Radio, and Adaptive Wireless Systems. Springer, London (2007)
7. Arslan, H. (ed.): Cognitive Radio, Software Defined Radio, and Adaptive wireless Systems. Springer, London (2007)
8. Benko, J., Cheong, Y., Cordeiro, C., Gao, W., Kim, C., Kim, H., Kuffner, S., Laskar, J., Liang, Y., et al.: A phy/mac proposal for ieee 802.22 wran systems. Part 1, 802–22 (2006)
9. Berlemann, L., Mangold, S.: Proposed enablers for realizing vertical spectrum sharing. Cognitive Radio and Dynamic Spectrum Access, pp. 145–187. Wiley, Chichester (2009)
10. Bian, K., Park, J.M.J.: Security vulnerabilities in ieee 802.22. In: Proceedings of the 4th Annual International Conference on Wireless Internet, pp. 9:1–9:9. WICON '08, ICST (Institute for Computer Sciences, Social-Informatics and Telecommunications Engineering), Brussels, Belgium (2008). http://dl.acm.org/citation.cfm?id=1554126.1554138
11. Cavalcanti, D., Ghosh, M.: Cognitive radio networks: Enabling new wireless broadband opportunities. In: 3rd International Conference on Cognitive Radio Oriented Wireless Networks and Communications, 2008, CrownCom 2008, pp. 1–6, IEEE (2008)
12. Chen, R., Park, J.M., Bian, K.: Robust distributed spectrum sensing in cognitive radio networks. In: INFOCOM 2008. The 27th Conference on Computer Communications. IEEE. pp. 1876–1884. IEEE (2008)
13. Commission, F.C., et al.: Second report and order and memorandum opinion and order in the matter of unlicensed operation in the tv broadcast bands additional spectrum for unlicensed devices below 900 mhz and in the 3 ghz band. Document 08 260 (2008)
14. Commission, F.C., et al.: Notice of proposed rulemaking, in the matter of unlicensed operation in the tv broadcast bands (et docket no. 04–186) and additional spectrum for unlicensed devices below 900 mhz and in the 3 ghz band (et docket no. 02–380), fcc 04–113 (2004)
15. Cordeiro, C., Challapali, K., Birru, D., Sai Shankar, N.: Ieee 802.22: The first worldwide wireless standard based on cognitive radios. In: New Frontiers in Dynamic Spectrum Access

Networks, 2005. DySPAN 2005. 2005 First IEEE International Symposium on, pp. 328–337. IEEE (2005)
16. Cordeiro, C., Challapali, K., Birru, D., Shankar, N.S., et al.: IEEE 802.22: an introduction to the first wireless standard based on cognitive radios. J. Commun. **1**(1), 38–47 (2006)
17. Cordeiro, C., Challapali, K., Ghosh, M.: Cognitive phy and mac layers for dynamic spectrum access and sharing of tv bands. In: Proceedings of the First International Workshop on Technology and Policy for Accessing Spectrum, p. 3, ACM (2006)
18. Gardellin, V., Das, S.K., Lenzini, L.: Self-coexistence in cellular cognitive radio networks based on the IEEE 802.22 standard. IEEE Wirel. Commun. **20**(2), 52–59 (2013)
19. Grønsund, P., Pawełczak, P., Park, J., Cabric, D.: "Sensing of wireless microphones in IEEE 802.22: A system level performance evaluation" Communications (ICC), 2013 IEEE International Conference on, IEEE (2013)
20. Hu, W., Gerla, M., Vlantis, G.A., Pottie, G.J.: Efficient, flexible, and scalable inter-network spectrum sharing and communications in cognitive ieee 802.22 networks. In: Cognitive Radio and Advanced Spectrum Management, 2008. CogART 2008. First International Workshop on. pp. 1–5. IEEE (2008)
21. Hwang, S.H., Um, J.S., Song, M.S., Kim, C.J., Park, H.R., Kim, Y.H.: Design and verification of ieee 802.22 wran physical layer. In: Cognitive Radio Oriented Wireless Networks and Communications, 2008. CrownCom 2008. 3rd International Conference on, pp. 1–6. IEEE (2008)
22. Kim, H., Shin, K.G.: In-band spectrum sensing in cognitive radio networks: energy detection or feature detection? In: Proceedings of the 14th ACM International Conference on Mobile Computing and Networking, pp. 14–25. ACM (2008)
23. Ko, G., Franklin, A.A., You, S.J., Pak, J.S., Song, M.S., Kim, C.J.: Channel management in IEEE 802.22 WRAN systems. IEEE Commun. Mag. **48**(9), 88–94 (2010)
24. Lei, Z., Chin, F., Wu, Y.: Beacon design and detection windows. In: IEEE International Conference on Communication Systems (ICCS), pp. 590–594. (2010)
25. Mahmoud, H., Yucek, T., Arslan, H.: OFDM for cognitive radio: merits and challenges. IEEE Wirel. Commun. **16**(2), 6–15 (2009)
26. Mitola III, J., Maguire Jr, G.Q.: Cognitive radio: making software radios more personal. IEEE Pers. Commun. **6**(4), 13–18 (1999)
27. Mody, A., Chouinard, G.: IEEE 802.22 wireless regional area networks: enabling rural broadband wireless access using cognitive radio technology (doc.: Ieee 802.22-10/0073r03). Tech. rep., June 2010 [Online]. http://www.ieee802.org/22
28. Mody, A., Reddy, R., Sherman, M., Kiernan, T., Shyy, D.: Security and the protocol reference model enhancements in ieee 802.22. IEEE doc pp. 802–22.
29. PHY, A.: Mac proposal for IEEE 802.22 WRAN systems part2: the cognitive mac (2006)
30. Shellhammer, S.J., Sadek, A.K., Zhang, W.: Technical challenges for cognitive radio in the tv white space spectrum. In: Information Theory and Applications Workshop, 2009. pp. 323–333. IEEE (2009)
31. Shellhammer, S.J., Tandra, R., Tomcik, J., et al.: Performance of power detector sensors of dtv signals in IEEE 802.22 wrans. In: Proceedings of the First International Workshop on Technology and Policy for Accessing Spectrum, p. 4. ACM (2006)
32. Shellhammer, S.J., et al.: Spectrum sensing in IEEE 802.22. IAPR Wksp. Cognitive Info. Processing pp. 9–10 (2008)
33. Stevenson, C., Chouinard, G., Lei, Z., Hu, W., Shellhammer, S., Caldwell, W.: IEEE 802.22: the first cognitive radio wireless regional area network standard. IEEE Commun. Mag. **47**(1), 130–138 (2009)
34. Sum, C.S., Villardi, G.P., Rahman, M.A., Baykas, T., Tran, H.N., Lan, Z., Sun, C., Alemseged, Y., Wang, J., Song, C., et al.: Cognitive communication in tv white spaces: An overview of regulations, standards, and technology [accepted from open call]. Communications Magazine, IEEE **51**(7), (2013)
35. Wyglinski, A.M., Nekovee, M., Hou, T. (eds.): Cognitive radio communications and networks: principles and practice. Academic Press, New York (2009)

36. Yu-chun, W., Haiguang, W., Zhang, P.: Protection of wireless microphones in ieee 802.22 cognitive radio network. In: IEEE InternationalConference on Communications Workshops, pp. 1–5. (2009)
37. Yucek, T., Arslan, H.: A survey of spectrum sensing algorithms for cognitive radio applications. IEEE Commun. Surv. Tutorials **11**(1), 116–130 (2009)

Chapter 2
A Review of Hardware Platforms for Whitespace Communication

Ojonav Hazarika and Amit Kumar Mishra

Abstract Designing a hardware platform for white space Technology is a complex task. The IEEE 802.22 standard for white space communications imposes tight specifications that require powerful CPUs and FPGAs as well as compliance with strict RF regulations. In addition, a commercial product needs to be produced at a low cost. In this chapter, we examine the specifications that directly impact the design of white space communications hardware. We then explore the various hardware platforms that are commercially available and identify their strengths and weaknesses. The hardware platforms that will be discussed here are the USRP N210, the Nuand BladeRF, Zepto SDR and the HackRF board. These platforms are costly considering the component cost of the hardware supplied and have been designed with different goals and hence don't meet the full white space communication requirements. However, these boards can be used as a yardstick for the design of a white space hardware and useful inferences can be drawn by observing their merits and de-merits.

The IEEE 802.22 specification for TV white space communications imposes tight specifications [1]. The 802.22 standard is based on OFDMA modulation for downstream and upstream links with the ability to channel bond adjacent TV channels should more bandwidth be required. The standard supports modulations schemes up to 64QAM, producing a data rate of 19 Mbps per 6MHz TV channel. The standard also makes use of a geo-location spectrum database or an Eigen Value method of Spectrum Sensing to detect other primary users.

These specifications require a platform with a powerful CPU and FPGA as well as compliance with strict RF power and spectral mask regulations. At the same time, a commercial product has to be low-cost. In this Chapter, we first examine the specifications that directly impact the design of white space communication hardware. We then explore the various hardware platforms that are commercially

O. Hazarika · A.K. Mishra (✉)
University of Cape Town, Cape Town, South Africa
e-mail: akmishra@ieee.org

A.K. Mishra and D.L. Johnson (eds.), *White Space Communication*,
Signals and Communication Technology, DOI 10.1007/978-3-319-08747-4_2

Table 2.1 Frequencies of operation: each of these channels have a bandwidth of 6 MHz

Device type	Frequency bands (MHz)
All TVBDs	512–608
	614–698
Only for communication between fixed TVBDs	54–60
	76–88
	174–216
	470–512

In total, there are 47 available channels of operation

available and identify their strengths and weaknesses. The hardware platforms that will be discussed here are the USRP N210, the Nuand BladeRF, Zepto SDR and the HackRF board [2–5]. These platforms are costly considering the component cost of the hardware supplied and have been designed with different goals and hence don't meet the full white space communication requirements. However, these boards can be used as a yardstick for the design of a white space hardware and useful inferences can be drawn by observing their merits and de-merits

2.1 Key Specifications

In this section, we shall briefly discuss the impact of some of the key Whitespace Communications specifications on the hardware design aspects. These specifications have been noted from the FCC and the IEEE 802.22 documents [1, 6].

2.1.1 Bandwidth of Operation

The first specification is the bandwidth of operation. As per the Whitespace regulations defined by FCC, the allowed channels have been shown in Table 2.1, [6]. The allowed frequency bands are part of the VHF and the UHF spectrum. This specification directly impacts the choice of the Radio Frequency hardware of the system. The choice of the RF hardware shall not only decide the operational bandwidth of the system but it will also play a part in deciding the price of the system.

2.1.2 Transmit Power

Another specification emerging from the FCC regulations is the maximum transmit power. It says that maximum Equivalent Isotropically Radiated Power (EIRP) of the system shall be less than 4 W (36 dBm). The choice of transmit power class of the system as shown in Table 2.2 shall decide the choice of the high power amplifier. General trend is that higher the power rating of the amplifier, higher is the price of the amplifying device.

Table 2.2 In-Band power: these specifications define the maximum limits of In-Band power levels and spectral densities radiated by the different classes of TVBDs

Device type	Typical conducted power output (dBm)	Typical antenna gain (dBi)	Absolute maximum EIRP (dBm)	Max power spectral density, PSD (in 100 KHz band) (in dBm EIRP)	Power control
Fixed-devices	30	6	36	12.6	Required
Personal/Portable device (operating @ adjacent channel to TV channels)	20	0	20	−1.4	Required
Sensing-only devices	–	–	–	−0.4	–
All other personal/portable devices	20	0	20	2.6	Required

2.1.3 Computational Specifications

The IEEE 802.22 WRAN architecture defines many computationally intensive tasks. These tasks require powerful processing units as well as onboard memory elements. Some of these tasks are as follows:

- OFDM and OFDMA: The specification defines 2,048 point FFTs for the transmit OFDM and receive OFDMA which are highly computationally intensive. The OFDM/OFDMA specifications are shown in Table 2.3.
- Spectrum Sensing: The spectrum sensing sensitivity specifications defined by the FCC are shown in Table 2.4. In order to sense the spectrum at such sensitivities, the system requires to implement efficient spectrum sensing algorithms. The spectrum sensing algorithms like the Eigen Value based sensing necessitate generation of huge covariance matrices and fast iterative calculations. Such tasks require sufficient memory as well as computational elements.
- Control and Management: Apart from the above resource intensive tasks, the control and the management plane defined in the Protocol Reference Model (PRM) of the IEEE 802.22 also requires storing and running of many routines at the same time.

Hence, the Hardware designer needs to ensure that necessary steps have been taken to provide enough memory and computational resources to perform the above mentioned tasks. The designer also needs to optimize the design of the radio frequency

Table 2.3 OFDM requirements

Property of OFDM	Value	Remark
Total number of sub-carriers (N_FFT)	2048	–
Number of guard sub-carriers (N_G)	368	(184,1,183)
Number of used sub-carriers (Nt = Nd + Np)	1680	–
Number of data sub-carriers (Nd)	1440	–
Number of pilot sub-carriers (Np)	240	–
Data sub-carriers/channel Pilot sub-carriers/channel	24 4	Total of 28 subcarriers/channel
Total number of channels	60	–
Length of cyclic prefix	74.7	in usecs (to compensate for unequal channel fadings)
Total size of the guard bands	1.08	in MHz

Table 2.4 Interference avoidance, spectrum sensing: sensitivity specifications

Signal type	Sensitivity-threshold (dBm)	Integration bandwidth	Detetctor type
ATSC, digital TV	−114	6 MHz	Average
NTSC, analog TV	−114	200 kHz	Average
Wireless microphone	−114	100 kHz	Average

hardware subsystems. The design decisions shall impact both the price and performance of the system. After having discussed some of the key specifications (from hardware perspective), we shall now move on to discuss some of the currently available SDR platforms.

2.2 USRP N210

The USRP(Universal Software Radio Peripheral) N210 is an SDR platform from Ettus Research, Fig. 2.1. Its a complete system which includes digital and RF subsystems allowing users to use this piece of hardware for various applications. Packed with high speed FPGA, dual ADC's, DAC's and Ethernet connections, its very powerful for data streaming to and from host processors. The USRP also provides seamless integration with the GNU radio which makes it an convenient platform for rapid prototyping [2].

The notable feature are:

- Xilinx Spartan 3A-DSP 3400 FPGA: Comprises of 54 K logic cells.
- Interfaces: Gigabit ethernet, 2 Gbps expansion interface, RF interfaces with SMA connectors etc.
- ADC/DACs: The USRP comes with high speed dual 100 Msps 14-bit ADCs and dual 400 Msps 16-bit DAC.
- Software compatibility: GNU Radio, LAB VIEW and Simulink.

Fig. 2.1 USRP N210—Ettus research. Photo courtesy—https://www.ettus.com/product/details/UN210-KIT

- Other features: DC—6 GHz operaration bandwidth, fully coherent MIMO capability, 2.5 ppm TCXO reference.

2.2.1 Hardware Architecture

The Hardware Architecture of the USRP is shown in Fig. 2.2. As we can see, the Spartan 3A-DSP FPGA forms the core of the USRP N210 system. The control and management is handled by a softcore microblaze processor. At the backend, it connected to one ethernet PHY and at the front end, its connected to the high-speed dual ADCs and DACs.The FPGA is also connected to a MIMO expansion header. Two USRP have to be connected to form a 2×2 MIMO configuration.

The USRP without connecting to the RF daughter board consumes around 8 Watts of power. With the WBX daughter card designed to work with the USRP, it can transmit a maximum RF power of 15 dBm. The WBX has a receive noise figure of 5 dB [7].

2.2.2 Merits

The merits of the USRP N210 are as follows:

1. The board has standard connectors and standard interfaces which makes it a ready to use commercial platform.
2. The processing bandwidth is 100 MHz which is very useful for spectrum sensing kind of applications.
3. Since, it integrates well with GNU Radio, fast prototyping capability of this board is a great advantage [8].
4. It does not have unnecessary extra-peripherals and hence, very application specific.

2.2.3 De-Merits

The USRP N210 has many de-merits:

1. Spartan 3A-DSP is a low end FPGA.
2. By implementing a softcore processor like the micro blaze consumes a third of its space leaving out very less space for other blocks.
3. Absense of on-board RAMs make it almost impossible to perform much signal processing on the board.
4. Priced at about 1,800 USD, it is not a very cheap platform.

2.3 NUAND Blade RF Board

The NUAND Blade RF board is an open source USB 3.0 software defined radio (SDR) board, Fig. 2.3. It contains a micro processor, an FPGA for configurable logic and the LMS6002D RF transceiver [3]. It has SMA connectors which have to connect to an RF front end. This board is capable for MIMO operation. The platform runs Linux, Windows, Mac and has GNU Radio software support [8].

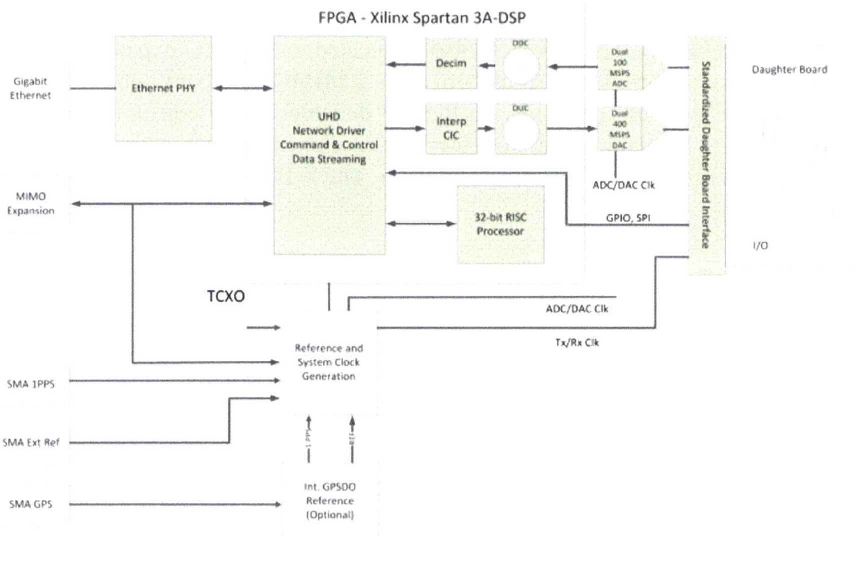

Fig. 2.2 USRP N210—hardware architecture. Photo courtesy—https://www.ettus.com/content/files/07495_Ettus_N200-210_DS_Flyer_HR_1.pdf

Fig. 2.3 NUAND—bladeRF board. Photo courtesy—http://nuand.com/

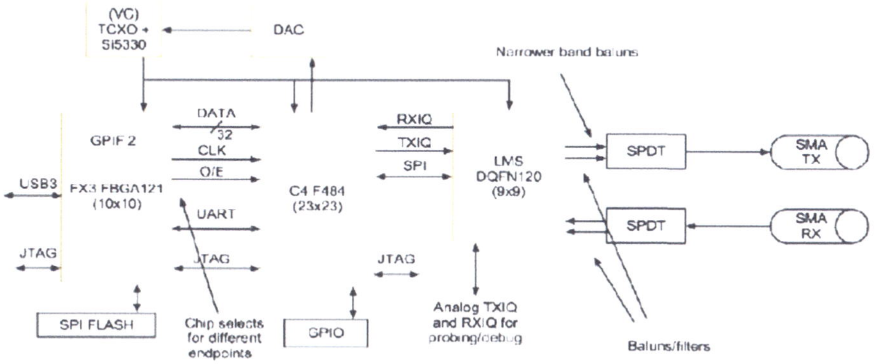

Fig. 2.4 NUAND—bladeRF hardware architecture. Photo courtesy—http://nuand.com/bladerf.pdf

2.3.1 Hardware Architecture

The bladeRF board has a processing core which is the ARM A-9 microprocessor, a programmable logic IC which is the Altera Cyclone-4 FPGA and the LMS6002D RF transceiver. The FPGA provides the interface between the ARM and the transceiver. Its RF section does not provide Fig. 2.4 any duplexing facilities. It just makes the transmit and the receive outputs available at two SMA connectors. This board can be powered by USB and has a 512 MB embedded SRAM. The transceiver is configured through the SPI interface from the Cyclone-4 FPGA [9]. The board comes with external JTAG interfaces for both the processor and the FPGA for the debug and configuration.

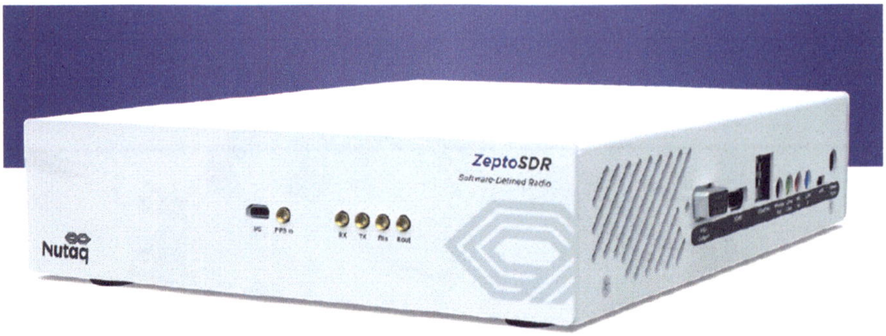

Fig. 2.5 Zepto SDR. Photo courtesy—http://www.nutaq.com/en/products/zeptosdr

2.3.2 Merits

The board has the following merits:

1. Priced at 650 USD, it is a cost effective board which has all elements for a radio frequency system functionality.
2. GNU Radio support [8].
3. With 115 KLE (K logic elements), it offers high amount of onboard programmable resources.
4. High Speed USB 3.0 functionality.
5. Small form factor of 5" by 3.5"

2.3.3 De-Merits

Inspite of the obvious merits, the NUAND bladeRF has the following de-merits:

1. Absence of ethernet functionality. In order to interface the board with packetized networks, an additional board has to be connected.
2. Absence of a duplexing system for transmit and receive ports.
3. Peak output power of 6 dBm is low.

2.4 Zepto SDR Board

The Zepto SDR board from Nutaq is a complete SDR product which houses two separate boards Fig. 2.5. One is the Zedboard from Digilent and the Radio420S board from Nutaq. The Zedboard and the Radio420S board are interfaced using the FMC connectors. While the Zedboard forms the digital processing core of the Zepto

SDR platform, the Radio420S board forms the air interface subsystem. Now, we shall discuss both the boards in greater detail [4].

2.4.1 ZedBoard

The ZedBoard is a very low cost Xilinx Zynq7000 development platform, Fig. 2.6 [10]. It is a very generic board which allows one to use it for a wide range applications. Due to its expandable features, the Zedboard is very convenient for rapid prototyping. The notable feature of the Zedboard are [11]:

- Xilinx XC7Z020-1CLG484CES Zynq-7000 AP SoC: It can be configured through QSPI flash, cascaded JTAG and SD card.
- Memory: 512 MB DDR3 and 126 Mb QSPI flash.
- Interfaces: USB JTAG, 10/100/1G Ethernet, USB OTG 2.0, SD Card, Digilent PMOD Headers, LPM FMC (FPGA Mezzanine card) header and so on.
- Display or Audio: The Zedboard comes with various display or audio connectors like HDMI, VGA, OLED display, audio line in, line out and microphone.

2.4.1.1 Hardware Architecture

The hardware architecture of the ZedBoard is shown in Fig. 2.7. Even though we shall discuss the Zynq-7000 SoC architecture in greater detail later, we would just like to mention that the SoC is partitioned into programmable logic (PL) and processing sysem (PS) elements. The processing sysem consists of the embedded dual core ARM Cortex A-9 processors with NEON floating point arithmatic logic units (ALUs). The programmable logic blocks have standard Xilinx FPGA elements like configurabe logic blocks (CLBs), DSP slices, RAM block and so on. There are also multiplexed IO pins which can be accessed either from the PS or the PL blocks. However, the MIOs refer to the PS pins and EMIO refer to the PL pins. As we can observe from Fig. 2.7, all the control, configuration and memory interfaces like USB UART, Ethernet PHYs, DDR3, SD card, clocks, resets etc are connected to the processing system or its MIOs. The application related peripherals like the FMC, general purpose IOs (GPIOs), audio and video interfaces are connected to the programmable logic [12].

The ZedBoard has a 10-Layer PCB stackup [13]. As we can see in Fig. 2.8, there are six signal layers. Three important aspects of the ZedBoard PCB design are noticeable from this stackup:

1. The two sets of internal signal layers are not standard striplines because each set has two signal layers placed adjacently to each other.
2. The internal signal layers are coupled to power planes as well instead of being coupled to just grounds.
3. Due to low layer count, each power supply does not have a single plane. Instead, planes have been split to accomodate for all the various power supplies.

As a result of these three design aspects, the cost of the ZedBoard PCB is very low.

2.4.1.2 Merits

Following are the merits of the ZedBoard:

1. Very Low Cost, less than 400 USD.
2. Very Generic and can cater to wide variety of applications due to its expandability and various features.
3. A convenient platform to learn and evaluate all capabilities of the Zynq7000 SoC.
4. Standard FMC interface for connection to daughter boards.

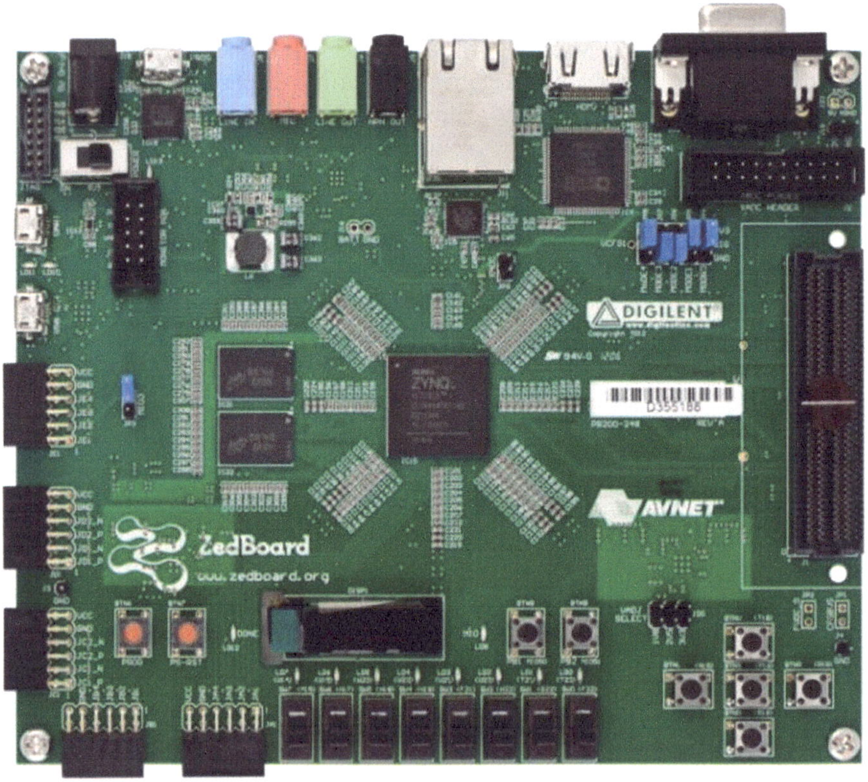

Fig. 2.6 Zedboard—hardware architecture. Photo courtesy—http://www.zedboard.org

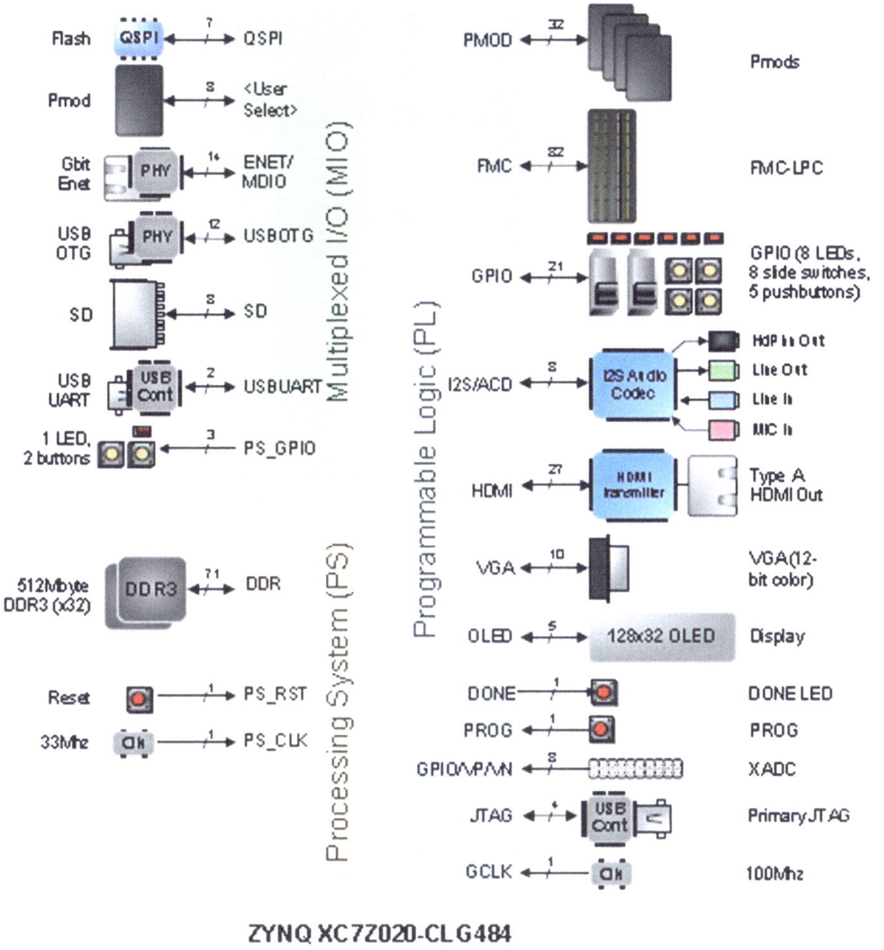

Fig. 2.7 Zedboard. Photo courtesy—ZedBoard (Zynq Evaluation and Development) hardware users guide

2.4.1.3 De-Merits

Following are the de-merits of the ZedBoard:

1. Not application specific as its meant to be an evaluation board.
2. Addional plugin boards have to be used if the final goal is to have a complete communication or radar system.
3. Due to many peripherals and connectors which consume a lot of space, the size of the board(5.3" by 6.3") is large when compared to more application specific modules.

Fig. 2.8 Zedboard—layer stackup. Photo courtesy—ZedBoard, power distribution and decoupling system

2.4.2 NUTAQ Radio420S Board

The Radio420S FPGA mezzanine card (FMC) is a software defined radio (SDR) RF transceiver module which uses the Lime Micro LMS6002D SoC, Fig. 2.9. It is a multi-mode module which supports time division duplex (TDD) as well as frequency division duplex (FDD) modes. The board has an operating frequency of 300 MHz–3 GHz with an instantaneous bandwidth of 1.5–28 MHz. The board has to be connected through an FMC connector to a digital motherboard for configuration and data transfer [14]. This board is aimed for communication applications like MIMO systems, cognitive radios, WiMAX, White Space, Wi-Fi, GSM, WCDMA and so on.

2.4.2.1 Hardware Architecture

The board has a simple hardware architecture. It contains the RF transceiver LMS6002D whose all the digital control and data line are connected to the low pin count (LPC) FMC connector. It has a selectable clock reference input. There are two RF output connectors, one for transmit and the other for receive. The outputs are connected through a set of baluns which convert the various differential transmit/receive outputs/inputs to 50 Ω single-ended outputs/inputs respectively [15].

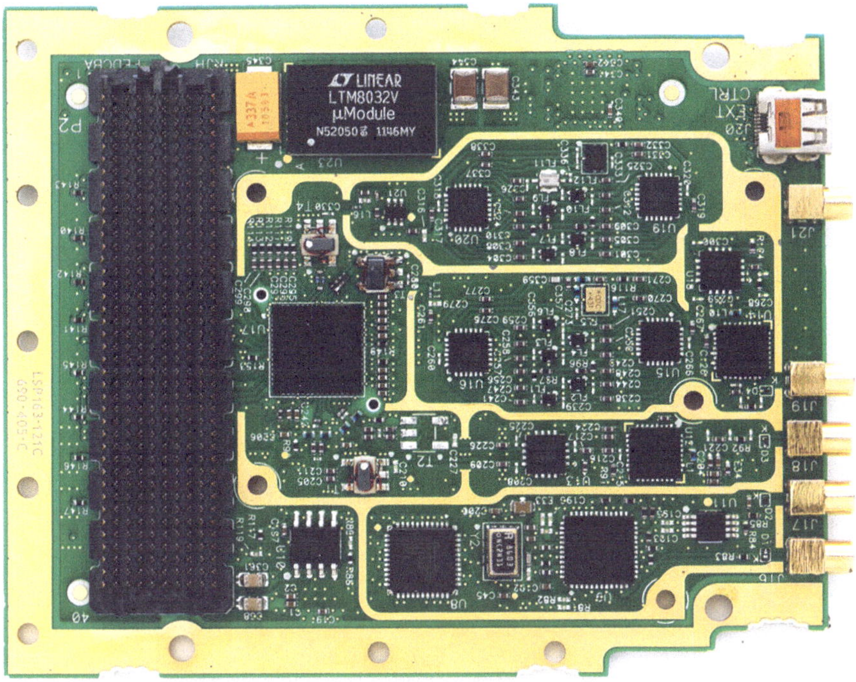

Fig. 2.9 NUTAQ—Radio420S. Photo courtesy—http://nutaq.com/en/products/view/+nutaq-radio420x

2.4.2.2 Merits

The merits of the NUTAQ Radio420X board are:

1. Simple design.
2. Utilizes all the features of the LMS6002D SoC.
3. Standard FMC connector for interfacing with motherboards.

2.4.2.3 De-Merits

1. Lacks proper frontend section. Even though its TDD and FDD compatible but implementation of those features would require a further RF frontend board.
2. Low output power for an RF module. The peak Output power is only 10 dBm.
3. The FMC connnector consumes a major portion of the board space.
4. Again like the Zedboard, its too generic and not a particular application or communication standard oriented board.

Fig. 2.10 HackRF SDR platform. Photo courtesy—http://www.rtl-sdr.com/hackrf-vs-bladerf-vs-usrp/

2.5 HackRF SDR Platform

The HackRF is a very new SDR platform platform which offers wideband performance at an extremely cheap cost, Fig. 2.10. The HackRF SDR platform consists of a CPLD device along with a wideband RF transceiver. This platform aims to provide widest band of operation at the cheapest price when compared to all other SDR platforms [5].

2.5.1 Hardware Architecture

The HackRF Hardware consists of a heterogeneous architecture with an ARM-based microcontroler and a CPLD which form the digital core of the platform. The microcontroler is the LPC43XX from NXP semiconductors [16]. It also contains a very wideband transceiver section capable of operating from 30 MHz to 6 GHz. The primary digital interface is the USB 2.0 and hence, has been designed primarily for use with a USB-attached host computer.

Table 2.5 Comparison

	USRP N210	Nuand blade RF	Zepto SDR	Hack RF
Operational bandwidth	DC—6 GHz	300 MHz–3 GHz	300 MHz–3 GHz	30 Mhz–6 GHz
Instantaneous bandwidth	100 MHz	28 MHz	28 MHz	20 MHz
Computational	Spartan-3A DSP FPGA	Altera Cyclone-4 FPGA	Zynq-7020	CPLD
Interfaces	USB 2.0	USB 3.0	2 Gbps Ethernet, HDMI, UART-USB, JTAG-USB, USB-OTG etc	USB 2.0
Sample size (ADC/DAC)	12 bit	12 bit	12 bit	8 bit
Price	High	Low	Highest	Lowest

2.5.2 Merits

The various positive points of the HackRF board are:

1. At 300 USD it aims to be the lowest priced SDR platform.
2. Very wide bandwidth of operation of 30 MHz–6 GHz.
3. Instantaneous bandwidth of 20 MHz.
4. The HackRF integrates with the GNU Radio framework.
5. The HackRF designs have been completely open-sourced.

2.5.3 De-Merits

Inspite of the many positives, it has many drawbacks. They are:

1. Absense of Ethernet functionality which is an integral part of a communication system.
2. For implementation of baseband elements, the HackRF only has a CPLD whose capabilities are very limited.
3. The digital IOs of the ADC/DACs are only 8-bit wide which means that the dynamic range of the platform would be low.
4. There are no onboard memory elements apart from the CPLD.
5. The development of this platform is still ongoing and hence, is not a matured product.

2.6 Chapter Summary

In this chapter, we discussed some implications of the Whitespace Technology specifications on the hardware design. Then, we identified and discussed four currently available SDR platforms. While the HackRF and the BladeRF are very low cost platforms, they are still in their development phases and hence, the products are not very matured yet. Also, they are not computationally very powerful and do not have much onboard memory elements. However, the USRP N210 and the Zepto SDR are commercial platforms and hence, very expensive. At 2295 USD, the Zepto SDR is a very powerful platform computationally. It incorporates the latest Zynq device from Xilinx and is packed with loads of other features along with 2 Gbps Ethernet facility. The USRP N210, on the other hand with a low-end FPGA and limited feature set is very expensive at 1,800 USD. Its high price is due to the fact that USRP series platforms are among the oldest commercially available SDR platforms leading to their immense popularity. It also has a very wideband performance when compared to the Zepto SDR and the Blade RF. However, from a Whitespace Technology perspective, this feature is not required as the operational bands are in the VHF and the UHF regions of the spectrum. To sum up, even though none of these platforms are optimized for Whitespace Communications, they can be used for prototyping and testing various functionalities while the hardware is being developed. The key features of the SDR platforms that we have discussed above have been summarized in Table 2.5.

References

1. IEEE standard for information technology: Telecommunications and information exchange between systems wireless regional area networks (WRAN): Specific requirements, 2011.
2. USRP networked series. https://www.ettus.com/product/category/USRP_Networked_Series
3. bladerf - the USB 3.0 superspeed software defined radio. http://nuand.com/
4. Zepto SDR.
5. HackRF, an open source SDR platform.
6. E. FCC, "Docket no. 08–260", Second Report and Order and Memorandum Opinion and Order, Nov, 2008.
7. USRP N200/N210 NETWORKED SERIES. https://www.ettus.com/content/files/07495_Ettus_N200-210_DS_Flyer_HR_1.pdf
8. GNU Radio. http://gnuradio.org/redmine/projects/gnuradio/wiki
9. bladeRF—Software Defined Radio. http://nuand.com/bladerf.pdf
10. Zedboard.org. http://www.zedboard.org/
11. Zynq-7000 EPP (XC7Z010 and XC7Z020) Data Sheet:DC and Switching Characteristics. http://www.xilinx.com/support/documentation/data_sheets/ds187-XC7Z010-XC7Z020-Data-Sheet.pdf
12. ZedBoard (Zynq Evaluation and Development) Hardware Users Guide. http://www.zedboard.org/sites/default/files/documentations/ZedBoard_HW_UG_v1_9.pdf
13. ZedBoard Power Distribution and Decoupling System.
14. Nutaq Radio420X. http://nutaq.com/en/products/view/+nutaq-radio420x
15. LMS6002 Multi-band multi-standard transceiver: ProductBrief. http://www.limemicro.com/download/Lime/ProductBrief.pdf
16. LPC4350/30/20/10—NXP Semiconductors.

Chapter 3
A Study on White and Gray Spaces in India

Kalpana Naidu, Yuva Kumar, B.M. Baveja, Rakesh Naik, Boli Sridhar, Shyam Ponnappa, Mohammed Zafar Ali Khan, S.N. Merchant and U.B. Desai

Abstract TV 'White Space' spectrum is characterized by large amounts of geographically available spectrum with excellent propagation characteristics, offering long range and exceptional building penetration compared to the spectrum used by WiFi and Bluetooth equipment. As a result TV white space spectrum is ideal for providing fixed broadband internet services to locations where the routing of cables or optical fibre is neither practical nor economical; and hence important from a developing world perspective. The prospects can be further broadend by considering 'gray space' spectrum which is characterized by spectrum that is not fully available in time or space. The amount of TV white space in the VHF and UHF TV band in South India from an analytical point of view is studied. The results are corroborated by extensive experimental data collected from various places in South India. We then present quantitative assessment and estimates for the gray spaces in GSM900 and GSM1800 bands for Hyderabad City. These results are indicative and allow for a whole lot of possibilities. We then describe a unique 'gray space' GSM voice call test bed setup at IIT Hyderabad. However, white space and gray space sharing have different policy and governance requirements. We conclude the chapter by discussing possible policy and governance structures from an Indian point of view.

Keywords White space · Gray space · Cognitive radio

These results were submitted to Deity, MCIT, GOI on 7th March 2011 and oral presentations were given at workshops, CORD 2012, 25th February 2012, at IIT Hyderabad and White Space Opportunities in India, 21st June 2013, Delhi.

K. Naidu · Y. Kumar · B.M. Baveja · R. Naik · B. Sridhar · S. Ponnappa ·
M. Zafar Ali Khan (✉) · S.N. Merchant · U.B. Desai
Department of Electrical Engineering, IIT Hyderabad, Hyderabad 502205, Andhra Pradesh, India
e-mail: zafar@iith.ac.in

S.N. Merchant
Department of Electrical Engineering, IIT Bombay, Bombay, India

A.K. Mishra and D.L. Johnson (eds.), *White Space Communication*,
Signals and Communication Technology, DOI 10.1007/978-3-319-08747-4_3

3.1 Introduction

Radio spectrum is a premium natural resource that has significant impact on wireless technology and social progress. In India, with 873.36 million mobile subscribers (which is much greater than the original predictions) [1], spectrum congestion and scarcity is a critical issue.

The ongoing switchover from analog terrestrial television (ATT) to digital terrestrial television (DTT) broadcasting will provide significant benefits to solve the spectrum congestion problem.

This is because while ATT broadcasting can deliver only a single TV program on each 8 MHz, to avoid interference ATT requires large geographic separation between transmitters that operate on the same frequency. These unused TV channels, designed for protecting adjacent TV transmissions from interference at a given location, are commonly referred to as TV White Space (TVWS) [2].

In contrast, DTT enables multiple TV programs to be carried on a single frequency channel which greatly increases spectrum utilization. Therefore, it is natural to expect that some TV spectrum will be released for new usage after the digital switchover popularly named as "Digital Dividend (DD)".

However, in India, analog transmission will continue along with digital transmission with no decided cutoff dates for switching off analog transmission. Additionally, India has limited bandwidth available in other internationally harmonised mobile bands such as the 900, 1,800 and 2,100 MHz bands. Therefore DD spectrum is attractive as fixed broadband services are limited in rural areas and DD spectrum might offer unprecedented broadband access capability for rural populations.

DD capacity has been analyzed in several developed countries [3–6]. Importantly, it has been shown that different propagation models significantly affect the estimated availability of the spectrum [6]. So an accurate local propagation model is important before making conclusions on spectrum availability.

Based on our spectrum measurement campaign, we have proposed a new path loss model for India [7, 8], that drastically reduces the mean squared error. Based on this model we calculate the TVWS in South India. Seperate plots for UHF and VHF are provided. Note that results from previous research on DD capacity estimations in other countries cannot be directly applied to India as the ATT and DTT coexistence problem is not considered (recall that India has no declared cutoff date for complete digital shift).

We then present our observation campaign for the GSM900 and GSM1800 bands and discuss the "gray spaces" available in both these bands. These results are then tallied with international observation campaigns in GSM band. Based on these we present results from testbed for delay constrained applications (voice) in the cellular band 'gray spaces'.

We then discuss the results (in the Indian scenario) from a regulatory point of view for both 'white spaces' and 'gray spaces'.

Table 3.1 Doordarshan TV transmitter classification

Band (Mhz)	No. of transmitters	Bandwidth (MHz)
VHF Band–I (54–68)	8	7
VHF Band-III (174–230)	1034	8
UHF Band-IV (470–590)	373	8

3.2 Frequency Allocation in India

Wireless licenses are allotted, in India, by the Wireless Planning and Coordination (WPC) wing of the Ministry of Communications and Information Technology and an overview is provided by the draft national frequency allocation plan (NFAP) published by the WPC [9]. However, detailed information is not publicly available. Significantly, the draft NFAP 2011 made a very encouraging statement of intent for the utilization of TVWS [9]. Also the following points from NFAP 2011 are of interest

- (Analog) TV broadcast transmitters operate only in the 470–590 MHz band in the UHF band
- Spectrum in the 470–890 MHz frequency band is earmarked for Fixed, Mobile and Broadcasting Services
- Digital broadcasting services including (digital TV broadcasting) will operate in the 585–698 MHz
- On completion of digitization, the 470–590 MHz in the UHF band, 54–68 MHz and 174– 230 MHz in VHF band; we presume, will be completely free. The important point is the actual date of complete digitization of terrestrial TV.

With regard to terrestrial TV, Doordarshan, a public service broadcaster, has exclusive use of terrestrial TV bands. Doordarshan today has more than 1415 terrestrial transmitters throughout India and transmits on two all-India channels DD National and DD News. The frequency band information of the various transmitters are given in Table 3.1. As a part of the digitization process, Doordarshan has four DTT units at present; one each in Delhi, Mumbai, Kolkata and Chennai, and plans to completely digitize its terrestrial transmission by end of 2017. From the tenders released by Doordarshan [10], it looks like India will have a total of 19 DTT 6 KW transmitters by 2014–2015. However, no finite time frame has been decided for stopping the analog transmission and it will continue along with the digital transmission till a substantial percentage of the population switches over to a digital mode of reception.

3.3 TV White Space Maps for South India

In this section we present TV white space Maps for South India. These maps are developed using the code provided by [11, 12] with some modifications (for allowing the propagation model of [7, 8]). Note that we need a more 'precise' definition of white space.

Following [13] 'white space' is the intersection of the spatial holes from 'pollution' and 'protection' viewpoints. The classical communication theory viewpoint or the 'pollution viewpoint takes into consideration the fact that even though a region is in use by a secondary device, the interference at the secondary receiver might be higher than the tolerable interference level. The 'protection' viewpoint stipulates that a secondary radio can only operate in locations where it cannot generate harmful interference to the primary system. Both viewpoints exclude regions around a transmitter and the actual size of the 'white space' is the intersection of the spatial holes from these two viewpoints.

As India does not have TV white space regulations, we use the regulations of the FCC (US). Also, as data on microphones is not available they are ignored. White Space Maps are very sensitive to propagation Models. Accordingly, we study the white spaces in India based on the the Hata Model. It has been shown in [7, 8] that the adaptive Pervez-Vega Model has significantly lower mean squared error as compared to the Hata Model and hence the 'White Space' maps obtained using this model are likely to be more reliable. Using the TV transmitter information provided by Doordarshan for South India,[1] we quantify TV white space in the UHF and VHF bands separately under the technical specifications given by the FCC for analog terrestrial TV in Sect. 3.3.1 and for Digital terrestrial TV in Sect. 3.3.2.

3.3.1 Analog TV White Space

From the technical details of Doodarshan terrestrial transmission centres, i.e., location, transmission power, range, frequency (VHF and UHF), we plotted the white space for particular region. VHF Band has a total of 9 channels (2 channels of 7 MHz in VHF I and 7 channels of 8 MHz in VHF III for a total bandwidth of 70 MHz) and UHF Band has a total of 15 channels for Analog TV (with a total bandwidth of 120 MHz).

- Figure 3.1 gives the white space availability using the FCC regulations in the VHF band. The lighter colour regions have more number of channels available, and dark (yellow) colour regions have less number of channels available. Observe from the figure that about 5–6 VHF channels (40–48 MHz) are available in about a third of South India with small areas of poor availability around cities like Hyderabad, Chennai, Pune and Mumbai.
- At least 2 channels are available in all of south India and leaving the areas of poor availability 5–6 channels are available in most of South India. This is in contrast with observations from studies in other countries (both developed and developing) where almost no space is available in the VHF band. India has on the average 40 MHz of TVWS in the VHF band.
- Figure 3.2 gives the white space availability using the FCC regulations in the UHF band. Again, the lighter colour regions have more number of channels available,

[1] This data is not publicly available in India and we could only obtain data for South India.

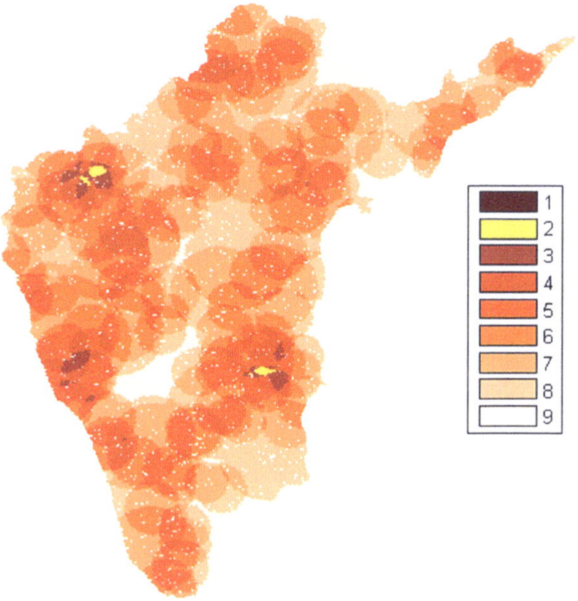

Fig. 3.1 White space plot for South India for 9 channels of VHF band

and dark (yellow) colour regions have less number of channels available. Observe from the figure that all the UHF channels (120 MHz) are available in most of South India with about four areas of lesser availability in Tamil Nadu and Kerala.
- Significantly, at least 9 channels (72 MHz) are available in all of South India.
- Overall, there is at least 86 MHz of TV white space available in South India. The average availability is much higher being around 140 MHz. Finally note that as digitization spreads the available white space will improve.

3.3.2 'Digital' White Space

Since India has a separate band allocated for DTT, we plot

- The white space available at present for the four installed DTT transmitters,
- The white space that will be available when 19[2] DTT transmitters will be installed (by 2017),
- The white space that will be available when the digitization is completed.

These plots give us an idea of how much white space is available in the DTT band as the digitization proceeds. Note that the proposed specifications for the 19 proposed

[2] This number has been obtained from the tenders floated by Prasar Bharati [10].

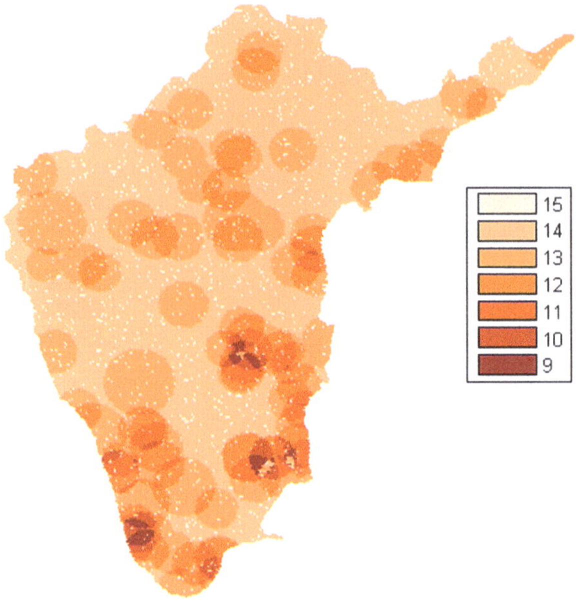

Fig. 3.2 White space plot for South India for 15 channels of UHF band

DTT transmitters is available [10] and we have plotted the 'Digital' White Space for the whole country in the first two plots and; for south India in the last plot. The digital plots are made assuming that the transmitters are FCC compliant for DVB-T2. Accordingly, in the code of [12], the minimum equivalent field strength at receiving place is set to 51 dB mu/m for minimum C/N = 20 dB. Observe from Fig. 3.3 that except regions around the four metros (Delhi, Mumbai, Kolkatta and Chennai) the entire country is a white space for all the 'digital' channels. Figure 3.4 gives the white space that will be available when all the additional 19 DTT transmitters are installed. Note again the entire country is a white space for all the 'digital' channels except for small regions around the 19 DTT transmitters. Also note that 11 channels will be available throughout India in the 'Digital' band. Figure 3.5 gives the white space, for South India, that will be available after completion of digitization. We have assumed that all the (present) 275 locations in South India have been converted to DTT with same antenna height but with a power of 6 KW. Channel allocation was done so as to minimize adjacent channel and co-channel interference and sorting latitude. Channel allocation starts with 21, next location will be allocated channel $max(x1, x2, \ldots, xn)+2$ if it is in same range of $x_1, x_2, x_3, \ldots, x_n$ channels. Observe that 6 channels will be available throughout India in the 'Digital' band.

In making of these charts we have assumed that except the power and sensitivity, all other parameters are the same as of the present 'analog' transmitters.

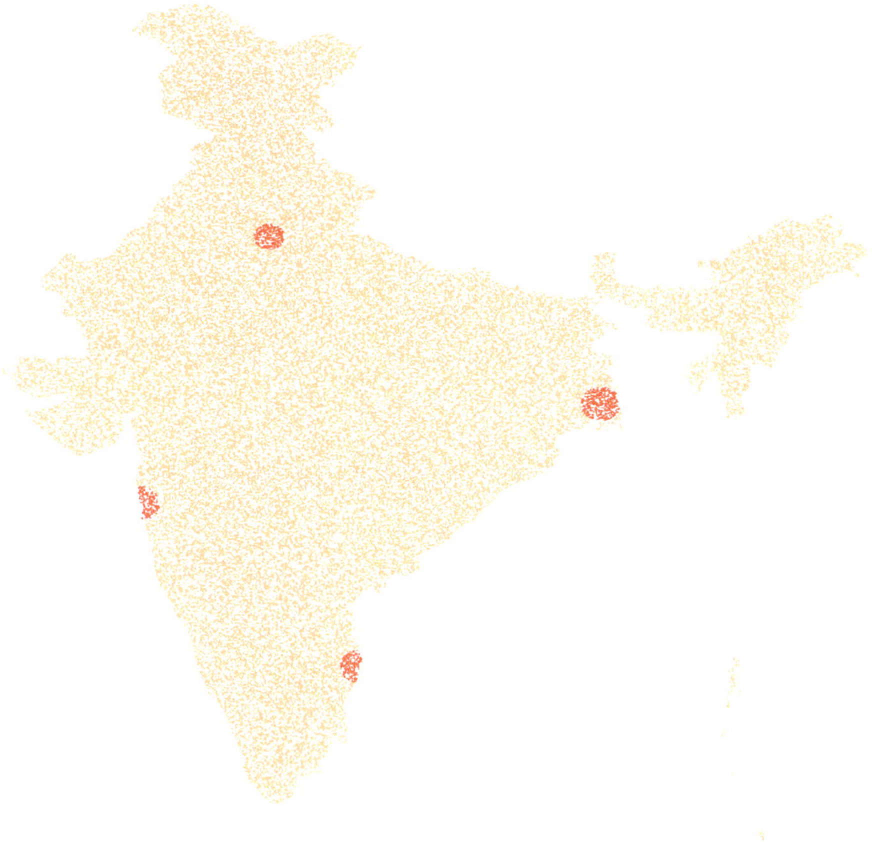

Fig. 3.3 Digital white space plot for India for 15 channels of UHF V band with present DTT transmitters

3.4 TV White Space Experimental Campaign

This section describes the steps followed during data collection and the description of the equipment used for White Space Experimental Campaign. The data collection tool used was an Anritsu Spectrum Analyzer MS2713E global positioning system (GPS) receiver set. Various applications of spectrum analyzer MS2713E include Spectrum Monitoring, Transmitter Verification, Interference Analysis of cellular system, Land Mobile Radio Systems, Wi-Fi Signal Strength Mapping etc. Here we describe measurement campaigns in two cities, Hyderabad and Chennai (for details of complete measurement campaign in Andhra Pradesh, please refer to [7]).

Power levels of Doordarshan (DoD) TV Transmitter in Hyderabad at different distances from the transmitter were measured using Anritsu's spectrum analyzer MS2713E. The latitude and longitude of Hyderabad is 17.2203100 N and

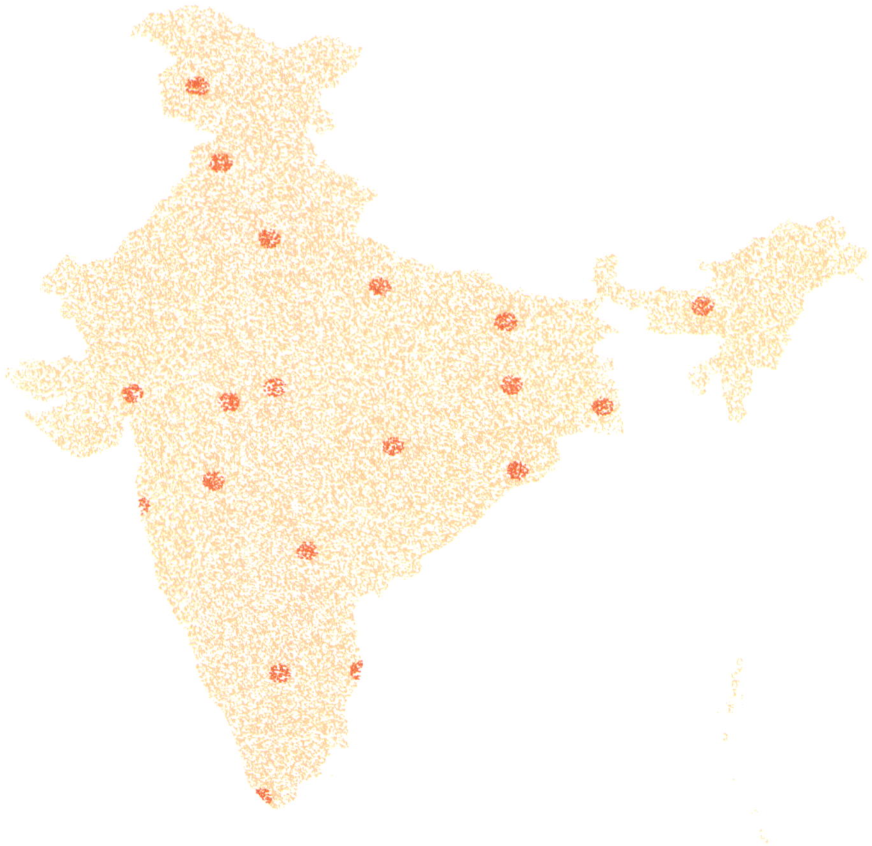

Fig. 3.4 Digital white space plot for India for 15 channels of UHF V band with 19 DTT transmitters

78.802700 E. The city is nestled on the Deccan Plateau and is positioned at a height of around 500 m from the sea level. Hyderabad is the capital and largest city of the southern Indian state of Andhra Pradesh. Occupying 650 km^2 (250 m^2) on the banks of the Musi River, it had a population of 6.8 million in 2011 with a metropolitan population of 7.75 million, making it India's fourth most populous city and sixth most populous urban agglomeration. Three routes were covered during the measurement campaign in Hyderabad city. Details of routes covered in Hyderabad city are given in Fig. 3.6 where the red, blue and green markers give the exact location of measurement, where the car was stopped. DoD National TV in Hyderabad transmits on channel 4-band 1 at 62.25 MHz frequency. DoD News TV in Hyderabad transmits on channel 12-band 3 at 224.25 MHz frequency. While transmission is taking place, anritsu spectrum analyzer was placed inside a car and driven along the mentioned routes. Field strength was measured continuously (every 5 s) and stored in an external pen drive for subsequent analysis.

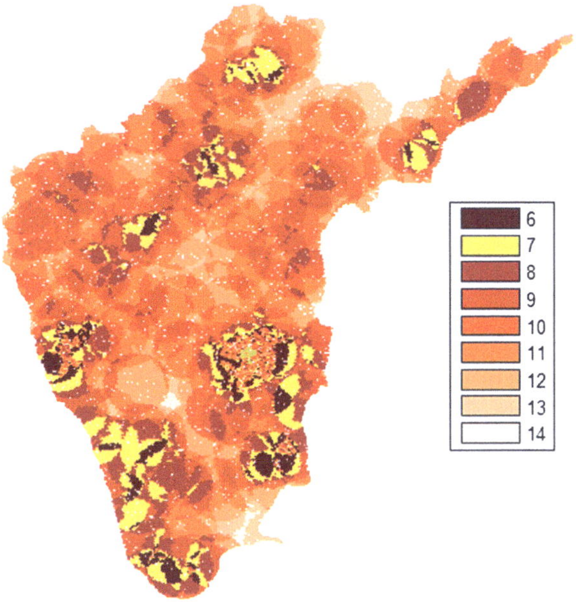

Fig. 3.5 Digital white space plot for South India for 15 channels of UHF V band after completion of digitization

Power levels of Doordarshan (DoD) TV Transmitter in Chennai at different distances from the transmitter were measured using anritsu spectrum analyzer MS2713 E. The latitude and longitude of Chennai city can be projected as 13.50000 N and 80.170000 E respectively. Chennai is the capital city of the Indian state of Tamil Nadu. Located on the Coromandel Coast on the Bay of Bengal, it is a major commercial, cultural, economic and educational center in South India. Five routes were covered during the measurement campaign in Chennai city. Details of routes covered in Chennai city are given in Fig. 3.7. DoD National TV in Chennai transmits on channel 5-band 3 at 175.23 MHz. DoD News TV in Chennai transmits on channel 7-band 3 at 189.26 MHz. DoD Podhigai TV in Chennai transmits on channel 11-band 3 at 217.25 MHz. While transmission is taking place, Anritsu's spectrum analyzer was placed inside a car and driven along the mentioned routes in Fig. 3.7. Field strength was measured continuously and stored in an external pen drive for subsequent analysis. From these received power levels, path loss was calculated. The sample measurement plots are shown in Figs. 3.8 and 3.9 for Hyderabad and Chennai respectively.From these measured pathloss exponent values it is found that the adaptive path loss model of Perez-Vega Zamanillo [7] has least average Root Mean Square Error (RMSE) of 11.12 dB which is least among the known path loss models for India. This model specifies the path loss as

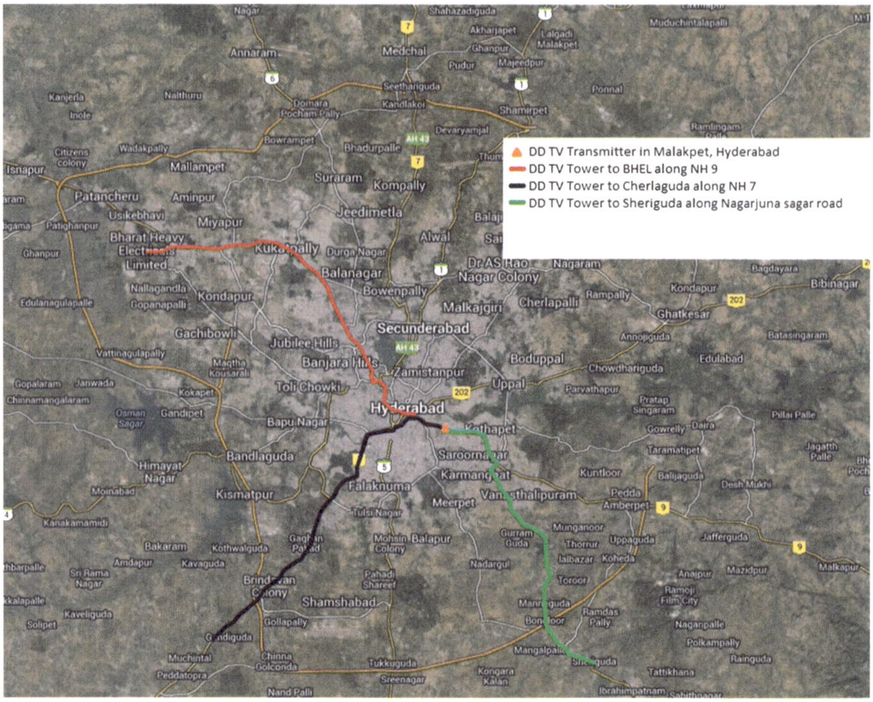

Fig. 3.6 Measurement routes in Hyderabad City

$$Y = a_0 + \sum_{k=1}^{26} a_k X_k \tag{3.1}$$

where a_k's are constants determined in [7] and given in Table 3.2, $X_1 = \log(f)$, $X_i = h^m d^n \log(d), i = 2, \ldots, 26$ and where $m = \left\lfloor \frac{k-2}{5} \right\rfloor$, $n = k - 1 \bmod 5$, h is the height of the transmitter in feet, f is the frequency of operation and d is distance between TX and RX in miles. To analyze the frequency usage, graphs are plotted between time and frequency.

The sensing results for IIT Hyderabad show that as expected 2 channels can be observed in the UHF band and 5 channels can be observed in the VHF band, corroborating the results of theoretical analysis.

3.5 Gray Spaces in GSM900 and GSM1800 in Hyderabad

In this section we present spectrum sensing results for the GSM band for both uplink and downlink in Hyderabad city for the two GSM bands used in India, GSM900

Fig. 3.7 Measurement routes in Chennai City

and GSM1800. We need a formal definition of gray space and accordingly define a 'gray' space as a white space for a limited time duration.

We have collected data in the different places of Hyderabad city continuously for three day periods using the afore-mentioned spectrum analyzer and plotted the 'gray' space using Anirtsu's Master Software Tools and MATLAB. Additionally, we have taken the GSM spectrum samples for every 500 m in Hyderabad city as shown in the Fig. 3.10 and estimated the GSM gray space using space-time interpolation. Fig. 3.11 shows the results of spectrum sensing at Tolichowki over a three-day period. The x-axis gives the frequency, y-axis the time and the color code gives the power levels of the signals. If the received power is less than −110 dBm then it is indicated by black, and by other colors if the received power is greater than −110 dBm. Fig. 3.12 gives the 3D plot of the same. Observe that both the plots indicate that there is significant amount of 'gray space' at Tolichowki, a prime location at Hyderabad.

Figure 3.13 gives a 3D plot for Gachibowli, another prime location. To formally analyze the gray space availability, we plot time percentage maps for each GSM

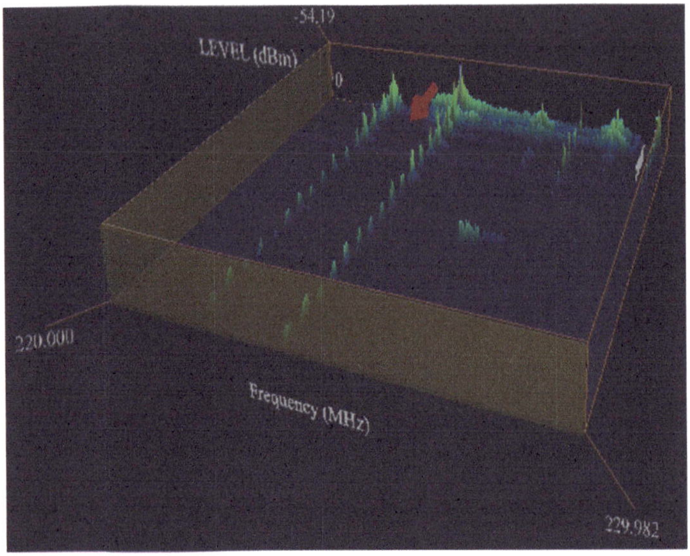

Fig. 3.8 Sensing at Hyderabad for a band of TV frequencies for 18 h

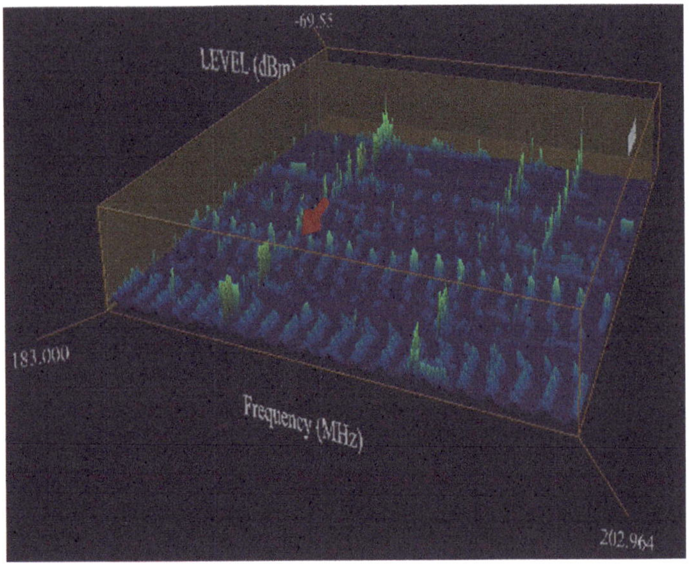

Fig. 3.9 Sensing at Chennai for a band of TV frequencies for 18 h

band where the signal strength is greater than −110 dBm. Figure 3.14 shows the time percentage plot for Tolichowki. Observe that there appears to be more 'gray' space available in UL as compared to DL. Also note that these percentage occupancy plots are conservative, as they do not take into consideration the TDM nature of GSM

Table 3.2 Values of coefficients in adaptive Perez-Vega Zamanillo model [7]

Coefficients	Value	Coefficients	Value	Coefficients	Value
a_0	$2.42e+2$	a_9	0.038	a_{18}	$-8.02e-06$
a_1	-17.80	a_{10}	-0.004	a_{19}	$7.54e-07$
a_2	-37.72	a_{11}	$1.25e-5$	a_{20}	$-2.27e-08$
a_3	17.16	a_{12}	-0.004	a_{21}	$4.30e-11$
a_4	-1.65	a_{13}	0.002	a_{22}	$-1.89e-08$
a_5	0.05	a_{14}	$-2.88e-4$	a_{23}	$6.83e-09$
a_6	$-5.90e-4$	a_{15}	$9.66e-6$	a_{24}	$-6.19e-10$
a_7	0.99	a_{16}	$-6.98e-8$	a_{25}	$1.54e-11$
a_8	-0.38	a_{17}	$2.180e-5$	a_{26}	$1.57e-13$

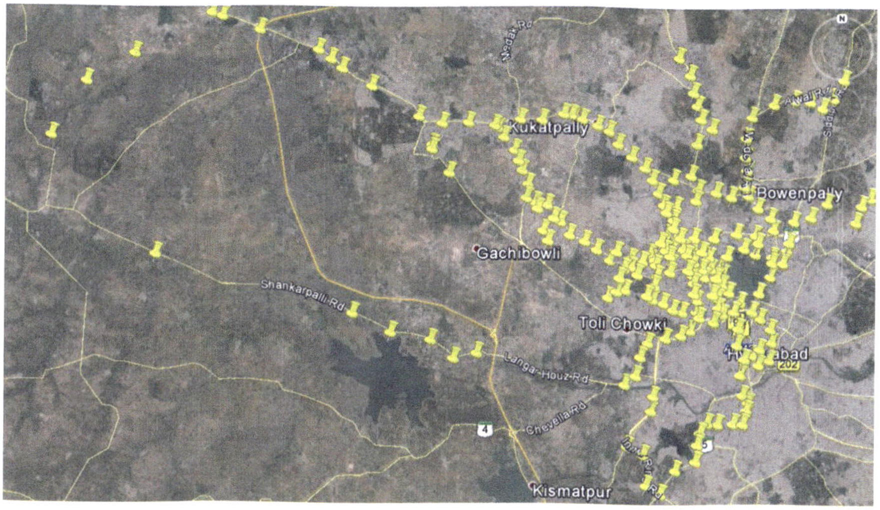

Fig. 3.10 Measurement locations in Hyderabad City for 'Gray Spaces' in GSM bands

900. So while a band is shown as occupied, only 1 in 8 possible TDM slots of the frequency bands may be occupied. A similar performance analysis was made for all the observation locations given in Fig. 3.10. The process was repeated for GSM1800. In order to get a complete picture from the sample points we use mesh interpolation of MATLAB. Figures 3.15, 3.16, and 3.17 give the 3D mesh interpolation plots of Hyderabad for various GSM sub-channels.

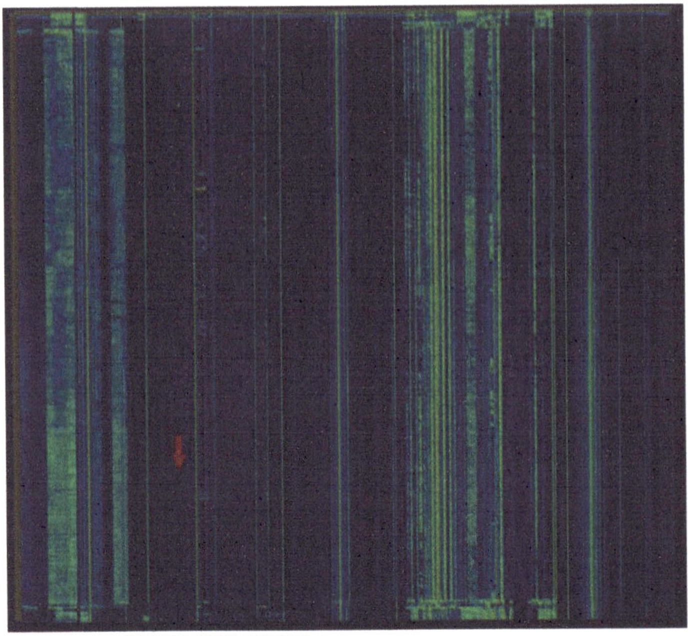

Fig. 3.11 GSM spectrum sensing for a 3 day period from 870 to 980 MHz at Tolichowki

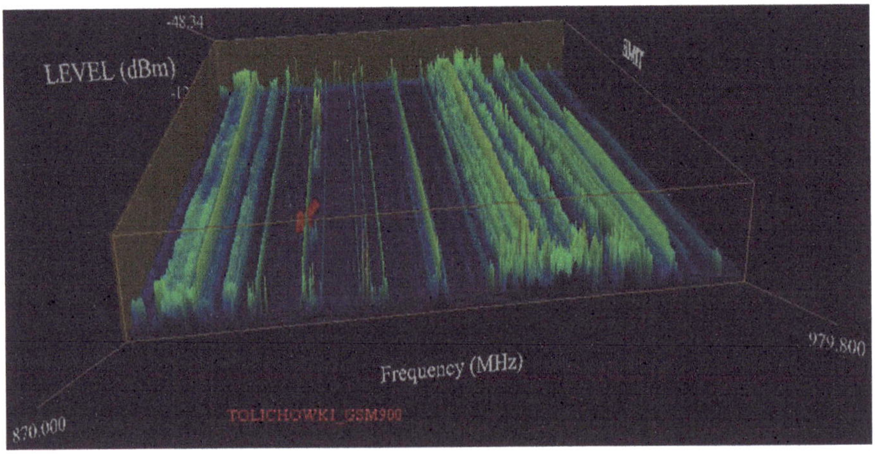

Fig. 3.12 3D plot for GSM spectrum sensing for a 3 day period from 870 to 980 MHz at Tolichowki

3.5.1 'Gray' Space Observations

This conservative measurement campaign demonstrates that there is plenty of opportunity to exploit the 'gray' spaces that are present in the GSM band in a mobile

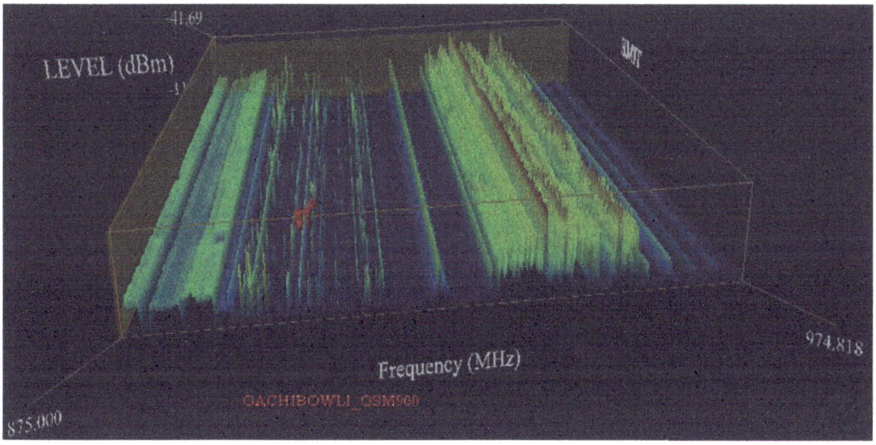

Fig. 3.13 3D plot for GSM spectrum sensing for a 3 day period from 870 to 980 MHz at Gacchibowli, Hyderabad

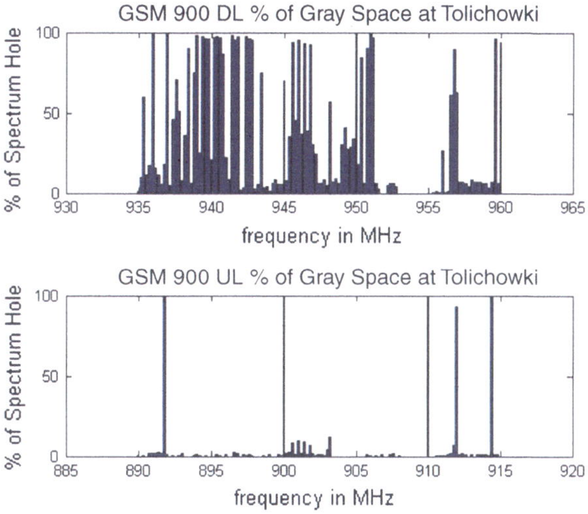

Fig. 3.14 Percentage occupancy plot for GSM spectrum sensing for a 3 day period from 870 to 980 MHz at Tolichowki, Hyderabad

intensive country like India. Also observe that during nights more spectrum bands are underutilized. Also, these results are in conformance with spectrum occupancy studies conducted in other places like New York and Chicago which present a spectrum occupancy of about 45 and 55 % respectively [18, 19] for GSM bands. The average occupancy of GSM band in Hyderabad, based on the above study is 50 %.

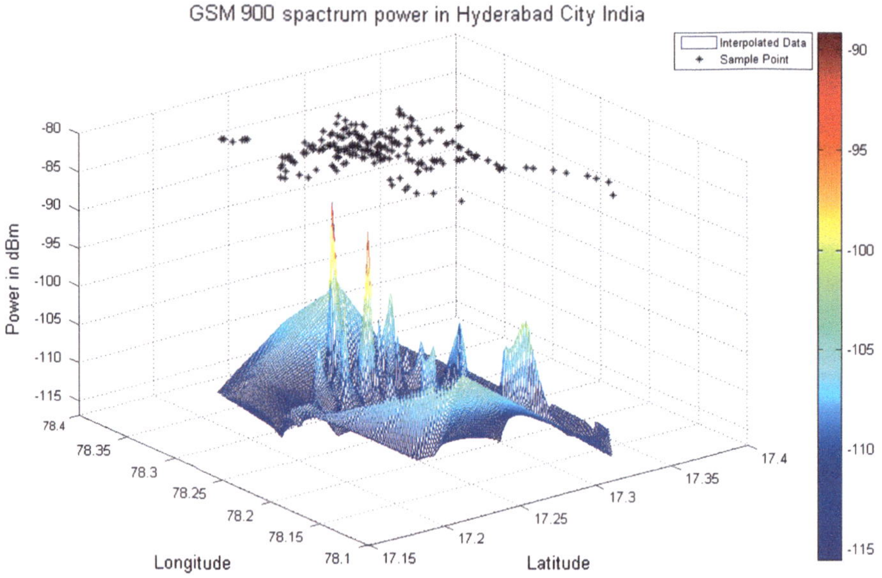

Fig. 3.15 3D mesh occupancy plot for GSM spectrum sensing for 900 MHz bands

3.6 GSM 'Gray Space' Voice Call Testbed

In the previous section we have observed that there is a strong possibility of effective utilization of gray spaces in the GSM band. In order to use Cognitive Radio (CR) communication technologies in this band, it is necessary to evaluate the practicality of algorithms in real time. Use of existing architectures will significantly reduce the the implementation time. Thanks to the availability of Software Defined Radio (SDR), it is now easy to modify radio capabilities via software and can be made as potential testbed. Open source hardware platforms like Ettus USRP, Rice WARP, Fairwaves UmTRX, OpenAirInterface can be used as off-the-shelf SDR [14].

Testbeds have been an active area of research for cognitive radio. Use of open source platforms as potential testbed components has also been investigated. Virginia Tech has worked with CORNET, Eurecom has OpenAirInterface [14]. Many cognitive radio testbeds have been developed in recent years for the usage of TVWS [15] and in WLAN [16]. But these works have mostly been done based on TV or ISM bands. Work on testbeds for GSM band has been presented in [17], but results in a call drop for the secondary user (SU) when the primary user (PU) starts transmitting. Here, we show that it is possible to successfully handover calls even when the PU starts transmitting.

For the testbed we use OpenBTS which is an open source GSM based architecture working with the existing SDRs like USRP and UmTRX [21]. Availability of OpenBTS makes implementation of CR capabilities in the GSM architecture simpler. To add-in the CR functionalities and make OpenBTS a potential testbed for CR

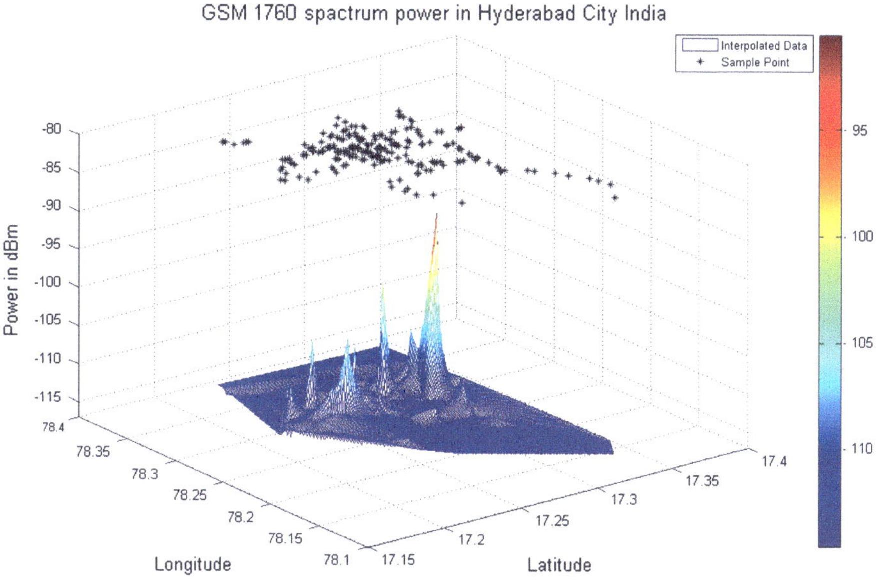

Fig. 3.16 3D mesh occupancy plot for GSM spectrum sensing for 1,780 MHz bands

research, we propose some changes in the source code of OpenBTS and addition of new custom built software modules to provide spectrum sensing and handoff management. With these additions, Cognitive GSM BTS (CGBTS) will be capable of simultaneously checking for free available frequencies and intimate the SU whenever the PU becomes active in order to initiate a handoff to another vacant frequency.

3.6.1 Testbed Overview

A pictorial representation of IITH Cognitive Radio GSM Testbed setup is shown in Fig. 3.18. In order to simulate the PU in the test bed, we use another OpenBTS transmission in the frequency used by the SU.

The Testbed setup uses 3 PCs and 5 USRP N210 equipped with WBX (RF) daughterboards and GPSDO installed in it. The software part consists of modified OpenBTS handover code and spectrum sensing application. The hardware usage is explained as follows:

- 1 PC equipped with two instances of modified OpenBTS handover version integrated with spectrum sensing application (for allowing handover from one frequency to another).
- 2 USRPs connected to this modified OpenBTS handover version will be acting as CGBTS. A Mobile (MS 1) will be connected to one of these BTSs. These along

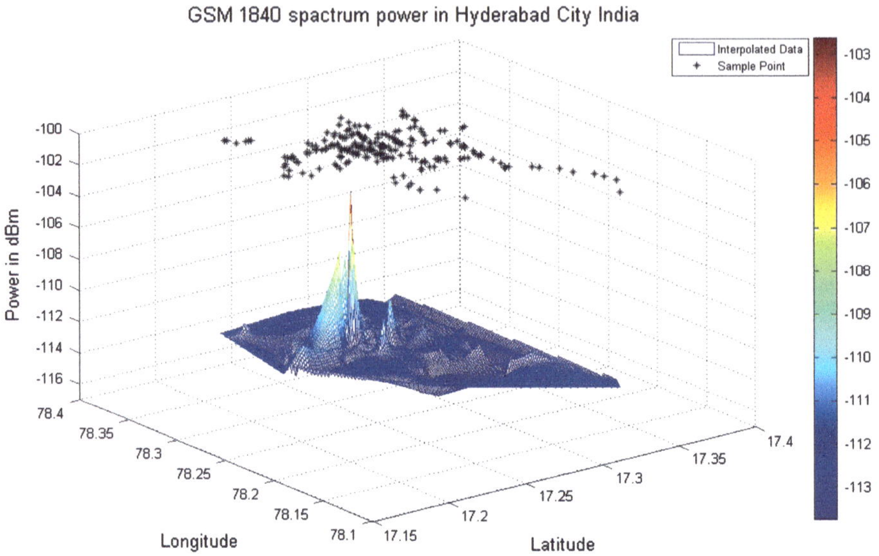

Fig. 3.17 3D mesh occupancy plot for GSM spectrum sensing for 1,840 MHz bands

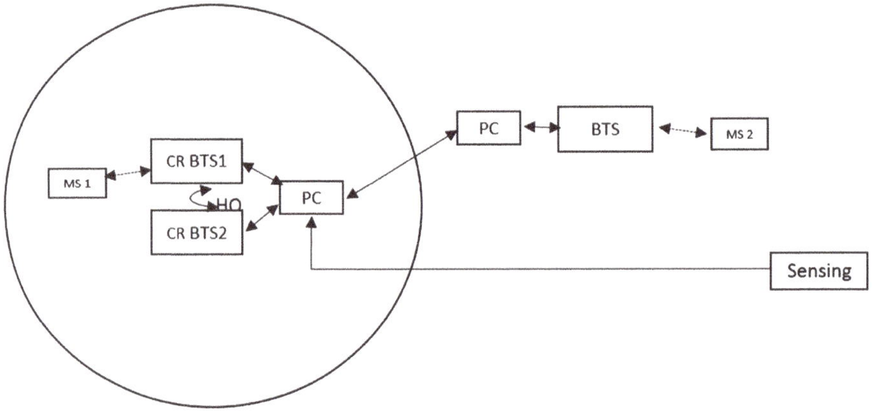

Fig. 3.18 Test bed pictorial representation

with the 1 PC equipped with two instance of modified OpenBTS that are referred to as CRBTS 1 and CRBTS 2 in Fig. 3.18. CRBTS 1 and CRBTS 2 together constitute the secondary user BTS.

- 1 USRP connected to a PC with a spectrum sensing application is placed outside the range of the CGBTS, will be acting as FFT-based spectrum analyzer and is shown as the sensing unit in Fig. 3.18.

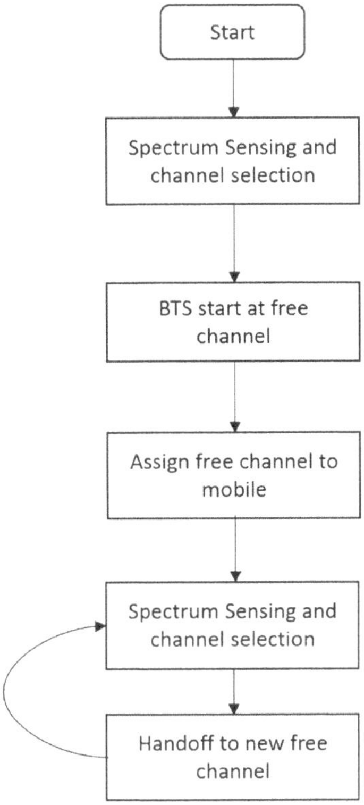

Fig. 3.19 Flowchart of CGBTS

- 1 PC equipped with OpenBTS public release connected with 1 USRP will be acting as external BTS for closing the call from MS 1. A second phone (MS 2) will be connected to this BTS.
- 1 PC equipped with OpenBTS public release and connected with 1 USRP for emulating the primary user PU (not shown in the pictorial diagram).

3.6.2 The Cognitive Functionality

An outline of the cognitive functionality in the OpenBTS is given in Fig. 3.19. CGBTS will start the spectrum sensing procedure. Once it has a list of free channels, it assigns the best of the available channels to the OpenBTS to start. The MS 1 call is connected using OpenBTS with another BTS. Spectrum sensing is done simultaneously by keeping the 'sensing block' outside the range of secondary BTS, so that the received signal power from CRBTS 1 and CRBTS 2 is less than −120 dBm at the sensing block. So even when call is in progress and a PU appears in the frequency used by the SU, sensing block can detect PU with minimal inerference.

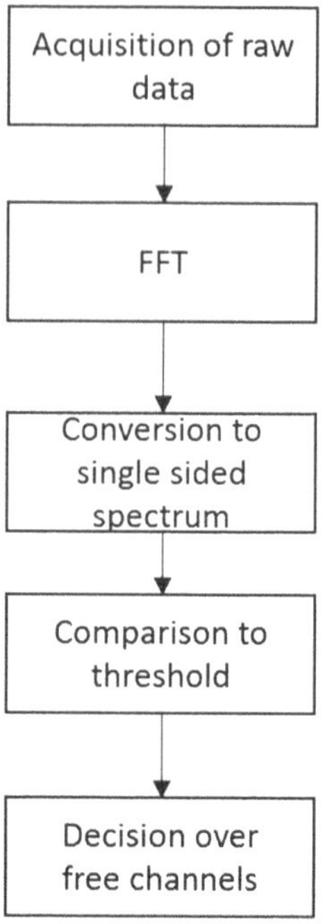

Fig. 3.20 Flowchart spectrum sensing

When a PU is detected, the sensing unit requests a handover in the OpenBTS (say CRBTS 1). Once the GSM Handover is initiated, MS provided measurement results are processed by CRBTS 2 on another (unoccupied) frequency, and the handover procedure is triggered by CRBTS 1. Note that handover is a standard protocol in GSM, initiated when the mobile user crosses over into the range (cell) of another base station without dropping the call. Here we use the same protocol to avoid dropping the call when the PU occupies its band. Subsequent to the handover of the MS to the next CRBTS with the available frequency, the first CRBTS is restarted with another available frequency to facilitate the next handover, thus maintaining the QoS for both the PU and the SU.

In doing so, the CRBTS allows the SU to make GSM calls using OpenBTS without dropping the call or degradation of service for either the PU or the SU.

3.6.3 Spectrum Sensing Algorithm

The spectrum sensing process for CRBTS is shown in a flow chart in Fig. 3.20. In order to get timely information on the availability of GSM subchannels, spectrum sensing is done based on FFT. The ADC is set to 25 MS/s 16 bit mode to stream IQ samples for processing. This allows us to sense up to 62 GSM subchannels in 12.5 MHz bandwidth simultaneously and decide upon these channels. We sense channels in GSM-900 band, it uses 890–915 MHz for uplink and 935–960 MHz for downlink, providing 124 RF channels each spaced at 200 KHz. Uplink and downlink uses duplex spacing of 45 MHz. Energy detection is used to determine free subchannels. Free channels are stored in an array (that is updated on the fly) for use when required.

3.6.4 Results and Observations

To check the working of the setup, we started the spectrum sensing algorithm and assigned the free channel to a CRBTS. The CRBTS1 was started and connected to MS1 and MS2 to the BTS. Once the call was established between MS1 and MS2, we started the OpenBTS which emulates the PU at MS1 frequency while call connection was still in progress. Spectrum sensing which was running simultaneously detected the PU and GSM handover of MS1 to CRBTS2 on an available channel without dropping the call. Complete handover timing was observed to be 2 s after the PU initiates a call. In this duration the PU and the SU interfere with each other until the handover is completed. The Frame Error Rate for a SU with a medium power PU (−90 dBm) varies from 3 to 1.28 % which is within the limits of voice call quality over VoIP [20].

This testbed provides a demonstration for realizing the efficiencies of next generation wireless systems without degrading PU and SU QoS and can be adopted without change in the GSM protocol. To the best of our knowledge, this is the first demonstration of CR for delay constrained applications. Note that in actual GSM base stations the switching time can be significantly reduced by using customized RF frontends. Further analysis of various other PU and SU parameters is ongoing.

3.7 TV White Space Policy Issues for India

In this section we discuss the policy issues of TV White Spaces in India. The major points of difference as compared to TVWS in other countries are as follows

- The TV band is wholly owned by the governmental broadcaster, Doordarshan with only two channels of transmission, in general.
- There is more than 72 MHz of bandwidth available in the analog band; 'Analog White Spaces' (in south India).
- Assuming conversion to DTT of all the transmitters, the available 'Digital White Space' is about 50 MHz in South India. Note that the band for DTT is 585–698 MHz with an overlap of only 5 MHz in the UHF band.
- Since Doordarshan has 4 DTT transmitters and 19 proposed to be set up by 2017 [10] both types of white spaces can coexist for at least another 10 years.

In this context, the sole government-owned broadcaster needs only two transmission channels, allowing for three to four programming channels. This will provide abundant bandwidth for its needs, while avoiding interference. Similarly, of the 13 UHF channels, 3–4 transmission channels will be more than adequate for Doordarshans digitized needs. So the first and most important point for the regulators and WPC is to look for more efficient channel allocation plans for both analog and digital terrestrial TV transmissions. This will result in a 'replanning' dividend and release at least 70 MHz in the VHF bands, 88 MHz in the UHF band V and 72 MHz in the UHF band VI. On completion of digitization another 32 MHz of 'Digital' dividend will be available. So a whopping 262 MHz will be available after completion of digitization, based on Doordarshans present requirements. Keeping another 8 channels for future expansion of broadcasting, approximately 200 MHz of prime bandwidth is available.

This bandwidth can either be reallocated or opened up for white space devices or a combination of both can be adapted. Note that the replanning of frequency will further increase the available white space. Conditional on the specifics of the replanning process, the 'replanning' and digital dividend in India should present at least three opportunities for WSD;

- Availability of at least 70 MHz of white spaces in the VHF bands
- Availability of at least 88 MHz of white spaces in the UHF V bands
- Availability of at least 105 MHz of white spaces in the UHF VI bands

These opportunities provide WSD proponents with enough incentives to engage the WPC in consultation processes that result in regulatory outcomes that:

1. Maximise the availability of whitespaces in both VHF and UHF bands
2. Licence these white spaces that are suitable for WSD without jeopardising ATT and DTT.

Given the large amount of unused spectrum in India and increasing need for spectrum for broadband, there is a strong possibility of reallocation of frequency bands away from broadcasting services. This will pose a significant challenge to WSD's in India.

Any reallocation of UHF spectrum away from the broadcaster, to other uses, would have major implications for WSD's because of the viability of enabling use of white spaces for commercial purposes. Also, a stringent reallocation of the UHF band may result in 'zero' tolerance for WSDs by the broadcaster.

3.8 Gray Space Policy Issues

From the gray space measurement campaign presented in Sect. 3.5 it is clear that there are sharing opportunities, other than TV white spaces, that can be exploited. However the regulatory requirements for gray space networks will be vastly different, as compared to white spaces. Example of primary-secondary gray space sharing mechanism has been already described in Sect. 3.5. Note that during handoff both primary and secondary systems interfere with each other. To avoid such a scenario primary and secondary may have to be interlinked, which implies that white space policy and governance rules will not be sufficient for gray space.

One important scenario for GSM 'gray' spaces is when the secondary users gain access to spectrum through the primary user and not through the regulator [22]. This scenario will immediately solve the interference issues involved as the primary can allot a band after keeping a lower transmit power constraint or and adjacent band with lower interference. This is in contrast with present TV white spaces systems, where secondary systems gain spectrum access from the regulator.

In the case where the regulator gives access to the gray spaces, it may have to introduce additional polices for the primary and maybe secondary, which is unattractive. The present TV white space policies of FCC and Ofcom require that the secondary users must periodically check a centralized database and terminate operation when and where required. An important aspect of gray space sharing is that gray space devices must rely on spectrum sensing as opposed to databases, due to the dynamic nature of the GSM gray spaces and the time constraints involved. Note that this itself will open up lots of problems, considering the fact that no sensing technique has been approved for white spaces. The problem is more confounded by the fact that sensing is more problematic for gray space sharing due to the risk of higher unintended interference and the lack of an obvious way to stop secondary transmissions when problems occur [22]. As such, due to lack of clarity, present U.S. policies prohibit TV band gray space sharing even though this band has larger time constraints. If the constraints on the spectrum sensing are too stringent this may lead to higher costs of gray space devices (GSD) making them unviable for developing economies.

In contrast the primary has detailed knowledge of itself which it can share with secondary users when required. Also the primary has the ability and reason to detect interference to itself faster, and the motivation to promptly demand that secondary systems stop causing harmful interference if it occurs. Such a mechanism will allow for low cost GSDs to function with not so stringent constraints for detection.

An alternative approach is common infrastructure, including shared use of the RAN (which implies sharing of spectrum as well), with central rather than distributed control. For example, a Next Generation Network that is run by a consortium of operators and the government, like Singapores OpenNet, led by a private sector minority partner (Axia in the case of Singapore). In such a scenario, the stringent requirements on detection can be relaxed considerably if not totally relaxed, as the operators can update a database on the spectrum availability.

Another aspect is that the primary's goal is not to improve utilization of spectrum resources from an overall perspective, unlike the regulator. As such better schemes to motivate the primary like payments for the right to secondary use must be introduced. This form of sharing already exists in some countries like the US [23, 24] which gives license-holders, in some bands, the right to allow access to secondary users and subsequently notify the FCC about these transactions. India could benefit vastly by allowing such a sharing scheme in GSM band, to start with. Additionally, the regulator (TRAI) needs to amend policies (with the agreement of the operators) that allows for a relaxation of QoS on handoff to permit shared gray space with a specific permissible degradation of service.

These policies can also be applied to TV gray spaces which are also significant in India (the DoD channels do not broadcast 24 h). As such, new telecom-related legal and business practices can and must emerge to allow for low transaction costs for gray space sharing.

3.9 Conclusions

In this chapter we have presented a three-year study of TV White spaces and GSM gray spaces in (South) India and Hyderabad. For South India, we have presented the white space available in the analog VHF and UHF bands. We have also studied the white spaces available (and that will be available) in the digital UHF bands. Results from extensive on going measurement campaigns have been reported. We have also shown the existence of significant 'gray' spaces in Hyderabad. We have identified the opportunities and the challenges for White Space and Gray Space Devices in India as a result of these studies. There is now a 'window of opportunity' for WPC to optimise the availability of white and gray spaces in India and establish regulatory arrangements, in consultation with White Space and Gray Space Devices operators, that will allow intensive use of these spaces in a symbiotic manner.

Acknowledgments The authors would like to thank Deity and DST for supporting this work under two grants to some of the authors.

References

1. Telecom Regulatory Authority of India, http://www.trai.gov.in
2. Ofcom: Digital Dividend Review, A statement on our approach towards awarding the digital dividend (2007)
3. Harrison, K., Mishra, S., Sahai, A.: How much white-space capacity is there? In: IEEE Symposium on New Frontiers in Dynamic Spectrum Access Networks (DySPAN), pp. 1–10. IEEE Press, New York (2010)
4. Nekovee, M.: Quantifying the availability of TV white spaces for cognitive radio operation in the UK. In: IEEE International Conference on Communications (ICC), pp. 1–5. IEEE Press, New York (2009)
5. Jntti, R., Kerttula, J., Koufos, K., and Ruttik, K.: Aggregate interference with FCC and ECC white space usage rules: case study in Finland. In: IEEE Symposium on New Frontiers in Dynamic Spectrum Access Networks (DySPAN), pp. 599–602. IEEE Press, New York (2011)
6. van de Beek, J., Riihijarvi, J., Achtzehn, A., Mahonen, P.: UHF white space in Europe—A quantitative study into the potential of the 470–790 MHz band. In: IEEE Symposium on New Frontiers in Dynamic Spectrum Access Networks (DySPAN), pp. 1–9. IEEE Press, New York (2011)
7. Sridhar, B., Khan, M.Z.A.: Comparison of propagation path models for UHF/VHF bands in India. In: IEEE TENS 2014, pp. 1–6. IEEE Press, New York (2014)
8. Sridhar, B., Khan, M.Z.A.: Adaptive Pervez-Vega model optimized for UHF/VHF bands in India. Technical report, IIT Hyderbad (2013)

9. Draft national frequency allocation plan: wireless planning and coordination wing. Ministry of Communications and Information Technology, Government of India (2011)
10. Specifications for supply of DVB-T2 TV transmiter systtem for setting up DTT at 19 locations. Prasar Bharati, Doordarshan (2010)
11. Mishra, S., Sahai, A.: How much white space is there?: Technical Report UCB/EECS-2009-3, EECS Department, UC Berkeley. http://www.eecs.berkeley.edu/Pubs/TechRpts/2009/EECS-2009-3.html. Jan. 2009
12. Anant Sahai's Code link, http://www.eecs.berkeley.edu/sahai/new_white_space_data_and_code.zip
13. Mishra, S.: Maximizing available spectrum for cognitive radios. PhD Thesis UC Berkeley. (2010)
14. Gustafsson, O., Amiri, K., Andersson, D., Blad, A., Bonnet, C., Cavallaro, J.R., Declerck, J., Dejonghe, A., Eliardsson, P., Glasse, M., Hayar, A., Hollevoet, L., Hunter, C., Joshi, M., Kaltenberger, F., Knopp, R., Le, K., Miljanic, Z., Murphy, P., Naessens, F., Nikaein, N., Nussbaum, D., Pacalet, R., Raghavan, P., Sabharwal, A., Sarode, O., Spasojevic, P., Sun, Y., Tullberg, H.M., Vander Aa, T., Van der Perre, L., Wetterwald, M., Wu, M.: Architectures for cognitive radio testbeds and demonstrators An overview. In: Proceedings of the Fifth International Conference on Cognitive Radio Oriented Wireless Networks and Communications (CROWNCOM), pp. 1–6. IEEE Press, New York (2010)
15. Franklin, A.A., Pak, JS., Jung, H., Kim, S.W., You, S.J., Um, J.S., Lim, S., Ko, G.Z., Hwang, S., Jeong, B.J., Song, M.S., Kim, C.J.: Cognitive radio test-bed based on ECMA-392 international standard. In: 7th International Symposium on Wireless Communication Systems, pp. 1026–1030. IEEE Press, New York (2010)
16. shizu, K., Murakami, H., Harada, H.: Cognitive wireless network infrastructure and restoration activities for the earthquake disaster. In: 14th International Symposium on Wireless Personal Multimedia Communications (WPMC), no, pp. 1–5. IEEE Press, New York (2011)
17. Pace, P., Loscri, V.: OpenBTS: a step forward in the cognitive direction. In: 21st International Conference on Computer Communications and Networks (ICCCN), pp. 1–6. IEEE Press, New York (2012)
18. Roberson, D.A.: Structural support for cognitive radio system deployment. In: 2nd International Conference on Cognitive Radio Oriented Wireless Networks and Communications, 2007. pp. 401–407. IEEE Press, New York (2012)
19. Patil, K., Prasad, R., Skouby, K.: A survey of worldwide spectrum occupancy measurement campaigns for cognitive radio. In: 2011 International Conference on Devices and Communications (ICDeCom), pp. 1–5. IEEE Press, New York (2011)
20. Validating voice over LTE end-to-end, Ericson Review http://www.ericsson.com/res/thecompany/docs/publications/ericsson_review/2012/er-volte-performance.pdf. (2012)
21. OpenBTS official page. http://openbts.org/
22. Jon, M.: Peha: spectrum sharing in the gray space. Telecommun. Policy **37**(2–3), 167–177 (2013)
23. Force Report, November 2002. Federal Communications Commission (FCC): Promoting efficient use of spectrum through elimination of barriers to the development of secondary markets. Report and order and further notice of proposed rulemaking. (2003) Retrieved from http://hraunfoss.fcc.gov/edocs_public/attachmatch/FCC-03-113A1.pdf
24. Federal Communications Commission (FCC): Promoting efficient use of spectrum through elimination of barriers to the development of secondary markets. (2004) Retrieved from http://hraunfoss.fcc.gov/edocs_public/attachmatch/FCC-04-167A1.PDFS

Chapter 4
The Television White Space Opportunity in Southern Africa: From Field Measurements to Quantifying White Spaces

Albert A. Lysko, Moshe T. Masonta and David Lloyd Johnson

Abstract The lack of sufficient fixed-line communication infrastructure in African rural areas has resulted in wireless communication being the only cost effective alternative solution for broadband connectivity. However, access to valuable spectrum—specifically sub-1 GHz spectrum—is mostly allocated to broadcasting or mobile telephony. The global digital switch over (DSO) of television (TV) broadcasting systems will see more sub-1 GHz TV band spectrum being made available for the digital dividend and also result in more TV white space (TVWS) spectrum. In order to ensure dynamic and efficient utilization of the TV white space spectrum, there is an increasing trend to use cognitive radiosystems that use geo-location spectrum databases and spectrum sensing as an enabling technology. In this paper, we overview the relevant signals and standards and present field measurement results showing the actual usage of TV bands before the DSO in selected urban and rural areas of Southern Africa. Measurements were conducted using low-cost and high-grade radio instruments. The low-cost spectrum analyser was built in-house using the Universal Software Radio Peripheral (USRP-2) and GNU Radio software. A metric to quantify available TV white space, based on the minimum acceptable field strength, is introduced and applied to quantify the availability of TV white space. Our results show medium spectrum usage in urban areas and very low spectrum usage in rural areas, making TVWS an attractive solution for rural broadband connectivity.

Keywords Measurements · Radio frequency · Southern Africa · Spectrum · Television white spaces

---

A.A. Lysko (✉) · D.L. Johnson
Council for Scientific and Industrial Research (CSIR) – Meraka, P O Box 395, 0001 Pretoria, South Africa
e-mail: alysko@csir.co.za

D.L. Johnson
e-mail: djohnson@csir.co.za

M.T. Masonta
Department of Electrical Engineering, Tshwane University of Technology, Pretoria, South Africa
e-mail: mmasonta@csir.co.za

A.K. Mishra and D.L. Johnson (eds.), *White Space Communication*,
Signals and Communication Technology, DOI 10.1007/978-3-319-08747-4_4

4.1 Introduction

White spaces refer to portions of licensed radio frequency (RF) spectrum band which are not utilized or are sporadically used at a given time in given geographical location. The amount of available white spaces depends on the incumbent, licensed or primary users (PUs) and can vary according to the PUs activities. Although white spaces can be found within any allocated spectrum band, the current focus is on the television (TV) band where there is ongoing migration from analogue to digital broadcasting or digital switch over (DSO). Thus, white spaces in TV bands are commonly referred to as television white spaces (TVWSs). The International Telecommunication Union (ITU) deadline for DSO in Region 1 countries (which corresponds to Europe, Russia, Africa and the Middle East) is set for June 2015 [1]. The DSO promises efficient spectrum use, increased competition, technology convergence and more spectrum for digital dividend (spectrum freed up in the digital migration process) as well as the exploitation of TVWS for broadband connectivity in rural areas. Unlike the Wi-Fi spectrum, which is above the 2 GHz frequency band, TVWS provides great propagation characteristics which can address Internet connectivity to sparsely populated and remote areas.

In 2008, the Federal Communications Commission (FCC) of the United States (US) announced a decision to open up TVWS for unlicensed or license-exempt use [2]. In order to make such a decision, extensive work commissioned by the FCC has been done in the US which includes spectrum occupancy measurements, consultations and TVWS trials. The same decision was also taken by the Office of Communications (Ofcom) in the United Kingdom [3], and this has created a large amount of interest from the research community on the exploitation of TVWS for broadband access. In Africa, notable activities on TVWS include the TVWS trials in South Africa [4–6], the development of a geo-location spectrum database (GSDB) [7] for the TV band and a number of spectrum occupancy measurement campaigns [8–11] with the aim of quantifying the amount of TVWS in both rural and urban areas. While there are no clear regulatory decisions on how to utilize TVWS in Africa, we believe that these efforts will play a crucial role in supporting the decision making process, especially post the DSO deadline.

In this paper, ultra-high frequency (UHF) TV band spectrum occupancy measurements results conducted in selected Southern African urban and rural areas are presented. This is a follow-up work that was first presented in [8], and now also includes results from the Cape Town area. Specifically, TV band occupancy measurements were conducted using a combination of the Meraka Cognitive Radio Platform and high-grade RF instruments in Southern Africa. In South Africa, spectrum occupancy measurements were conducted in Pretoria and Cape Town in Gauteng (representing a typical urban environment) and in Philipstown in the Northern Cape Province (representing a typical rural environment). In Zambia, spectrum occupancy measurements were conducted in Macha, a rural area in Southern Zambia. As one would expect, our occupancy measurements found that over 40 % of TVWS is typically available in urban areas,whereas rural areas have over 90 % of TVWSs available for broadband connectivity usage.

The remainder of this chapter is organized as follows: Sect. 4.2 provides the TVWS context in South Africa. Section 4.3 presents a detailed methodology used for estimating TVWSs in Southern Africa using the Meraka Cognitive Radio Platform. The detailed methodology used for estimating TVWSs in Cape Town is discussed in Sect. 4.4. Section 4.5 presents the measurement results in order to quantify the amount of TVWS availability in Southern Africa. Some remarks on the optimizing speed and accuracy of measurements are provided in Sect. 4.5. Section 4.7 concludes the chapter.

4.2 TV White Space Context in South Africa

This section presents the TVWS context in the South African environment. Like other countries, South Africa is undergoing a national DSO process and is expected to complete this process by June 2015—the deadline set for ITU-R region 1 countries.

4.2.1 Overview of the South African TV Broadcasting Network

South Africa has over 4,000 registry entries in the list of TV programs being transmitted. The location of TV transmitters, as per the listing by the Independent Communications Authority of South Africa (ICASA), is shown in Fig. 4.1. The figure shows an overview of the distribution of TV transmitters around South Africa. It is easy to see that the highest density of the transmitters corresponds to the most populated regions, such as Gauteng, and Western Cape. It is also possible to see that the original

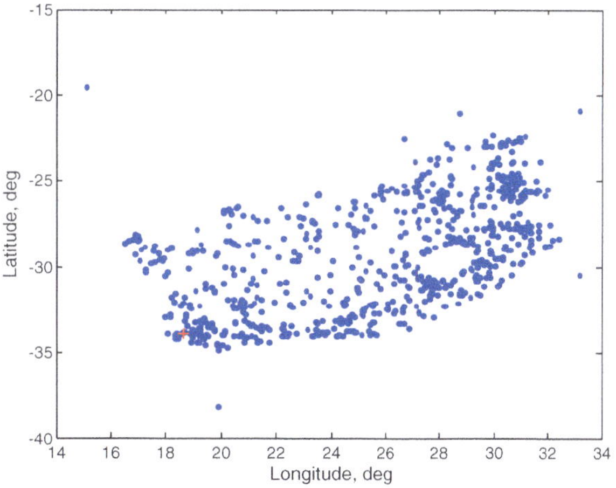

Fig. 4.1 Location of TV transmitters around South Africa. TV transmitters are shown with *blue* dots

data set includes some outliers and, if used for mission critical applications, such as a national geo-location database [7], it must therefore be treated with caution.

On October 3, 2012, South African Communications Minister Dina Pule launched a demonstration of South Africa's digital terrestrial television (DTT) technology in the Northern Cape, ahead of the planned December launch of the country's migration from analogue to digital TV broadcasting. In 2012/2013, South Africa had both analogue and digital broadcasting available in many areas. Analogue broadcasting was available nation-wide, whereas digital broadcasting was introduced in a testing phasein limited areas, prior to the official dual illumination period [12]. Thus, in many locations, especially in the more densely populated areas like Gauteng and Western Cape, TV broadcasting was de-facto in dual illumination mode.

4.2.2 Expected Spectral Masks and Minimum Signal Strength Values for TV Broadcasting in South Africa

In South Africa, analogue transmitters use PAL system I with 8 MHz channel bandwidth similar to what is used in the UK. The digital transmitters use terrestrial digital video broadcasting (DVB-T2) and the DVB hand-held option (DVB-H). This Section deals with describing the signals in the frequency domain and using the information available from the standards and legal documents to derive the metrics to be used as a threshold to identify and quantify the amount of white space.

4.2.2.1 Analogue Broadcasting: PAL-I

With reference to [13], the key parameters of a PAL-I signal are captured in Table 4.1. A schematic representation of the components present in PAL-I spectrum are shown in Fig. 4.2a.

A sample of a measured analogue signal's spectrum is shown in Fig. 4.2b. It is possible to see the following distinct components: video intensity (luma) carrier surrounded with luma spectrum, color representation carrier, audio carrier and NICAM (Near Instantaneous Companded Audio Multiplex) [16]. The NICAM is an optional component, and may be present in some of the spectra measured. The chroma component is optional, as well, but, in the modern world, is present in practically all analogue TV transmissions.

References [17, 18] point to the following information regarding the relevant specifications of a digital method to transmit sound, NICAM:

- Modulation: differentially encoded Quaternary Phase-Shift Keying (QPSK), with bit rate 728 kbit/s $\pm$ 1 part/million; one packet sent every millisecond;
- Carrier frequency of 6,552 MHz $\pm$ 1 part/million above the vision carrier frequency;
- The power ratio between the peak vision carrier and the modulated digital signal is approximately 100:1 (20 dB) for systems B, G, H and I;

Table 4.1 Key parameters of PAL-I signal, as per [14–16]

#	Parameter	Value and unit	Rec. ITU BT/EN
1	Nominal radio-frequency channel bandwidth	8 MHz	1701
2	Sound with respect to video carrier	+5.9996 ± 0.0005 MHz	1701
3	Nearest edge of channel relative to vision carrier	1.25 MHz	1701
4	Nominal width of the main sideband	5.5 MHz	1701
5	Nominal width of the vestigial sideband (VSB)	1.25 MHz	1701
6	Minimum attenuation of vestigial sideband	20 dB at −3.0 MHz and	
		30 dB at −4.43 MHz[a]	1701
7	Chrominance sub-carrier frequency	4 433 618.75 Hz ± 1 Hz	1700
8	Type of chrominance sub-carrier modulation	Suppressed-carrier amplitude-modulation of two sub-carriers in quadrature	1700
9	Power ratio between the peak vision carrier and the analogue FM sound carrier	Approximately 10:1 or 10 dB[b]	300163

[a] Additional restrictions may apply for a co-channel transmittance of a DVB-T signal
[b] However, in the UK with system I this ratio has since been relaxed to 20:1 or 13 dB, bringing the UK in line with Systems B and G within Europe

Note In the case of cable distribution with adjacent channel operation, it is recommended to use a power ratio of approximately 300:1 (25 dB) for PAL system I.

Minimum Field Strength: PAL-I

ITU-R [19] states that the median field strength values for which protection against interference is planned should not be lower than the values shown in Fig. 4.3 and Table 4.2.

The minimum field strength can be used as the threshold to determine whether a channel with analogue PAL-I signal is usable for viewing.

Typical TV Receiver Parameters: PAL-I

Recommendation [20] provides the values for antenna gain and cable loss, as shown in Table 4.3.

The section on calculation of minimum field strength for DVB-T 8 MHz systems in [21] also uses the same antenna gain values for the respective bands, although it

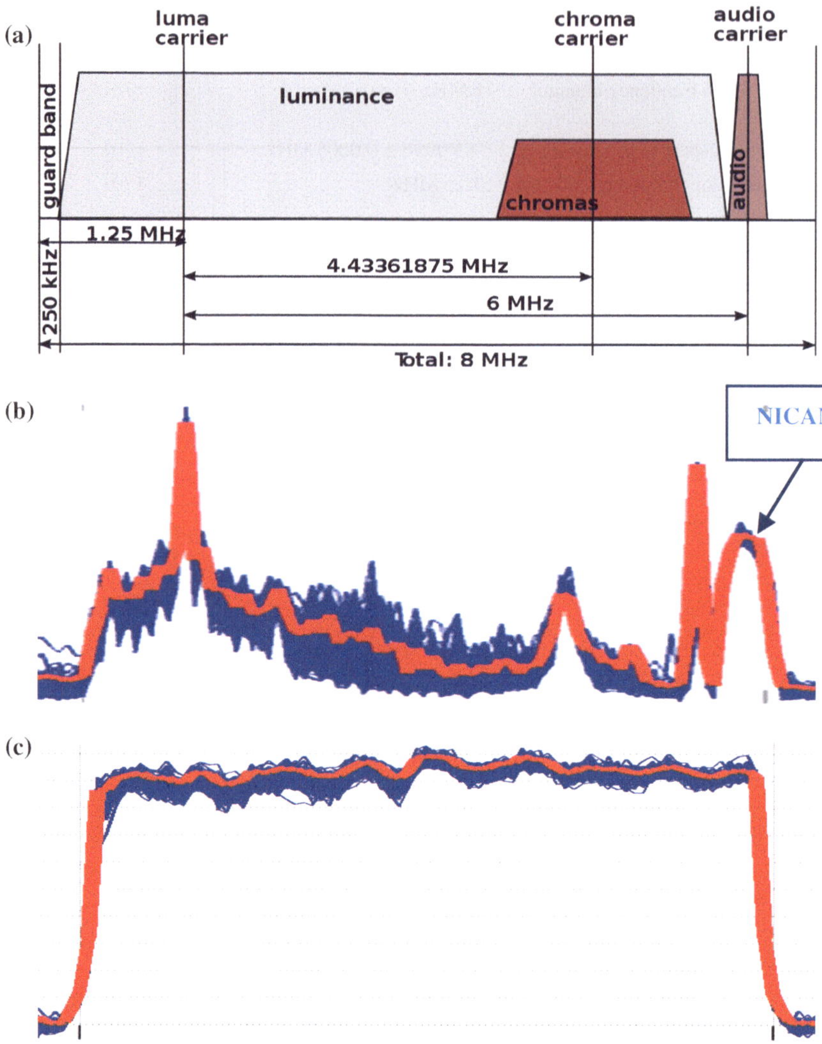

Fig. 4.2 **a** TV signal's spectrum for PAL system I (Courtesy of http://en.wikipedia.org/wiki/PAL); **b** sample of real world PAL-I spectrum with NICAM; **c** sample of real world digital spectrum DVB-T2

uses 5 dB cable loss (referred to as *feeder loss*) instead of 4.5 dB. As shown in [21], when considering DVB-T2 8 MHz systems operating at 650 MHz, an antenna gain of 11 dB and feeder loss of 4 dB for the "fixed" roof top scenario is used. In [22] they focus on DVB-T2, at 650 MHz, and also use the same parameters.

Note on interpretation of antenna gain:

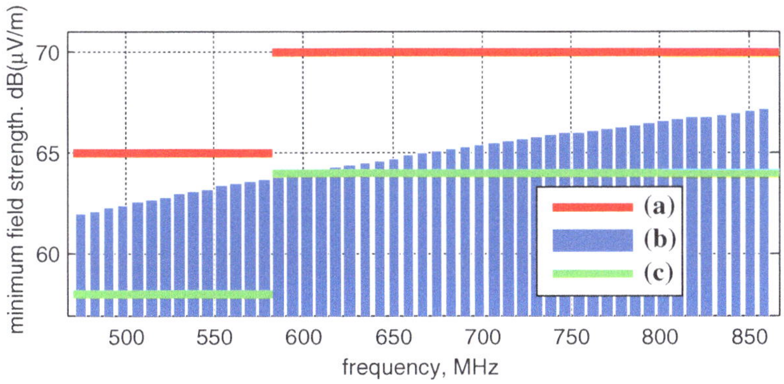

Fig. 4.3 The median and minimum field strength values permissible with and in absence of interference [19]. The bars were calculated based on the formula E_{min} (dB(μ V/m)) = 62 + 20 log(*f*/474). The (*a*)–(*c*) refer to the respective lines in Table 4.2

Table 4.2 Field strength values (in dB(μ V/m)) recommended for planning analogue TV broadcasting in UHF [19] These values refer to the field strength at a height of 10 m above ground level

Band			IV	V
Frequency range, MHz			470–582	582–960[a]
	Scenario			
Medium/	(a)	Urban area	65	70
Minimum value,	(b)	Urban area with no interference	62 (at 474 MHz)	67 (at 842 MHz)
dB(μ V/m)	(c)	Area with better receivers and antennas	58	64
	(d)	Area with better receivers and antennas and no interference	52	58

The grey block values are calculated using the formula E_{min} (dB(μ V/m)) = 62 + 20 log(*f*/474), where *f* is frequency in MHz; with the result shown in Fig. 4.5. The lines (a) and (c) refer to median field strength values. The lines (b) and (d) refer to minimum field strength values

[a] The upper frequency of the band has been changed to 862 MHz, and is planned to be moved to 790 MHz, and possibly even lower

The antenna gain is interpreted to be in dBd (with respect to a half-wave dipole) rather than dBi (with respect to an isotropic radiator, where the two are related as G(dBi) = G(dBd) + 2.15 dB).

The grounds for this are as follows:

- Recommendation [20] lists the dipole conversion factor.
- Recommendation [23] lists the values in dBd. The expressions also include a 1.64 correction factor corresponding to a 2.15 dB difference.
- Recommendation [21] lists the values in dBd.
- Report [24] lists the values in dBd, and also proposes an additional correctional factor to take into account the increase in the antenna gain with frequency:

Table 4.3 Typical minimum signal strength, antenna gain, feeder loss per [20]

Band	IV	V
Frequency range, MHz	470–582	582–960[a]
Minimum field strength, dB(μ V/m)	62	67
Antenna gain, dBd	10	12
Cable loss, dB	3	4.5
Dipole conversion factor[(b)], dB	20.5	25

[a] The upper frequency of the band has been changed to 862 MHz, and is planned to be moved to 790 MHz, and possibly even lower
[(b)] $20 \log 2\pi/\lambda$

Corr = 10log10(FA/FR), where FA is the actual frequency being considered and FR is the relevant reference frequency.

Suggested Parameters for Modelling a Typical Fixed Roof Top Antenna Installation

Based on the references indicated and discussions made in the previous section, the following is considered as a model for the fixed rooftop antenna installation:

- Antenna gain, $G = 10 + 8\ \log10(f/474)$,
- Feeder loss, $L = 3 + 6\ \log10(f/474)$,

where f is the frequency in the middle of a channel, in MHz.

The model ensures that the parameters specified by Table 4.3 are satisfied. The resultant curves for the parameters are shown in Fig. 4.4. It is clear that the maximum difference between the difference of gain and loss for the proposed model and [20] is within 0.5 dB. The difference to the other above-mentioned recommendations and to [1] is within 1 dB.

Considerations for TV Receivers

As per [20]:

- "In order to obtain the protection ratios given in [25] the minimum field-strength values given in [19], and meet other frequency planning constraints, the" noise-limited sensitivity of at least −58 dBm in UHF band "for reference receivers for different transmission systems should be met".
- A principal receiver in UK has the noise-limited sensitivity of −65 dBm in the UHF band.

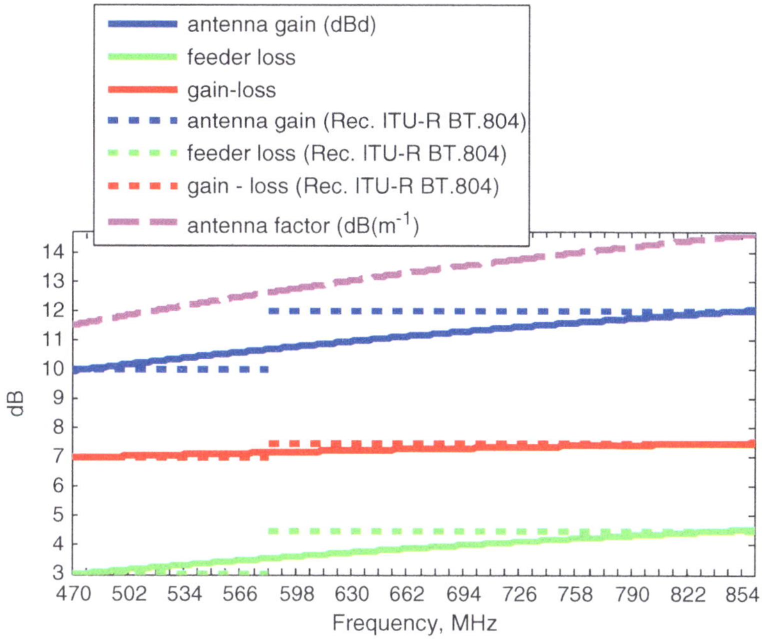

Fig. 4.4 Model curves for the selected specifications for a fixed rooftop antenna installation. The *upper* scale shows the TV channel number

Estimation of the Minimum Acceptable Power Level

The minimum usable power level is estimated from the minimum field strength given in Sect. 2.2.1.1. The calculations are done in the following sequence:

1. Unit-less antenna gain G is converted into antenna aperture area $A = G\lambda^2/4/\pi$ $[m^2]$, where λ is wavelength;
2. The incident flux S is computed as $S = E_{min}{}^2/120/\pi$ $[\mathrm{W{\cdot}m^{-2}}]$, where E_{min} $[\mathrm{V{\cdot}m^{-1}}]$ is the minimum field strength.
3. Incident power P_i is computed as $P_i = S{\cdot}A$ [W], and may then be converted into dBW or dBm.

The result of this calculation, based on the model discussed in Sect. 2.2.1.3 is shown in Fig. 4.5. The values shown represent full power (i.e. within the defined signal bandwidth, e.g. 300 kHz around the video carrier for PALI, rather than within the spectrum analyzer's resolution bandwidth).

Considering that the typical uncertainty in the field measurements is on the order of a few dB, it is possible to say that the minimum required power level for the power incident onto a fixed rooftop TV antenna for PAL-I is about **−56 dBm**.

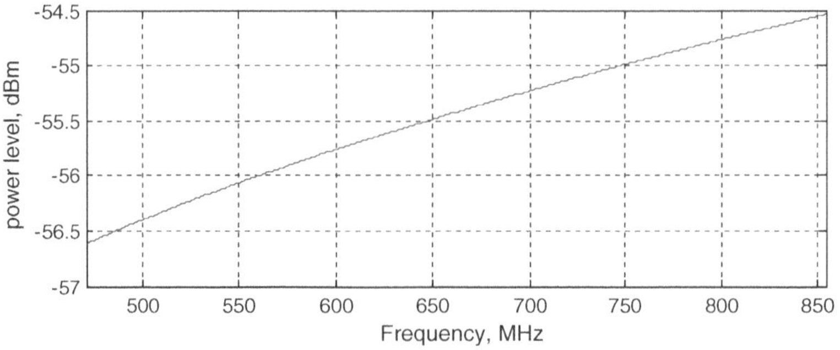

Fig. 4.5 Minimum incident power level at the antenna, corresponding to the minimum field strength for PAL-I reception

4.2.2.2 Digital Broadcasting: DVB-T2

DVB-T2 is based on orthogonal frequency division multiplexing (OFDM). A sample of a measured digital signal's spectrum is shown in Fig. 4.2c. The spectrum is composed of a multitude of tightly packed subcarriers appearing as a continuous, nearly rectangular block. The bandwidth of the "standard" DVB-T2 signals is 7.61 MHz [22, 26]. In South Africa, DVB-T2 is said to currently run in 256 QAM, code rate 3/5, PP4, 32k extended mode, with 1/16 guard interval, resulting in a slightly wider bandwidth of 7.77 MHz.

Minimum Field Strength: DVB-T2

The sample scenarios considered in [22] reveal a minimum equivalent field strength at receiving location at 650 MHz of 45.3 dB(μ V/m) for a fixed scenario, 50.2 dB(μ V/m) for a portable outdoor/urban scenario, and 42.5 dB(μ V/m) for a mobile/rural scenario etc. Just like for the analogue signal, these values are extrapolated over the UHF band by using a logarithmic formula. The main difference is that the mobile and portable etc. scenarios were introduced and are treated differently to a fixed scenario:

for fixed reception: $F_{s1} = F_s + 20\ \log 10(f_1/f)$, and

for portable and mobile reception: $F_{s1} = F_s + 30\ \log 10(f_1/f)$,

where F_{s1} is the minimum median field strength for a frequency f_1 using the value of the field strength F_s for the frequency f given, applicable within the Bands IV/V.

Models for Antenna Installation and Resultant Minimum Acceptable Power Level

The same model is used as in Sect. 2.2.1.3.

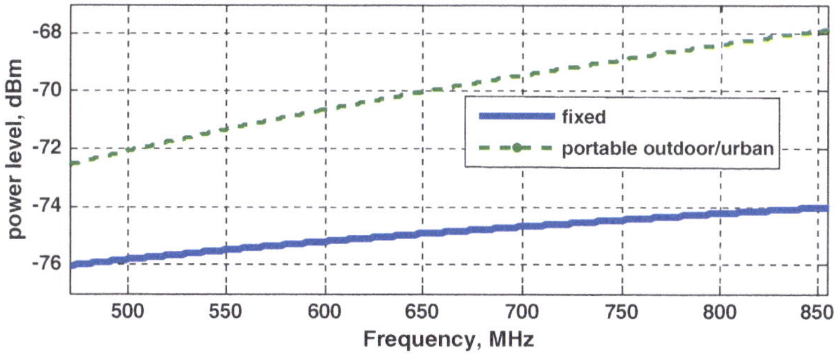

Fig. 4.6 Minimum incident power level at the antenna, corresponding to the minimum field strength for DVB-T2 reception under the "fixed" and "portable outdoor/urban" scenarios

Repeating the same type of calculations as in the case of analogue signals, it is possible to obtain the minimum required incident power level for fixed scenario DTT with a "standard profile" DVB-T2.The result is shown in Fig. 4.6. As it follows from the figure, the minimum required power level for the power incident onto a fixed rooftop TV antenna for DTT is about **−75 dBm**. This figure is expected to be slightly stricter than the one required for the South African mode of DVB-T2 (because the bandwidth of the South African signal is slightly wider).

It may need to be taken into account that a margin additional to this minimum required power level is normally necessary to ensure protection of the primary users, i.e. TV broadcasting.

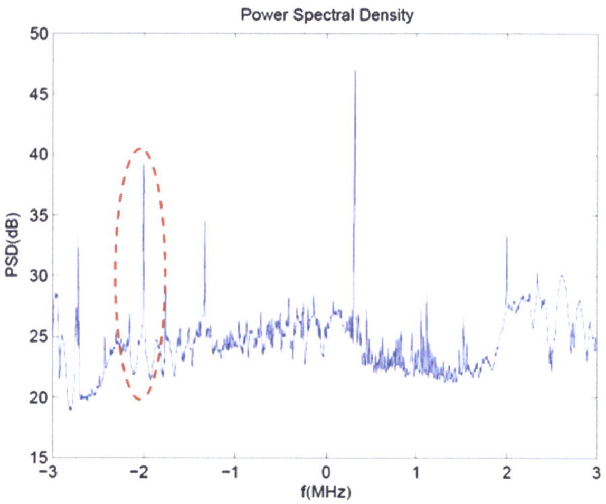

Fig. 4.7 Received periodogram of a typical TV channel with a wireless microphone signal and narrowband interference signals. The channel centre frequency is converted to the baseband, and the scale of the y-axis is not calibrated. The wireless microphone signal is at f = −2 MHz (highlighted), and the other spikes are from unknown emissions [30]

4.2.2.3 Note on Wireless Microphones

Wireless microphones that are used for Public Address systems often also operate in the UHF band and are treated as primary users. Wireless microphones are capable of emitting bandwidths from 50 kHz to 600 kHz [27]. A sample of the spectrum emissions produced by FM wireless microphone is shown in Fig. 4.7. It is clear that this type of transmitteris difficult to detect (e.g. the measurements to capture the spectrums shown were carried with the resolution bandwidth of 1.8 kHz). A number of spectrum measurement scan also be observed in [28]. The approach currently used by the FCC is to create dedicated channels for wireless microphones and store these in the geo-location spectrum database [29].

The spectral profiles of the expected types of signals have now been considered to a degree sufficient for visual recognition from a spectrum scan made with a typical set up consisting of a spectrum analyzer connected to an antenna. The next step is to take a deeper look at some of the options possible for making spectral scans.

4.3 White Space Mesurement Configuration Using the Meraka Cognitive Radio Platform

The aim of our measurements was to scan the VHF/UHF spectrum bands, i.e. from about 50 MHz–1 GHz, using a low-cost spectrum analyser based on the Ettus USRP2. We conducted the frequency scan in two different places within South Africa. One of these measurements was carried out in Pretoria (an urban area in South Africa) and the other one in Philipstown (a rural area in the Northern Cape Province).

Solutions that provide acceptable spectrum scan results using low-cost equipment are crucial if scientists in African countries, that can't afford high-grade costly scientific equipment, are to contribute to the global picture of available white space spectrum. The Ettus USRP-based spectrum analyser presented here as well as other solutions such as those based on the ASCII-32 device [10] used in Malawi are crucial to fulfill this aim.

4.3.1 Known TV Transmitters Around the Test Sites

The list of TV transmitters in Pretoria and Philipstown are shown in Tables 4.4 and 4.5, respectively. These tables were generated using data from an updated terrestrial broadcasting frequency plan of 2013 published in government gazette no 36321 by Independent Communications Authority of South Africa (ICASA) on 2nd April 2013 [31]. According to ICASA's 2013 terrestrial broadcasting frequency plan, there were no live DTT transmissions in Pretoria and Philipstown.

Table 4.4 Known broadcasting transmitters around Pretoria East

Location and coordinates	Service	Status	Ch. no	Frequency (MHz)	ERP (kW)
Pretoria (S 25° 41′ 21″ E 27° 29′ 02″)	M-Net	Operational	21	470	84.6
	CSN	Operational	25	502	28.2
	e.tv	Operational	29	534	138
Note	*SABC 1, 2 and 3 broadcast from the VHF channels 8, 5 and 11 (all at 100 kW ERP) from the Pretoria transmitter*				
Menlo Park (S 25° 46′ 16″ E 28° 16′ 05″)	SABC 2	Operational	53	726	0.04
	SABC 1	Operational	57	758	0.04
	SABC 3	Operational	65	822	0.04
	M-Net	Operational	61	790	0.04
	CSN	Operational	44	654	0.04
	e.tv	Operational	48	686	0.04
Sunnyside (S 25° 45′ 58″ E 28° 12′ 21″)	SABC 2	Operational	55	742	1
	SABC 1	Operational	63	806	1
	SABC 3	Operational	59	774	1
	M-Net	Operational	67	833	1
	CSN	Operational	46	670	1
	e.tv	Operational	38	606	1
	TSHW	Operational	27	518	1
Pretoria-North (S 25° 41′ 29″ E 28° 10′ 02″)	SABC 2	Operational	40	622	0.05
	SABC 1	Operational	52	718	0.05
	SABC 3	Operational	46	670	0.05
	M-Net	Operational	50	702	0.125
	CSN	Operational	56	750	0.12
	e.tv	Operational	37	598	0.05

Table 4.5 Known TV broadcast transmitters around Philipstown

Service	Status	Ch. no	Frequency (MHz)	ERP (kW)
SABC 2	Operational	26	510	0.004
SABC 1	Operational	22	478	0.004
SABC 3	Operational	30	542	0.004
e.tv	Operational	34	574	0.004

4.3.2 Meraka Cognitive Radio Platform

The Meraka Cognitive Radio Platform (MCRP), operating as a low-cost spectrum analyser, was used to conduct RF spectrum measurements in both urban and rural areas of South Africa. The MCRP is shown in Fig. 4.8. The platform consists of four

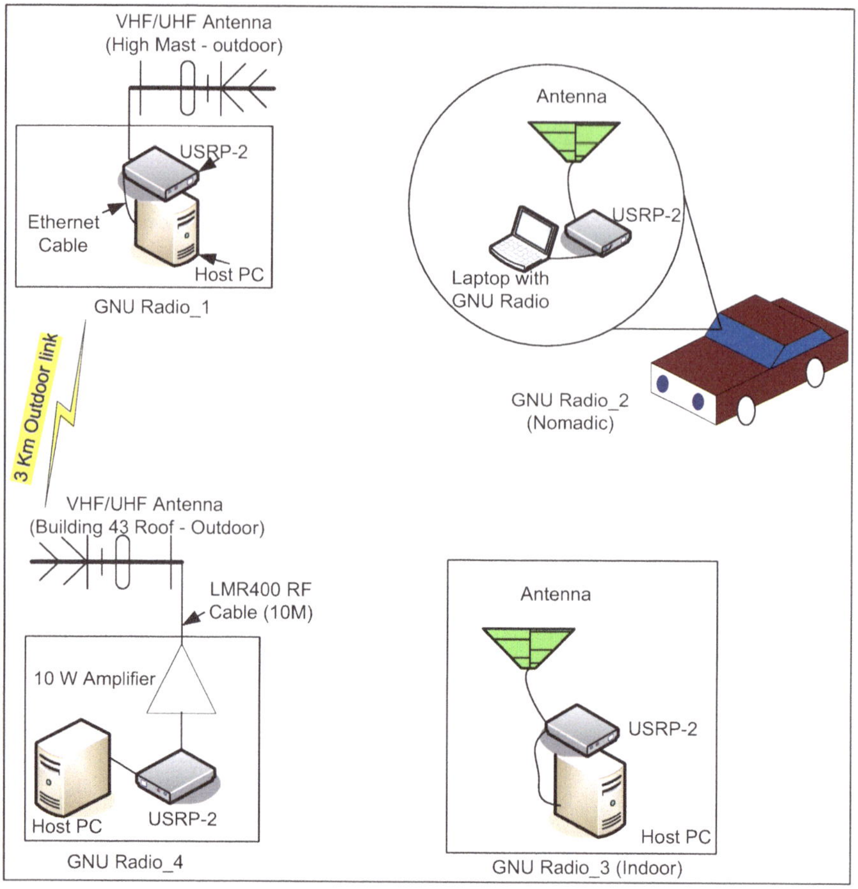

Fig. 4.8 The Meraka Cognitive Radio Platform [8]

nodes, and each node is connected to the Internet using the Ethernet cable (when in a fixed environment). One of these nodes can operate as a hand-held or standalone receiver or transmitter using a laptop with GNU Radio software. A single fixed node is built up of three major hardware components: (i) a high speed computer (powered by 2.60 GHz Dual Core Intel Pentium Processor, 2 GB memory and 500 GB hard-drive), (ii) version two of the Universal Software Radio Peripheral (USRP-2) package (with a single WBX daughter-board) and (iii) an antenna.

The USRP-2 is composed of a motherboard and one or few daughter-boards. The USRP-2 is a flexible Software Defined Radio (SDR) device developed by Ettus Research LLC [32]. The SDR is a radio communication system where components that would have typically been implemented in hardware are implemented using software. The motherboard performs baseband processing while the daughter-boards provide the RF front-end part of the radio. The MCRP is built on WBX daughter-boards with the transceiver covering the 50 MHz–2.2 GHz frequency range.

The MRCP is suitable for conducting experimental research studies in wireless communications. Typical functions include spectrum monitoring, a low-cost spectrum analyser and two way communication between nodes. Depending on the nature of an experiment, a 1 W amplifier can be used to boost the signal during the transmission (at unlicensed bands). An amplifier proved to be important for long distance transmission experiments since the USRP-2 is limited to 100 mW output power. As shown in the Fig. 4.8, two nodes are used for outdoor 3 km link where either node can receive or transmit in a selected frequency channel.

During the field spectrum measurements exercise, one fixed node was used for Pretoria measurements and the other nomadic node was used as a handheld spectrum analyser for Philipstown measurements. For fixed installations, a high gain Ellies VHF/UHF Combo antenna with 15 elements was used and for nomadic measurements an Ettus log-periodic 6 dBi PCB-based antenna was used. Detailed setups for each measurement are discussed in the following subsections.

4.3.2.1 Pretoria Measurements

For the Pretoria scans, an 8 dBi log periodic high gain Ellies antenna was used on a rooftop. We used a directional antenna that connected directly to the fixed MCRN equipment as shown in Fig. 4.9. Measurements were collected in one direction since the antenna was fixed. This placed the antenna at around 5 m above the ground. In Pretoria, an outdoor CR node was used for the spectrum scan, which ran for more than a week. Figure 4.10 shows a map of Pretoria containing the point were measurements were carried out and the four locations where known TV broadcasting transmitters are located.

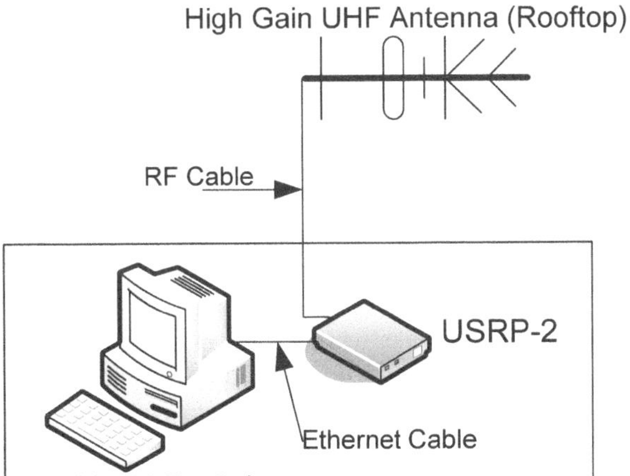

Fig. 4.9 Setup for Pretoria RF Measurements

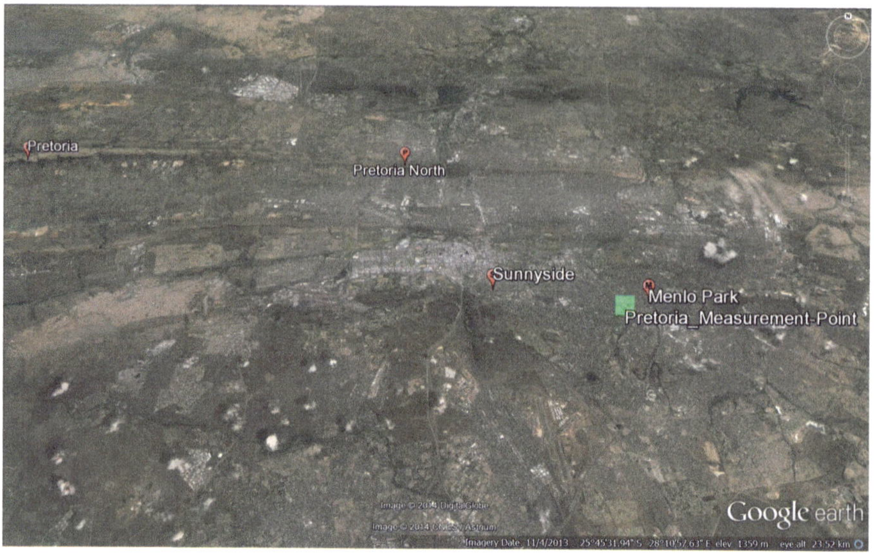

Fig. 4.10 Map of Pretoria showing known TV broadcasting transmitters and the location where measurements were collected

4.3.2.2 Philipstown

In Philipstown, measurements were collected at the location denoted by the following coordinates: S 30° 26′ 09.05″, E 24° 28′ 22.42″ E. The known TV broadcasting transmitter in Philipstown is located within a kilometer from the point where measurements were done, as shown in Fig. 4.11. Nomadic or handheld equipment was used for the spectrum scans which were collected for a few minutes in three different directions at 90° to each other.

4.3.2.3 Macha Measurements

In Macha, measurements were conducted at the location: S 16° 25′ 31.63″, E 26° 47′ 00.81″. The nearest known TV transmitter to Macha is located in Pemba, which is about 60 Km away from the point where frequency measurements were conducted as shown in Fig. 4.12. In Zambia, there are eight private broadcasters and 1 public broadcaster, Zambia National Broadcasting Corporation (ZNBC). ZNBC is authorized to broadcast nationwide, while others are restricted to a defined transmission coverage area of around (100–150 km) in radius [33]. In order to expand TV coverage in rural areas, the ZNBC signal is transmitted via satellite and then received in a number of rural districts where it is re-distributed using local terrestrial TV transmitters. By 2010, it was estimated that about 65 % of Zambian population were able to receive the ZNBC terrestrial TV signal [33]. Figure 4.13 shows how nomadic equipment was

Fig. 4.11 Map of Philipstown showing known TV broadcasting transmitters and location where measurements were collected

Fig. 4.12 Map of Macha, Zambia showing location were measurements were taken and location of TV transmitter

setup for rural spectrum measurements. Philipstown and Macha measurements were taken at human height level (1.5 m) with handheld EttusUHF antenna, LP0410 Log Periodic antenna (400 MHz–1 GHz) with 5–6 dBi gain. In all measurements where the MCRP was used, multiple consecutive scans were done using 800 kHz bandwidth and Fast Fourier Transform (FFT) size of 2,042. The data was post-processed and FFT bins were averaged to 25 kHz buckets.

4.4 White Space Measurement Configuration for the TV White Space Trial InTygerberg Near Cape Town

As a part of initiation of the TV White Space trial in Cape Town [4], measurements were made to estimate availability of the spectrum in the Ultra-High Frequency (UHF) band around Tygerberg, Cape Town, South Africa Fig. 4.13.

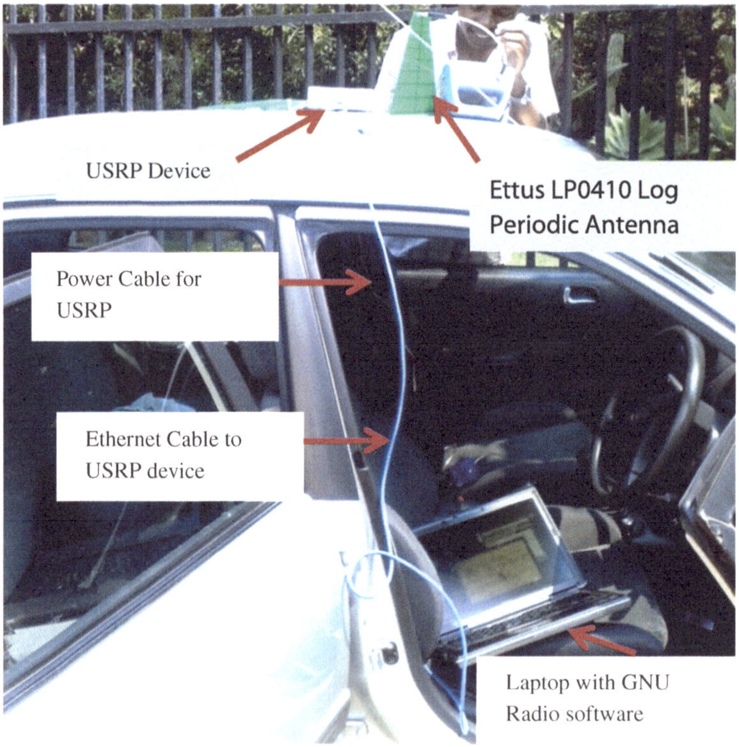

Fig. 4.13 Setup of nomadic node for outdoor remote feald scan

4.4.1 Known TV Transmitters Around the Test Site

The list of TV transmitters in Tygerberg is shown in Table 4.6. However, as it will be clear from the measurement results, there are also TV signals received in Tygerberg, which do not seem to originate from these transmitters.

The following should be considered when considering the data presented in Table 4.6:

- The Tygerberg site is located at S 33° 52′ 31″, E 18° 35′ 44″ at height 75 m, and an omnidirectional antenna with vertical polarization was used for measurements.
- Channel 38 shown in Table 4.6 was historical allocated to Cape Community TV but the current active allocation is for DTT MUX 1.
- Channel 28 is said to be allocated to mobile DTV, channels 65 and 66 to NEOTEL, and the frequency range 820–846 MHz to Studio Transmitter Links (STL). The frequencies for STL and NEOTEL overlap.

It should also be noted that Digital Dividend 1 will affect frequencies above 790 MHz. In addition, there are provisional plans to implement Digital Dividend 2 affecting the frequencies above 694 MHz [31].

Figure 4.14 shows the indicative information about the TV transmitters near the four test sites. The transmitters shown were selected using the following criteria: distance less than 300 km, operational status being "Active" and the signal strength at the location of Tygerberg hospital estimated to be more than −115 dBm. The estimation was based on the effective radiated power (ERP) and free space propagation loss [9] and did not include the influence of the actual landscape or radiation pattern.

Figure 4.15 shows the location where spectrum measurements were conducted in the Cape Town area. Information on the Cape Town measurement sites and some brief measurement details are provided in Table 4.7.

Table 4.6 TV programs in from the transmitter in Tygerberg

Service	Status	Ch. no	Frequency, MHz	ERP, kW
SABC2	Operational	22	479.25	2
SABC1	Operational	26	511.25	2
MNET	Operational	30	543.25	1
SABC3	Operational	34	575.25	2
DTT MUX1	Operational	38	610	2
CSN	Operational	42	639.25	1
ETV	Operational	46	671.25	2
DTT MUX2	Operational	50	706	2

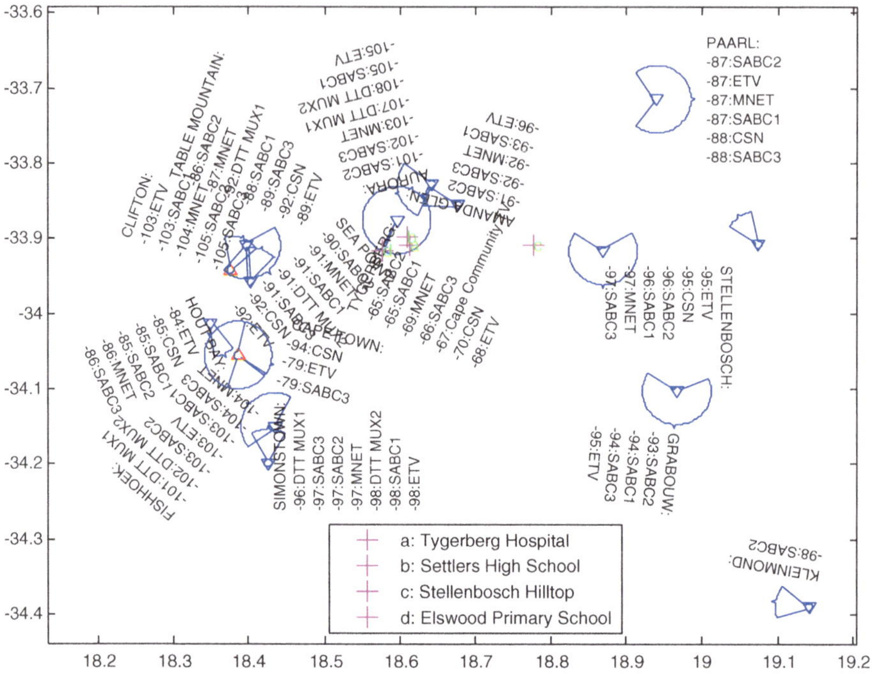

Fig. 4.14 Indication of the location and specifications of nearby TV transmitters and the directions of their radiation. The small spike on the *blue* antenna radiation pattern *curves* shows the direction of the main beam. The width of the *round* portion shows the beam width of the antenna beam. The *omnidirectional* patterns are shown as *circles* with an arbitrary direction of the main beam. The text includes the name of the TV station, approximate value of the signal received at the Tygetberg hospital (in dBm; estimated using free space loss formula), and the name of the service. The *blue flat-top* and *red flat-bottom triangles* correspond to the *vertical* and *horizontal* polarizations, respectively. The locations of the four test sites are also shown, as *magenta* crosses

4.4.2 Measurement Set-Ups

In Cape Town, some of the measurements were done using a set up schematically illustrated in Fig. 4.16.

The signal measured by the instrument equals

$$P_{rx} = P_i + G - L,$$

where G is antenna gain, L is the loss in the cable, P_i is the incident power, and all quantities are in dB. It is assumed that the connectors do influence the results in a negligible manner, e.g. compared to the uncertainty in the measurement due to the fading.

Fig. 4.15 Location of the test sites in Cape Town area

Table 4.7 Information on the measurement sites and brief measurement details

Site	GPS	Comment
Tygerberg Hospital	−33.90799, 18.61252	24 h run; 150+ scans of full UHF band
Settlers High School	−33.89764, 18.60821	Short run
Stellenbosch Hilltop	−33.90778, 18.77782 476 m altitude	Short run; had to switch from FSH6 to PR100 due to interference; directional information was obtained
Elswood Primary School	−33.91632, 18.57716 49 m altitude	Short run; Two antenna elements fall off down during the measurement due to wind; obtained 3+2 scans

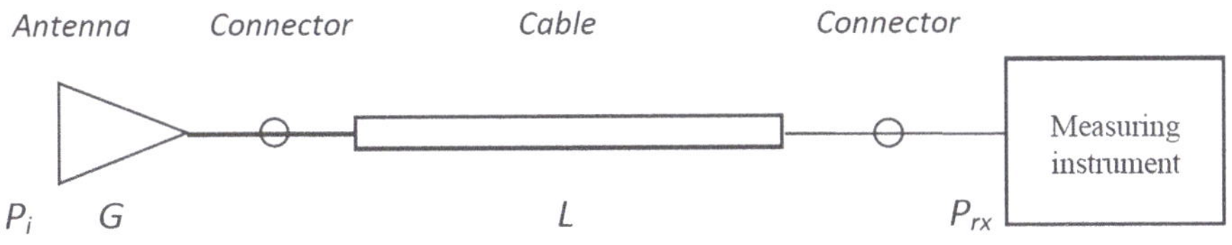

Fig. 4.16 Signal path in the measurement chain: from incident power P_i, through the antenna gain G and cable loss L, to the signal measured by the instrument, P_{rx}

4.4.2.1 Equipment Used

The 24 h long series of measurements on top of the Tygerberg hospital were performed using an R&S ESVD test receiver with a R&S HK033 omnidirectional coaxial dipole antenna.

The scans were done from 470–854 MHz (i.e. 384 MHz or 48 television channels). The R&S ESVD was used as the measuring instrument. Due to the limited memory of instrument, the whole band (470–854 MHz) was split into sub-bands measured one by one, increasing the frequency. The key specifications for the measurements are given below, in the form of human-readable commands sent to the instrument:

SCAN:FREQUENCY:STEPSIZE 100 kHz;
SCAN:RECEIVER:DETECTOR PEAK;
SCAN:RECEIVER:BANDWIDTH:IF 10 kHz;
SCAN:RECEIVER:MEASUREMENT:TIME 100 ms;
SCAN:RECEIVER:ATTENUATION:AUTO On;
SCAN:RECEIVER:ATTENUATION:MODE LowNoise;
SCAN:RECEIVER:Range 60 dB;
SCAN:RECEIVER:PreAMPLIFIER ON;

A few initial scans were performed with the preamplifier off (approximately 10 dB gain difference). This was been taken into account in processing the data acquired.

Note on Timing: *Total data collected covers a period of 24 h 20 min. The time-dependent plots and statistical characterisation use 23 h 2 min selected from the middle of this data set (approximately, from 29-Aug-2012 16:26:01 to 30-Aug-2012 15:18:52). The selection was done mainly to simplify the data processing (as one hour of data was lost due to a power failure at the measurement site).*

One full frequency scan from 470–854 MHz took on average around 9 min. Each such scan was composed of scanning through 10 sub-bands.

The rest of the measurements were done using R&S FSH6 portable spectrum analyzer, and R&S PR100 portable receiver. The antennas were always set to receive vertical polarization.

4.4.2.2 Correction Factors

Before the signals are measured by the instruments, they are received by an antenna, and passed through an RF cable. As these devices influence the results of the measurements, all of them require characterization.

The losses in the cables have been measured and the result is shown in Fig. 4.17: The curves denoted with "filtered" are the ones applied to the spectrum measurements. They were been obtained by applying a running average filter to the actually measured values. The filtering has been done to reduce the measurement noise.

The antenna gain for R&S antenna HK033 is specified to be 2 dBi $\pm$ 1.5 dB.As discussed in Sect. 4.4.2.3, the other antenna used in some of the measurements, i.e. the

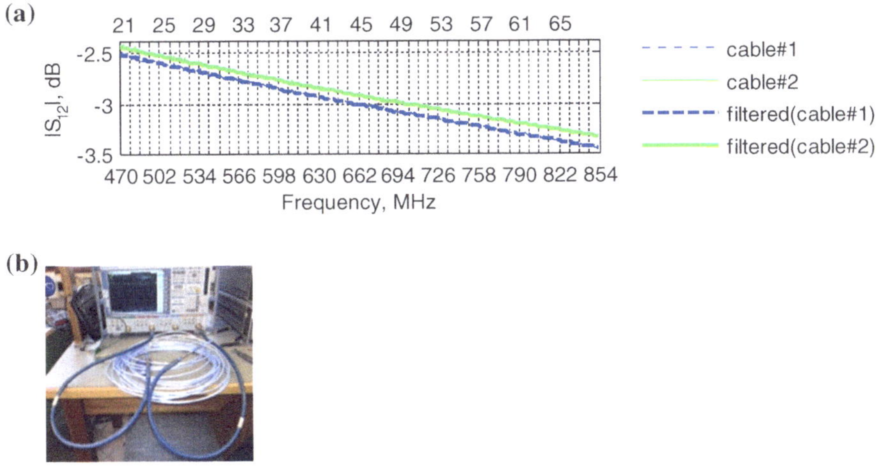

Fig. 4.17 **a** Measured loss in the cables $|S_{12}|$, and **b** measurement set up

antennas ST4, has a deep null covering several TV channels, significantly reducing the sensitivity in those bands (around 622–672 MHz, corresponding to TV channels 40–46, where the gain drops to as low as −20 dBi, thus reducing the sensitivity by 15–20 dB). This is applicable to the measurements done at the two schools.

It may be noted that the antenna gain nearly compensates for the loss in the cable.

4.4.2.3 Note on the Challenges with Using the Antenna AntennaCraft ST4

In order to ensure validity and tractability of measurements, the antennas used for the spectrum scans have been characterized. The wire discone antenna ST4 made by AntennaCraft was found to have expected features in its gain pattern. In particular, it was found that even for the horizontal plane radiation pattern, the results indicate a significant drop in the antenna gain, most pronounced at the frequency bands centered around 540–690 MHz. This was first identified from a numerical model using WIPL-D software [34, 35], and then confirmed with the open range antenna gain pattern measurements. A summary of the results of this study are shown in Fig. 4.18, where it is easy to observe a significant drop in the sensitivity of the system using ST4.

As some of the measurements shown in Fig. 4.18, were made using this antenna, the appropriate corrections have been introduced. Unfortunately, due to the low sensitivity of the system around the two identified frequency bands, the corrections also, in effect, raise the measurement noise floor around those areas.

Further examination indicated that the influence of wind or inaccurate vertical positioning of the antenna could be a significant contributing factor as well. It was found that the antenna gain variation in the vertical (elevation) plane may cause significant and frequency dependent variation in signal strength received from the

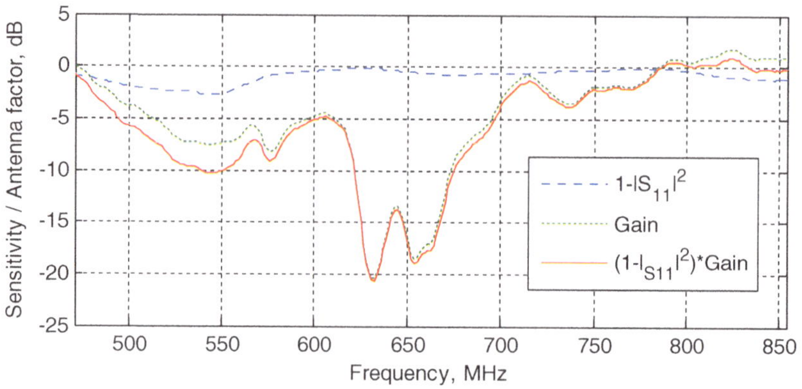

Fig. 4.18 Change in the partial system sensitivity function (shown with *red solid line*) due to the behaviour of the Gain and mismatch factor $(1 - |S11|^2)$ of the wire discone antenna ST4

antenna. For example, a −15 to 15 deg movements due to wind can cause up to 25 dB variation in the read level of the incident signal (from −20 to + 5 dBi).

The above-mentioned findings indicate a need for strict requirements on the necessity to ensure a rigid mounting of the antenna as well as a good alignment to the vertical orientation.

4.4.3 Measurement Results

The location (via Google Earth) of the measurement set up and the pictures from setting up the measurements are shown in Fig. 4.19a–d.

As it can be seen from 4.19b, there should be very little blockage to the measurements. The only blockage is due to the two towers on the sides of the building. The choice of not using one of those towers (which would have been a more ideal location) was dictated by the safety considerations of transporting the heavy equipment across the roof.

4.5 Quantifying the Amount of TVWS

This section presents the results for all the spectrum measurements conducted using the two equipment discussed in the previous sections.

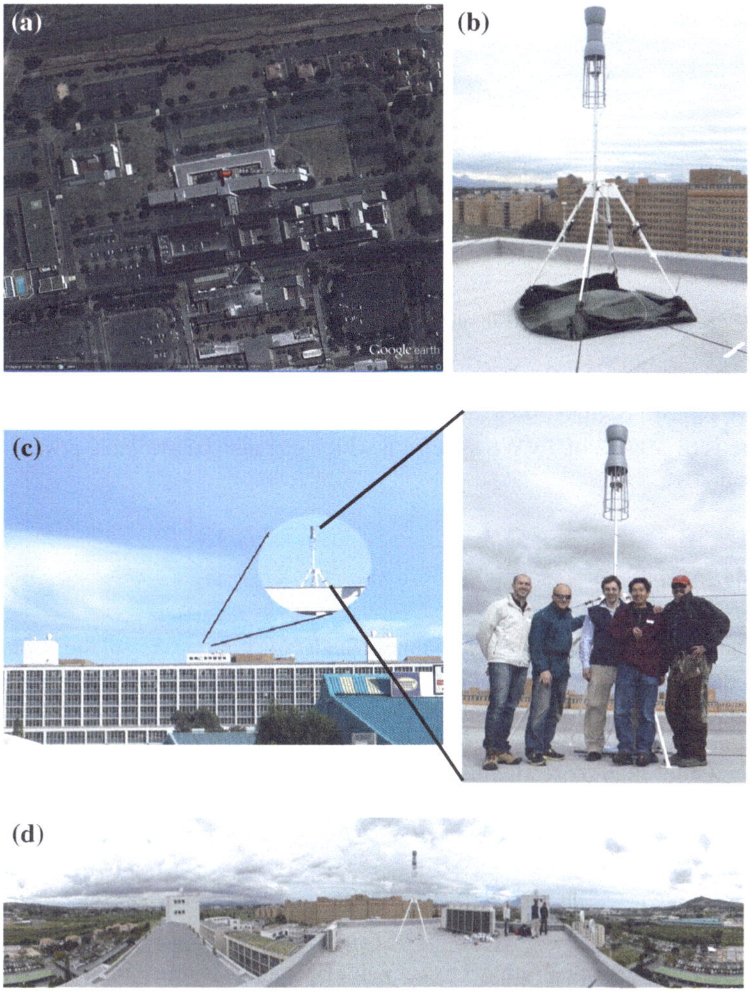

Fig. 4.19 **a** *Top* view at the measurement location; **b** measurement set up composed of the R&S HK033 antenna mounted on a tripod and connected to a computer controlled receiver (under the rain-proof cover); **c** side view at the measurement set up; **d** panoramic view from the location of the set up

4.5.1 Philipstown

Figure 4.20 shows the TV UHF band scan results performed in Philipstown. According to the national terrestrial broadcasting frequency plan of 2013 [31], there are only 4 TV stations in the area which occupy channels 22 (at 478 MHz), 26 (at 510 MHz), 30 (at 542 MHz) and 34 (at 574 MHz). However, due to the very low transmit power, only one channel (#34) was detected by equipment. With only four channels being

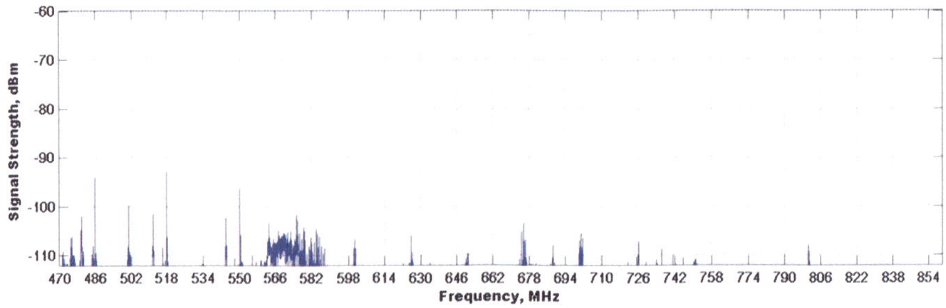

Fig. 4.20 Spectrum activity results in rural Philipstown, South Africa (relative vertical scale)

allocated, out of possible 29 channels (between 470 and 694 MHz band), it is clear that there is over 80 % of TVWS spectrum which can also be available post the DSO process in 2015.

4.5.2 Macha

In Zambia, ZNBC has only one TV channel which covers over 65 % of the population. Most of the private broadcasting channels are not covering the rural or remote areas. While we could not find the actual channels used at Pemba TV transmitter, which is the nearest transmitter to Macha, our results shows some activities between 470 and 590 MHz. Other than those channels, TVWS availability in Macha rural area is over 90 % as shown in Fig. 4.21.

4.5.3 Pretoria

Unlike in rural areas, spectrum occupancy in urban areas is much higher mainly due to a high number of TV transmitters needed to provide quality viewing for the greater population. Although some transmitters used low power, there is no channel reuse

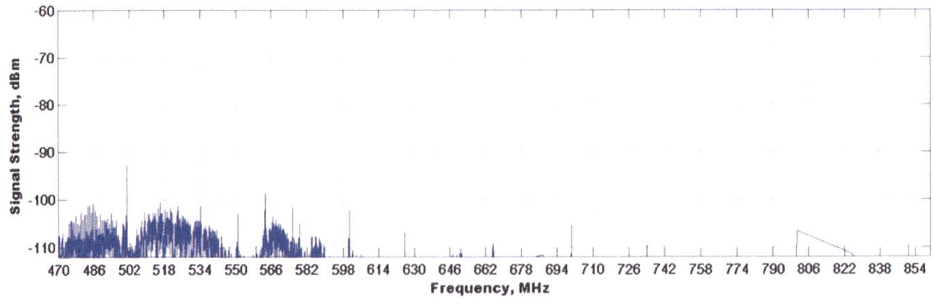

Fig. 4.21 Spectrum activity results in rural Macha, Zambia (relative vertical scale)

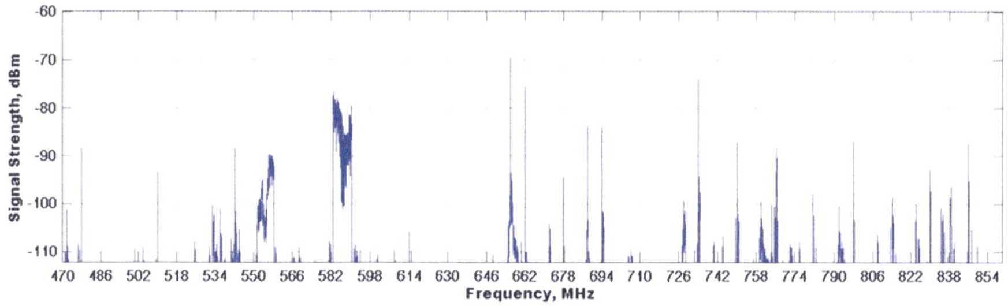

Fig. 4.22 Spectrum activity results in Pretoria West (relative vertical scale)

within the Pretoria area as shown in Table 4.4. Out of possible 29 UHF channels, a total of 22 channels were used for TV broadcast around Pretoria (there are also three high power VHF transmitters). As shown in Fig. 4.22, there are many activities which were detected during our spectrum measurements. It is important to note that measurements in Pretoria lasted for more than 24 h since the equipment was fixed with a roof-top antenna. Based on our measurements and number of channels already used for broadcasting, TVWS availability in Pretoria can be estimated to be over 40 %. However, this number can even be greater, depending on the location, especially since most of the existing transmitters are using low power.

4.5.4 Tygerberg

As a part of initiating the TV White Space trials in Cape Town, more specifically in Tygerberg, an urban area, the spectrum availability was measured at four locations. Most of the discussions are derived from the measurements made on the roof of Tygerberg hospital, where the spectrum scan was run for 24 h and used only calibrated professional grade equipment.

4.5.4.1 Samples of Spectrum Scans Sade at the Tygerberg Hospital Roof

This subsection discusses the raw results of the measurements.

Note on Data Presentation

The plots shown in the next several pages show the frequency spectrum use in the following form:

- The **upper plot** displays all the received spectra measurements (in blue) as well as the maximums over the 3-hourly averaged values (in red). The latter plot was obtained by computing averages over all values of time with a running window of 3 h, at each frequency; after this, a maximum was found of these values, again individual for each frequency. Thus computed average value shows the presence of systematically available signals in the bands versus more sporadic signals.

- *For the 3 h averaged values*: It may be noted that if TV station broadcasts only for very short time (and the transmitter is transmitting during the time only), e.g. only 1 h over any given 3 h time interval, the value will be displayed with a magnitude reduced by the ratio of the duration of the program to the 3 h reference frame. For example, a 45 min long transmission will have 1/4th of the true value and thus the displayed magnitude which is 6 dB lower then actually present. Such programs may be identified by considering this upper plot against the 3rd plots with time representation (waterfall plot). In addition, one may expect a peak at that frequency in the 2nd plot, as well.
- It was possible to increase the noise floor, by at least 10 dB, but at an expense of setting a narrower IF bandwidth and thus much longer scanning time—a much longer scanning time would result in the possibility of missing shorter transmissions.
- The top of the plot shows the TV channel numbers.
- Around the top of the plot, there are also crosses (x) to indicate the frequencies registered for some TV transmitters which could be visible in the measurements (e.g. as full-fledged profiles of signals or just weak signals).

  It may be noted that the reference frequency (stated in the ICASA documents for a particular TV transmitter) for the fully visible profiles of analogue signal correspond to the visual carrier and are usually 1.25 MHz to the right from the start of the particular TV channel (e.g. TV channel 21 starts at 470 MHz, and so the visual carrier may be expected to be around 471.25 MHz, possibly with a small offset used to minimize interference in large analogue TV broadcasting networks).

  It may be noted that the reference frequency for a digital signal usually corresponds to the middle of the TV channel.

  The associated texts tell the estimated signal strength value (in dBm) as well as the name of the station, polarization (V for vertical and H for horizontal), and the program it transmits. As it has been mentioned in the earlier Sect. 4.4.1 on the current frequency allocations around the area of interest, the estimated received power is calculated using the propagation loss derived from the free space loss formula. This simple formula disregards the antenna's radiation pattern and landscape elevation profile. In addition, the computed values relate to the total power of TV signal, whilst the measured values are based on the IF bandwidth of 10 kHz (this gives over 10 dB difference for analogue TV signal's video carrier occupying at least 100–500 kHz, and even more for digital signals). Thus, the numbers shown in this text can only be used as a guideline, and cannot be compared to most of the measurements directly.

• The **2nd** plot displays the uncertainty in the measured mean value. This was computed as a standard deviation from the raw data, at each frequency, using the time series. A high value, e.g. above 5 dB, may indicate presence of significant occasional transmissions made or presence of quiet periods.

- The **3rd** plot displays the time variation of the signals at each frequency, shown as a temperature plot. As in the previous plots, the horizontal axis shows the frequency. The vertical axis shows the time of the day (from the beginning of the measurement, before 6 PM until the end of the measurement, after 3PM on the next day). The colour of the data indicates the signal strength observed during that time at a specific frequency, e.g. red corresponds to strong signals and blue to weak signals. The colour scale is shown under this plot. It is possible to see that some signals are present on continuous basis (e.g. much of the signals in 22nd and 26th channels), whilst some signals are transmitted occasionally (e.g. signal visible in the portions of channels 27 and 31).
  - It may be noted that the colour scale on this plot has been selected to saturate strong signals (as they are already visible well), and highlight medium range signals.
  - The plot uses approximation of the spectral content being constant within one step of time ($\sim$9 min).
- The **4th** plot shows occupancy of each individual frequency resolved, indicating how much use a particular channel is estimated to be used, in percents, over the entire measurement period. This quantify is computed as a ratio of the number of times the signal exceeds a pre-defined threshold to the total number of samples (154). Three thresholds were used: −95 dBm, −90 dBm, and −80 dBm.

Sample of scans—channels 27–32

The scan results shown in Fig. 4.23 indicate that TV channels 27, 29, 31 and 32 were not actively used, channel 28 is used for digital transmission (for mobile communication needs) and channel 30 (likely, MNET) has a weak (likely unusable) TV signal. The occasional peaks present in the channels 27, 29, 31 and 32 are likely to be due to narrow-band analogue transmissions. It is possible that the signals in the channels 29 and 31 are paired.

It is also possible to see that the analogue TV transmission in channel 26 (shown only partly) has NICAM sound, in addition to the normal audio component.

Sample of scans—channels 33–38

In Fig. 4.24, the plots indicates that channels 34 (SABC3) and 38 (DTT MUX 1 or Cape TV) are definitely used, by analogue and digital transmissions, respectively. The channels 33 and 35 seem to be mostly available. The channels 36 and 37 have low level signals/traces of what appears to be TV signal(s) from far away transmitter(s).

It may also be noted that the analogue TV transmission in the channel 34 does not have NICAM sound.

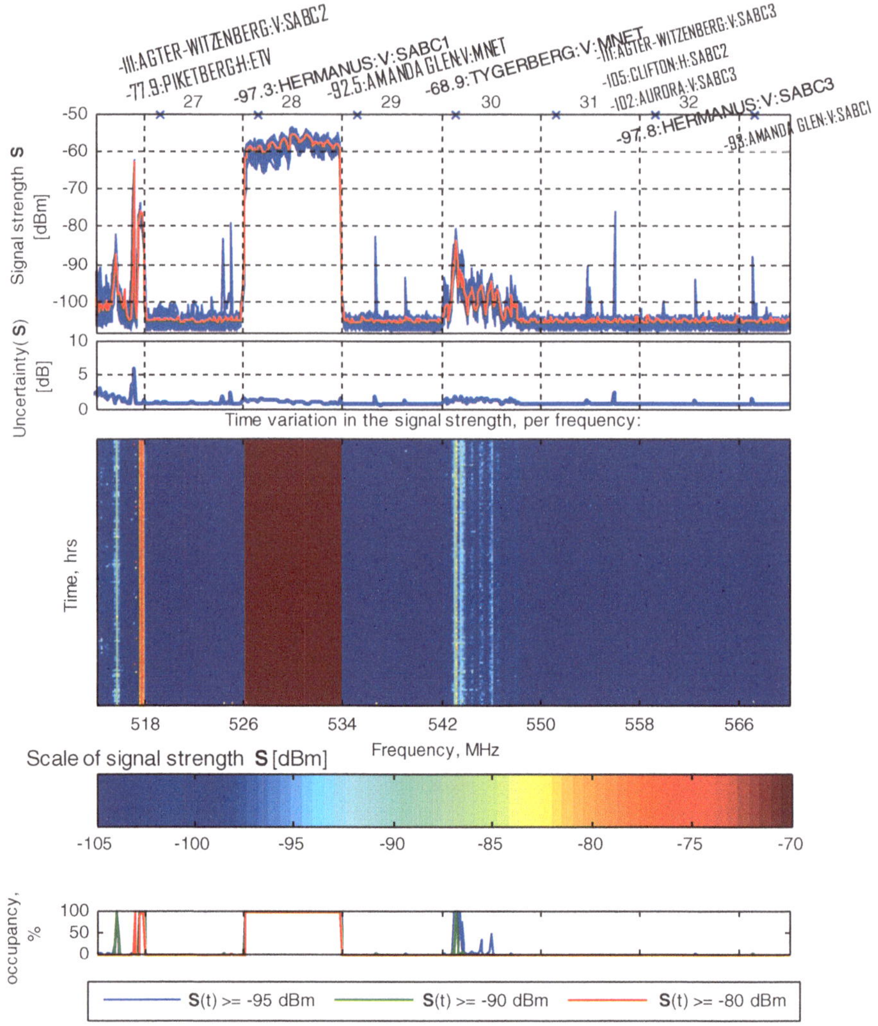

Fig. 4.23 Spectrum activity results for Tygerberg hospital roof: TV channels 27–32

The spectrum activity plot in the upper part of Fig. 4.24 may be compared to an equivalent plot based on measurements made in May 2013, approximately 8 months after the first round of the measurements. The latter plot is shown in Fig. 4.25. The interpretation of the data in this plot requires taking the following into account:

- Unlike the much longer measurements used for Fig. 4.24, this new set of data is based on two rounds of just 3 frequency scans over a total period of less than 30 min.

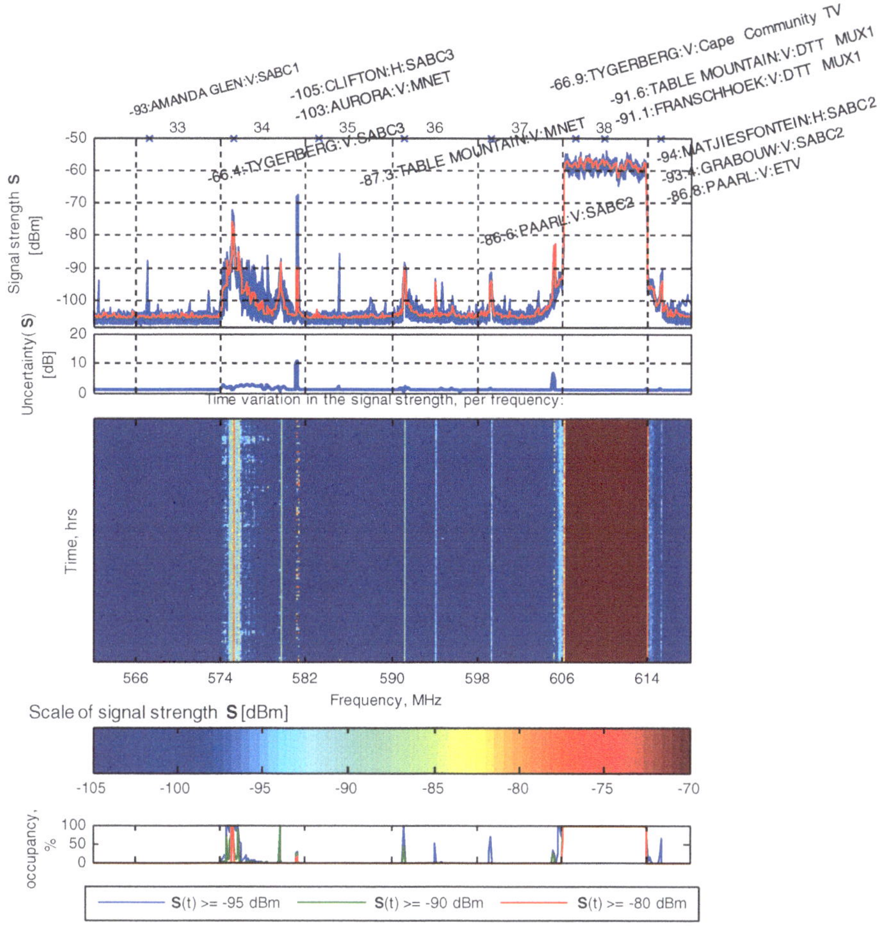

Fig. 4.24 Spectrum activity results for Tygerberg hospital roof: TV channels 33–38 in Aug 2012

- The plot displays all the received spectra measurements, i.e. 3 scans made over 27 min long period (in green) as well as the two averages over the duration of the scans (in red).
  - Each spectrum scan has been converted from the received voltage in dBμ V into received power in dBm and then into the strength of the incident electric field in dB(μ V/m), as discussed in Sect. 4.6.1.
  - One curve corresponds to the mean value of the scans for the scenario when a transmitter(TR) is on and transmitting.
  - Another curve corresponds to the mean value for the TR being off and silent.
  - Mean values are based on the three data points, for most of the measurements.
  - The top of the plot shows the TV channel numbers, as well as lists some of the TV transmissions that may be seen in the measurement. The propagation

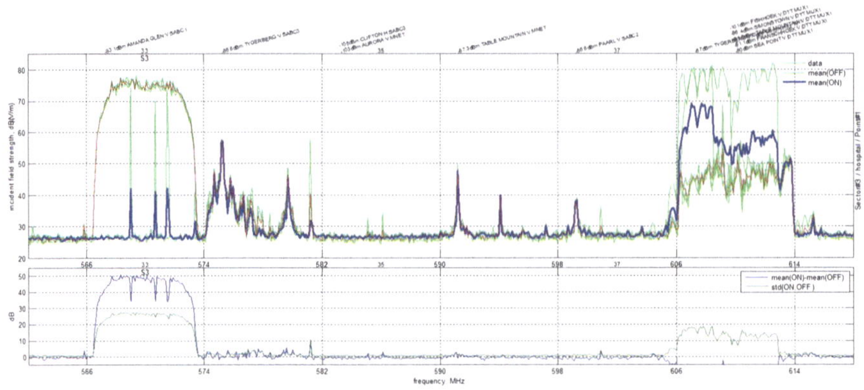

Fig. 4.25 Spectrum activity results for Tygerberg hospital roof: TV channels 33–38, in May 2013

loss was estimated using the free space loss formula disregarding the antenna's radiation pattern and elevation profile. Thus, the numbers can only be used as a guideline, and cannot be compared to most of the measurements directly.

Comparing the plots in Figs. 4.24 and 4.25, one may find that the plots are effectively the same for the contributions due to the most of TV broadcasting channels (except the channel 38). They remain essentially the same. In May 2013, the channel 38 was being tested by SENTECH and so it was periodically switched on and off. Thus, the scanned values were different.

The spectrum in the channel 33 is an overlap of 6 scans, of which 3 are for a recently introduced transmitter being on and other 3 for the same transmitter being off. It is possible to see that when the transmitter is off, there are some other signals (3 peaks) present in the band. It was checked against the ICASA frequency allocation tables and no legal allocations (except for the TV broadcasting) were found for this channeling this geographic area.

4.5.4.2 Summary of the Results

The results are summarized in Fig. 4.26. The four upper plots show the individual spectrum activity per location. The lowest plot shows an overlap of those plots, to highlight the good correlation between the scans obtained around Tygerberg area.

The good correlation between the results measured in different parts of Tygerberg and even from the hilltop in Stellenbosch indicated that the shadowing effects did not disturb the measurements and the spectrum scans obtained may be used as a reliable source of data to estimate spectrum availability.

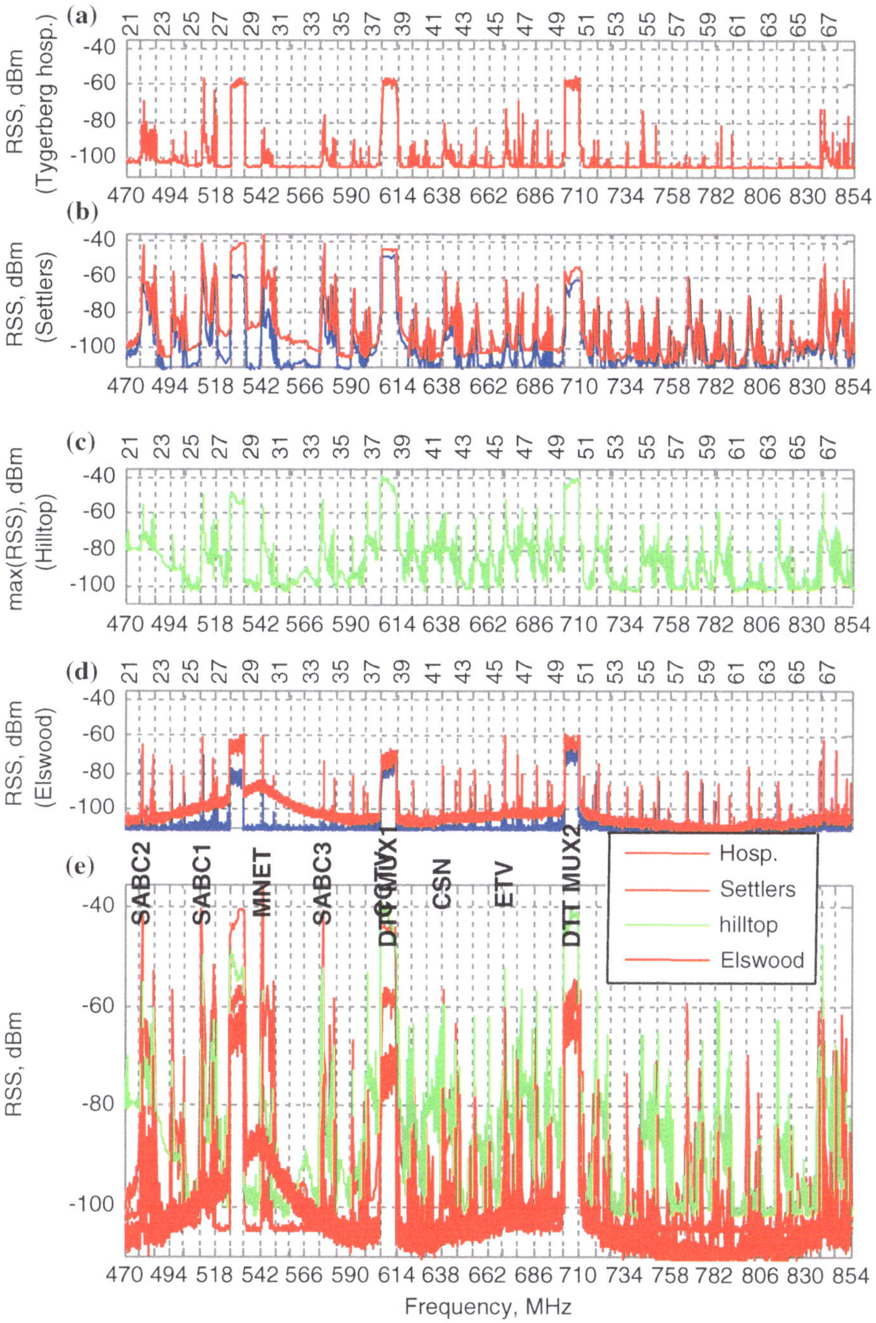

Fig. 4.26 Spectrum activity comparisons for four locations near Tygerberg, Cape Town, South Africa. The four *upper* plots correspond to the measurements done **a** at the roof of Tygerberg hospital, **b** at Settlers High School, **c** at a hilltop near Stellenbosch, **d** at Elswood Primary School, **e** overlaps of plots **a**–**d**

4.5.5 Notes on Criteria for Evaluation of Spectrum Availability

Determination of availability of a channel may be made using a number of various criteria. Disregarding the more advanced and accurate but computationally very demanding approaches, such as the ones based on singular value decomposition, the following discussions may be made.

In Fig. 4.27, each plot shows a histogram illustrating the occupancy of a channel. The horizontal scale is for the signal strength and maps strengths from −115 dBm to −50 dBm (the values are not shown in order to keep the picture clearer). The vertical axis is in logarithmic scale, showing the number of times the signal in the band had this (corresponding to the respective value on the horizontal axis) level. This presentation permits illustrative interpretation of the availability. For example, the channels 21, 29, 53 and 61 have only low level present in them, and never encountered a high level signal. This may be interpreted as availability of these channels. On the contrary, the channel 22 has nearly uniform occurrence of low and high signal levels, and can thus be considered as busy.

The simplest way to estimate the availability is by using a threshold of a predefined level. Several thresholds were tested. One was based on the maximum value of the signal observed in a given TV channel. Another threshold was based on the mean value of the signal observed in a given TV channel. Also, for a reference purpose, a metric based on the minimum value of the signal observed in a given channel was also applied. The results, based on the Tygerberg hospital roof measurements, are shown in Figs. 4.28 and 4.29.

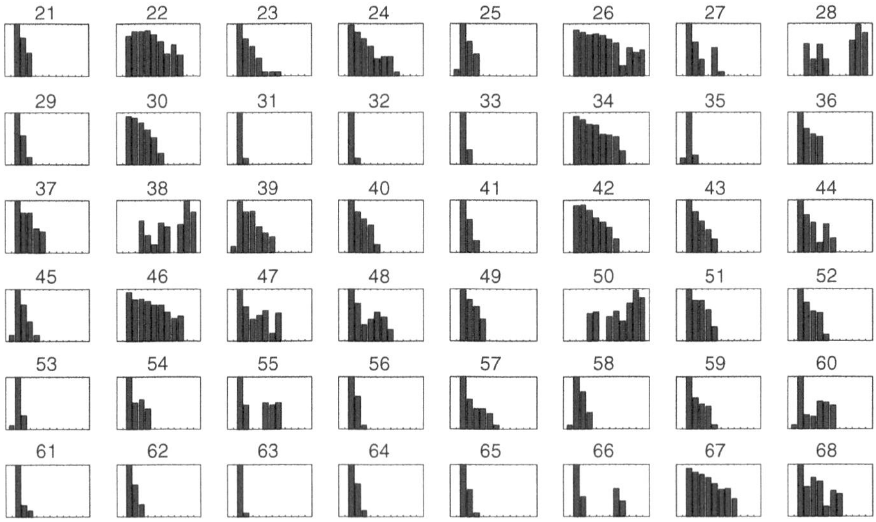

Fig. 4.27 Distributions of occupancy, per TV channel

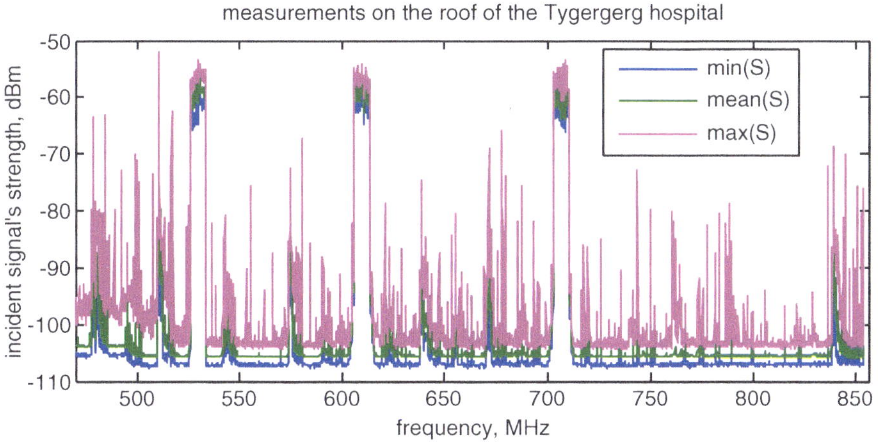

Fig. 4.28 Signal strength versus frequency (via different metrics: minimum, average and maximum levels of the signal detected over time in each frequency band resolved by the receiver)

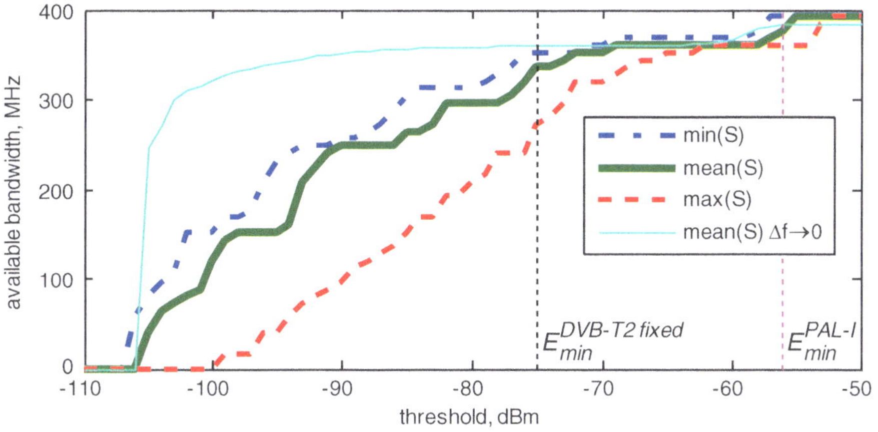

Fig. 4.29 Availability of spectrum in TV bands around Tygerberg hospital in August 2012, estimated using different threshold types, based on the minimum, mean and maximum values of the signal observed at each 8 MHz wide TV channel (except for the curve 'mean(S) Δf→0' calculated using 100 kHz wide channel width)

Figure 4.28 illustrates the effect of different metrics used for the processing the spectral signature in time domain. It affects the perceived noise floor as well as the perceived maximum level of the signal. It may also be noticed that the quantity of the effect differs for the analogue and digital signals, likely due to the way the receiver measures the signal using its swept narrowband filters.

Figure 4.29 shows the availability of the spectrum calculated from the data shown in Fig. 4.28. There are several observations and conclusions which may be made from the plot in Fig. 4.29:

- The ‘minimum acceptable signal strength’ indicators derived in Sects. 2.2.1.5 and 2.2.2.2 for PAL-I and DVB-T2, respectively, yield huge amounts of the white space available, on the order of 336–376 MHz, or 86–96 % (the first number is given for DVB-T2 fixed scenario and the other number is giver for the PAL-I fixed scenario). Even with an additional margin of 10 dB, there is still over 43–92 % of the TV spectrum that is available.
- Introduction of digital terrestrial television reduces the amount of white space available, because of the better sensitivity afforded by the digital receivers. However, the quantity of the spectrum available for local reuse will still be significant.
- The curve ‘mean(S) $\Delta f \rightarrow 0$’ shows the availability of the spectrum, when the width of a channel is very small (here, it is equal to the frequency step of the receiver, or 100 kHz). This curve may be interpreted as an upper limit for the amount of spectrum without much obvious activities (and also including the gaps in the spectrum signatures from the valid users).
  - This curve is well above the equivalent curve ‘mean(S)’ for 8 MHz wide channel, but the difference is reduced with the growth in the level of the threshold.
  - Around the noise floor, the curves converge due to high uncertainty in the measured values.
- Using the worst case scenario of a threshold, i.e. the maximum value of the signal observed in a TV channel (curve ‘max(S)’) still shows a very large amount of white space.

Comparing the relatively densely occupied spectrum shown in Fig. 4.28 against the measurements done in more rural areas, like those from Sects. 4.5.1 and 4.5.2, reveals an abundance of TV white space spectrum relative to urban areas such as Cape Town.

4.6 Notes for Optimizing Measurements

As discussed in 4.4.2.3, it may be important to ensure the correct choice of antenna and the antenna mounting.

Depending on the instrumentation available, one may need to calibrate the measurement set up against a reference. For example, the commonly accepted correct method of measuring power is using RMS (root mean square) detector. Some, especially older, instruments may not have a RMS detector available. If so, one may need to calibrate the instrument against another, reference instrument. A comparison of the measurement results using different spectrum analyzers is shown in Fig. 4.30. It should however be noted that the calibration factor may depend on the type of signal being measured.

Considering that the spectrum scans normally consume significant time, it may be desirable to optimise the overall measurement time by reducing the measurement time set in the measuring instrument. This is especially applicable to the older instruments. In order to minimize the measurement time whilst keeping sufficient accuracy of signal presentation, a study was done on the effect of measurement (integration)

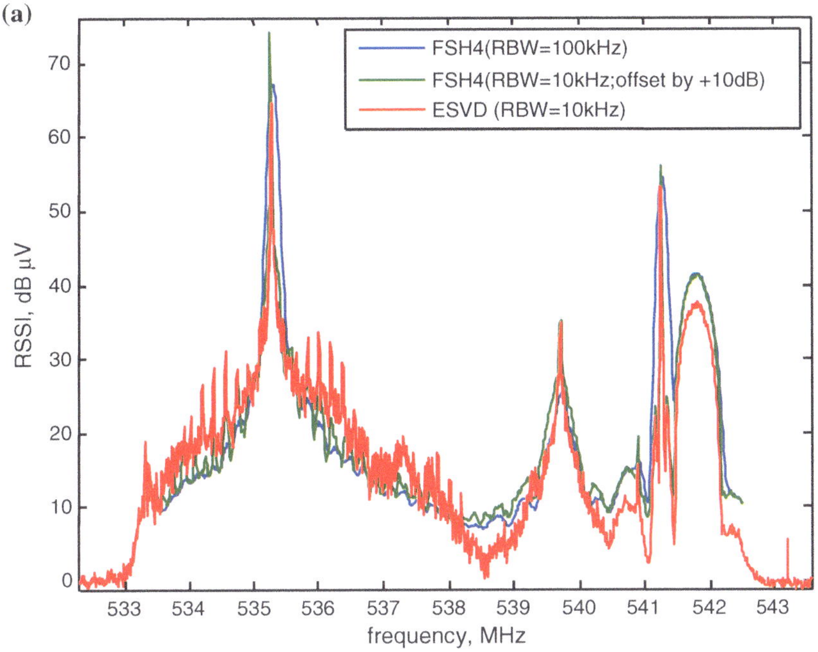

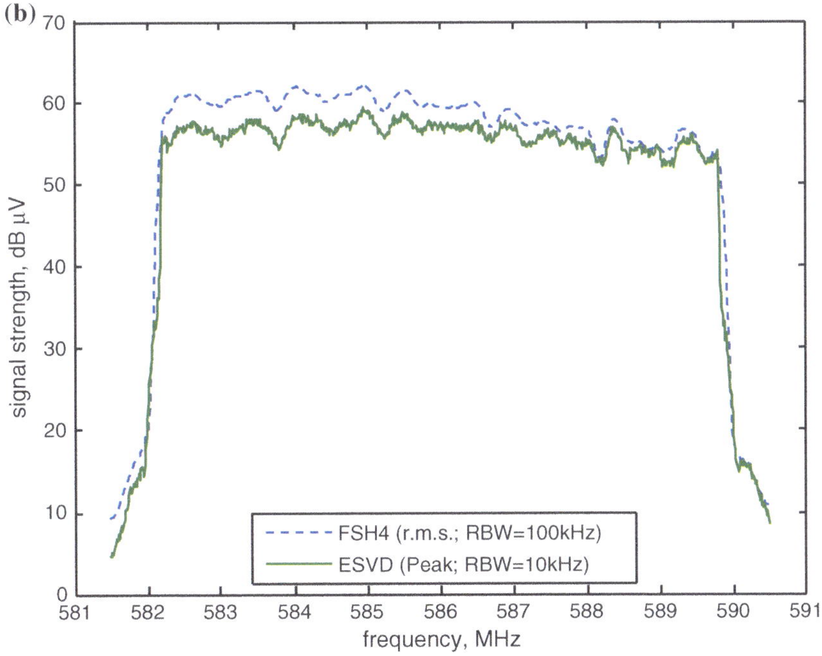

Fig. 4.30 Spectrum of **a** analogue PAL-I and **b** digital broadcasting signals, as measured by FSH4 with RMS detector and resolution bandwidth (RBW) of 100 kHz, and ESDV with Peak detector and RBW of 10 kHz

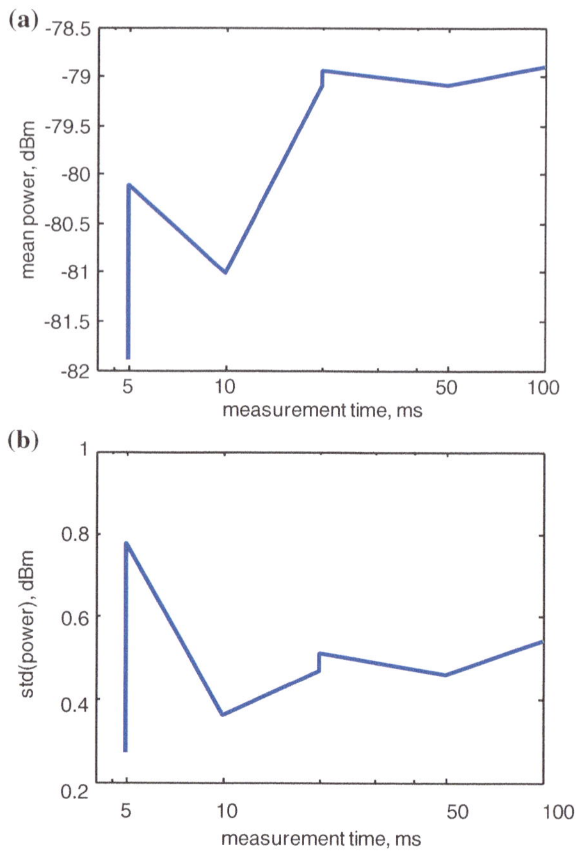

Fig. 4.31 Average total power (**a**) and its standard deviation (**b**) versus measurement time

time on the result of the measurement with R&S ESVD test receiver using the PEAK detector. A live digital terrestrial television (DTT) broadcast was used as a reference signal, whilst it was measured using different values of detector *measurement time*. The plots for the variation in the average power measured using Peak detector and respective standard deviation in the data sets are show in Fig. 4.31. The standard deviation remains about the same (determined by the source DTV signal), whilst the measured average of the peak grows slowly and stabilizes at 20 ms. It may be concluded that 20 ms, is sufficient time for an accurate measurement (within the uncertainty of the measurement).

4.6.1 Correspondence Between Incident Power and Incident Field Strength

The results stated for the Tygerberg measurements have been presented in terms of the incident power, in dBm. This quantity reflects the power of a signal incident from the air onto the antenna and, as per Sect. 4.4.2, is calculated by subtracting the gain of the antenna and adding the loss in the cable to the values measured by a spectrum analyser.

It may however be noted that the broadcasting regulations specify the minimum acceptable value of the field strength rather than the power. As an equivalent alternative to the equation shown in Sect. 4.4.2, the field incident onto an antenna [dB(μ V/m)] may be computed as a sum of the voltage measured by the receiver [dB(μ V)], antenna factor AF [dB(m^{-1})] and losses in the cable [dB]:

$$\mathrm{E_i} = \mathrm{V_{rx}} + AF \ + \mathrm{L}, .$$

where voltage measured by the receiver is related (for a 50 Ω transmission line) to the power measured by the same receiver as

$$\mathrm{V_{rx}} = P_{rx} + 107dB.$$

The antenna factor may be computed as a function of wavelength λ [m] and unitless antenna gain G_u as

$$\mathrm{AF}_u = \frac{9.73}{\sqrt{\mathrm{G}_u}}, \quad \mathrm{AF}\,(\mathrm{dB}) = 20 \cdot \log_{10}\left(\frac{9.73}{\sqrt{\mathrm{G}_u}}\right), \quad G\,(\mathrm{dB}) = 10 \cdot \log_{10}(\mathrm{G}_u)$$

As a note, one may easily notice that the uncertainty in the value of the gain, given in dB, will translate into the same uncertainty in the antenna factor and field strength.

These formulas may be for example applied to the antenna used for the measurements on the roof of Tygerberg hospital. Considering that the antenna gain for R&S antenna HK033 is specified to be 2 dBi ± 1.5 dB, the antenna factor may be calculated and the result is plotted in Fig. 4.32.

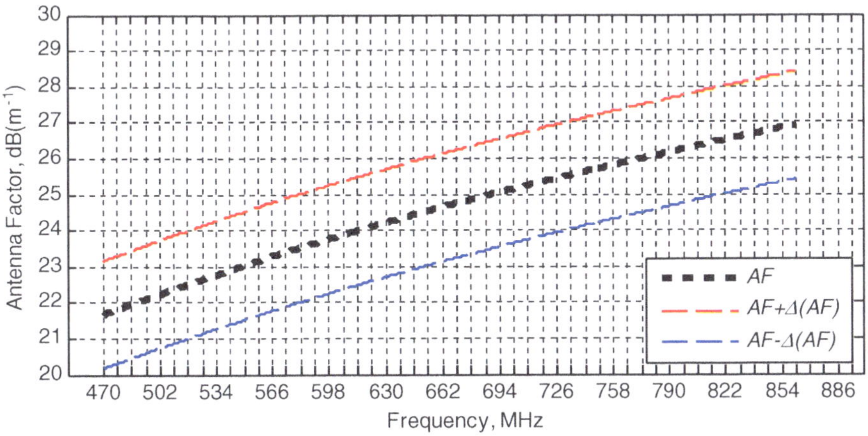

Fig. 4.32 Antenna factor computed for the antenna R&S HK033, together with the *upper* and *lower* bounds based on the specification of the uncertainty in the gain

4.7 Conclusion

In this chapter we have introduced a standards-based metric for quantifying the amount of available TV white space. Spectrum scans reveal an abundance of TV white space bandwidth, especially in rural areas, that may be used for broadband or machine-to-machine (M2M) communications in Africa. The urban areas may offer at least 40 % of the UHF spectrum to low RF power TVWS applications. The rural areas are likely to offer more than 90 % of the UHF spectrum. The spectrum scan measurements were carried out during the current dual-illumination period and even more TVWS will be released after the global digital switch over (DSO). This study serves as a starting point towards informing the development of fully operational TV white space networks for broadband or M2M connectivity. It is likely that rural areas will be a specific target of TV white space broadband networks as they provide excellent propagation characteristics for reaching sparsely populated regions with large distances between population settlements. There are ongoing research efforts towards understanding the TVWS ecosystem in South Africa through TVWS trials and a planned countrywide study of TVWS availability and these studies will ultimately culminate in new regulation on spectrum use in TV white spaces in South Africa -most likely in early 2015.

References

1. ITU, Final acts of the regional radiocommunication conference - GE06, International Telecommunications Union - Radio Sector, Switzerland, Geneva (2006)
2. FCC, Second report and order and memorandum opinion and order in the matter of: unlicensed operation in the tv broadcast bands, ET Docket No. 08–260, Federal Communications Commision, Washington, DC, United States (2008)
3. Ofcom, TV white spaces: approach to coexistence - technical analysis, Office of Communciations, London, United Kingdom (2013)
4. TENET, The Cape Town TV white spaces trial, Available: http://www.tenet.ac.za/tvws. (2013) Accessed 14 March 2014
5. Mikeka, C., Thodi, M., Mlatho, J., Pinifolo, J., Kondwani, D., Momba, L., Zennaro, M., Moret, A.: Malawi television white spaces (TVWS) pilot network performace analysis. J. Wirel. Networking Commun. **4**(1), 26–32 (2014)
6. Haji, M. A.: Licensing of TV white space networks in Kenya, Geneva, Switzerland: Presented at ITU-R SG1 Workshop: Spectrum management issues on the use of white spaces by cognitive radio systems (2014)
7. Mfupe, L., Montsi, L., Mzyece, M., Mekuria, F.: Enabling dynamic spectrum access through location aware spectrum databases. In: IEEE Africon, Mauritius, 9–12 Sept 2013
8. Masonta, M., Johnson, D. L., Mzyece, M.: The White Space Opportunity in Southern Africa: Measurements with Meraka Cognitive Radio Platform, Springer Lecture Notes, Popescu-Zeletin et al. (Eds.), vol. 92, Part 1, pp. 6–73 (2012)
9. Lysko, A. A., Masonta, M., Johnson, D. L., Venter, H.: FSL based estimation of white space availability in UHF TV bands in Bergvliet, South Africa, in SATNAC (2012)

10. Zennaro, M., Pietrosemoli, E., Mlatho, J., Thodi, M., Mikeka, C.: An assessment study on white spaces in Malawi using affordable tools. In: IEEE Global Humanitarian Technology Conference, Seattle, USA (2012)
11. Kagarura, G.M., Okello, D., Akol, R.N.: Evaluation of spectrum occupancy: a case for cognitive radio in Uganda. In: IEEE 9th International Conference on Mobile Ad-hoc and Sensor Networks, Dalian, China (2013)
12. Masonta, M., Mekuria, F., Mzyece, M.: Analysis of ICASA broadcasting frequency plan for possible use of TV white space for broadband access. In: Southern Africa Telecommunication Networks and Applications Conference (SATNAC), Stellenbosch, South Africa (2013)
13. ITU-R, Conventional analogue television systems, Recommendation ITU-R BT.470-7, Geneva (2005)
14. ITU-R, Characteristics of radiated signals of conventional analogue television systems, Recommendation ITU-R BT.1701-1, Geneva (2005)
15. ITU-R, Characteristics of composite video signals for conventional analogue television systems, Recommendation ITU-R BT.1700, Geneva (2005)
16. ETSI, Television systems; NICAM 728: transmission of two-channel digital sound with terrestrial television systems B, G, H, I, K1 and L", ETSI EN 300 163 V1.2.1 (1998–2003) (1998)
17. ITU-R, Television systems; NICAM 728: transmission of two-channel digital sound with terrestrial television systems B, G, H, I, K1 and L, ETSI EN 300 163 V1.2.1 (1998)
18. ITU-R, Transmission of multisound in terrestrial television systems PAL B, D1, G, H and I, and SECAM D, K, K1 and L", Recommendation ITU-R BS.707-4, Geneva (1998)
19. ITU-R, Minimum field strengths for which protection may be sought in planning an analogue terrestrial television service, Recommendation ITU-R BT.417-5, Geneva (2002)
20. ITU-R, Characteristics of TV receivers essential for frequency planning with PAL/SECAM/NTSC television systems, Recommendation ITU-R BT.804, Geneva (1992)
21. ITU, Planning criteria, including protection ratios, for second generation of digital terrestrial television broadcasting systems in the VHF/UHF bands, Recommendation BT.2033-0, Geneva (2013)
22. EBU, Frequency & network planning aspects of DVB-T2, Tech 3348 r3, Geneva (2013)
23. ITU-R, Planning criteria, including protection ratios, for digital terrestrial television services in the VHF/UHF bands, Recommendation ITU-R BT.1368-10, Gevena (2013)
24. ITU-R, Frequency and network planning aspects of DVB-T2, Report ITU-R BT.2254, Geneva (2012)
25. ITU-R, Radio-frequency protection ratios for AM vestigial sideband terrestrial television systems interfered with by unwanted analogue vision signals and their associated sound signals, Recommendation ITU-R BT.655-7, Geneva (2004)
26. Fischer, W.: Digital Television: A Practical Guide for Engineers. Springer, Berlin (2004)
27. ETSI, Electromagnetic compatibility and radio spectrum matters (ERM); wireless microphones in the 25 MHz to 3 GHz fre-quency range; Part 1: technical characteristics and methods of measurement, ETSI EN 300 422–1 V1.3.2 (2008–03) (2008)
28. Ofcom, Spectrum efficiency of wireless microphones (Final Report) 2202/DWM/R/2/2.0, 2010. Available: http://stakeholders.ofcom.org.uk/binaries/research/technology-research/sewm/finalreport.pdf (2010) Accessed 19 May 2014
29. Shellhammer, S., Sadek, A.K., Zhang, W.: Technical challenges for cognitive radio in the TV white space spectrum. In: Information Theory and Application (ITA) Workshop, San Diego, CA, USA (2009)
30. Sun, H., Zhang, T., Zhang, W.: Separating the wheat from the chaff: sensing wireless microphones in TVWS, arXiv:1205.2141v1 [cs.IT] (2012)
31. ICASA, Final terrestrial broadcasting frequency plan 2013, Government Gazette. 36321, South Africa (2013)
32. Ettus Research, USRP family of products, Ettus REsearch, 2014. Available: http://home.ettus.com/ Accessed 21 May 2014
33. Habeenzu, S.: Zambia ICT sector performance revier 2009/2010: towards evidence-based ICT policy and Regulation. Research ICT Africa **2**(17), 1–38 (2010)

34. Kolundzija, B., Ognjanovic, J.S., Sakar, T.K.: WIPL-D: Electromagnetic modeling of composite metallic and dielectric structures. Software and User's Manual, Artech House Publishers, Boston (2000)
35. Lysko, A.A.: New mom code incorporating multiple domain basis functions. In: XXXth URSI General Assembly and Scientific Symposium (URSI GASS 2011), Istanbul, Turkey (2011)

Chapter 5
Spectrum Sensing for Cognitive Radio Networks

Srinu Sesham and Samrat L. Sabat

Abstract Spectrum sensing is a primary and an essential function in cognitive radio (CR) for dynamic spectrum access. Cognitive user should detect the whitespaces or vacant frequency bands as quickly as possible without causing significant interference to the primary user (PU) communication. This requires efficient signal processing algorithms. Hence, research into novel sensing techniques is being aggressively engaged in both academia and industry. In this chapter, we aim to detect low signal-to-noise ratio (SNR) signal under probable channel impediments. In practice, detection performance (detection probability (P_d) and false alarm probability (P_f)) of a single cognitive user is often compromised by multipath fading and shadowing issues in the channel. Multinode/cooperative detection is an effective method to minimize the impact of these issues and to improve the detection performance by exploiting spatial diversity. In the literature, various spectrum sensing methods are presented for cognitive radio networks. Each and every one has its own advantages and compromises. According to the IEEE 802.22 WRAN standards for sensing TV whitespaces, the channel detection time and channel move time should be ≤ 2 s. In addition, the false alarm and detection probabilities should be <0.1 and >0.9, respectively. In this chapter, energy detection is considered for whitespace detection due to its ability to scan the spectrum quickly. Moreover, weighted gain combining (WGC) and equal gain combining (EGC) methods are presented to improve the cooperative sensing performance.

S. Sesham (✉)
University of Cape Town, Cape Town, South Africa
e-mail: seshamsrinu83@gmail.com

S.L. Sabat
University of Hyderabad, Hyderabad, India

A.K. Mishra and D.L. Johnson (eds.), *White Space Communication*,
Signals and Communication Technology, DOI 10.1007/978-3-319-08747-4_5

5.1 Introduction

5.1.1 Utilization of Radio Spectrum

The Radio spectrum is an indispensable natural resource for evolution of future generation wireless systems. Moreover, it is a very costly and limited natural resource. According to the recent statistical studies and reports published by Federal Communications Commission (FCC), fixed-allocation of spectrum bands to specific wireless communication applications lead to poor utilization of spectrum interms of different dimensions such as frequency, time, and geographical space [15]. Figure 5.1 illustrates the spectrum occupancy (utilization) in each band averaged over seven locations over the range of frequency bands from 30 MHz–2.9 GHz. It is reported that, significant amount of spectrum is available for dynamic spectrum access and the maximum spectrum occupancy range is in between 5.2 and 13.1 % at certain frequency bands [31]. The temporal and spatial utilization range is 15–85 % approximately [2]. The figure reveals that, certain portion of the radio spectrum is sparsely used and significant amount of the radio spectrum is unutilized [2]. It indicates that

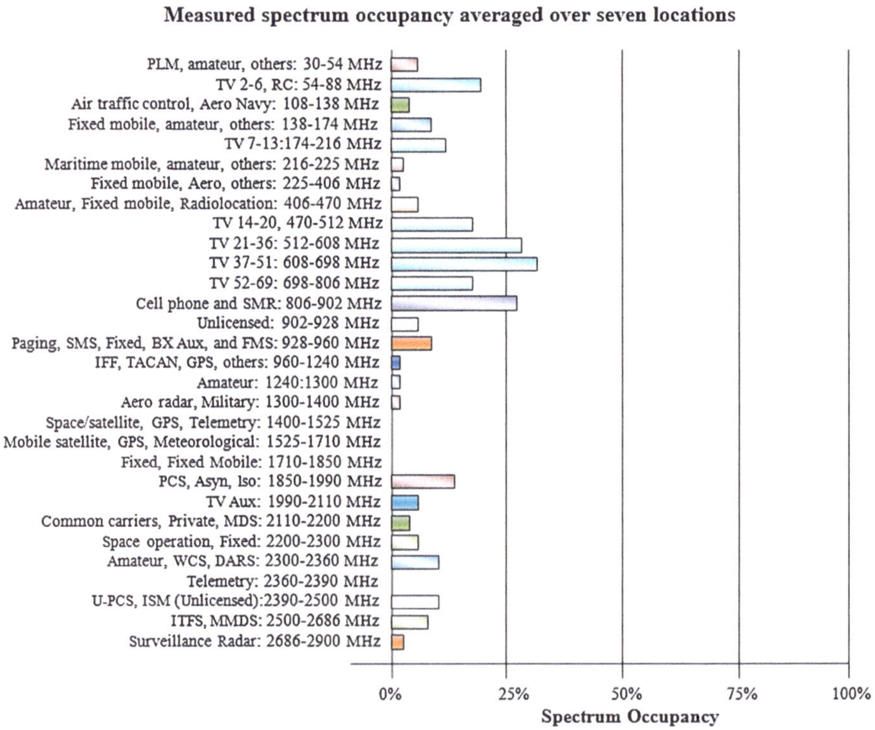

Fig. 5.1 Spectrum occupancy in RF bands averaged over seven locations [31]

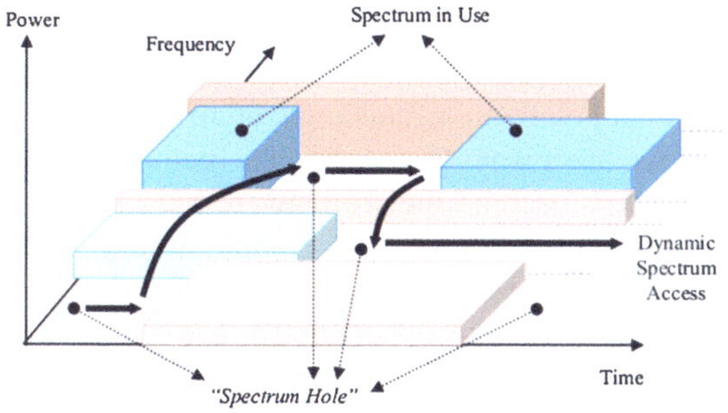

Fig. 5.2 Spectrum hole and dynamic spectrum access [25]

scarcity of spectrum resources is not due to fundamental lack of radio spectrum resources, but due to inefficient spectrum allocation policy or under utilization of allocated spectrum. In order to solve the impending spectral scarcity and under utilization of allocated spectrum, cognitive radio (CR) technology has been proposed [33]. Cognitive radio refers to "a wireless architecture in which a communication system does not operate in a fixed assigned band, but rather searches and finds an appropriate band for its communication", as shown in Fig. 5.2.

To improve the spectrum utilization efficiency, cognitive/secondary user (SU) system essentially requires dynamic spectrum access. To achieve this, cognitive user should be able to scan the primary/licensed user's frequency band as quickly as possible without interference to the licensed user communication. This requires efficient spectrum sensing technique. Hence, research into novel techniques for efficient spectrum utilization is being aggressively engaged in both academia and industry. According to the IEEE 802.22 WRAN standards for sensing TV bands, channel detection time and channel allocation time must ≤ 2 s, and the detection probability and false alarm probability as 90 and 10 % respectively [7]. Different signal processing techniques are being used for spectrum sensing. Still there is a need to develop an efficient sensing algorithm which tends to meet the cognitive radio standards.

5.1.2 Whitespace Detection

Spectrum sensing is an essential function in CR system to know the radio spectrum usage or presence of licensed user in a particular frequency band. A conventional approach to classify spectrum usage in space is divided into three types: black spaces, gray spaces, and white spaces [9]. Black spaces are occupied by high power local interferences, gray spaces are occupied by low power interferences, while white spaces are free of any interference excluding ambient noise. The white and gray spaces are the spectrum opportunities or spectrum holes, which can be used by the cognitive users. Hence, spectrum sensing can also be termed as whitespace detection.

However, this conventional approach of spectrum opportunities only exploits three dimensions of the spectrum: frequency, time, and space. The additional dimensions can be utilized such as code, polarization, and angle of arrival [57]. Spectrum awareness can be obtained in two other ways by using active method or passive method. In the active methods, the radios become spectrum aware by detecting and estimating the spectrum. Active methods have advantages compared to passive methods with respect to immense applications, low initial and maintenance cost. In passive methods, the information regarding the unoccupied spectrum is provided to the secondary users with the use of database and beacons [19]. Passive methods need support from the PUs who is under no obligation to change their operation to assist the secondary user network. Therefore, in practice, passive methods are difficult to implement. Hence, we focus on active method for sensing the spectrum using cognitive radio technology.

Generally, spectrum sensing techniques can mainly be divided into transmitter based signal detection, receiver based signal detection, and interference-based detection [53]. In case of the receiver based detection method, the local oscillator (LO) block in RF front-end of the licensed user receiver emits the leakage power that inturn is useful to know the activity of licensed user in a band of interest. This can be realized by keeping few low cost wireless sensors near to the primary user receiver. The sensors measure the LO leakage power to determine the channel status. Based on the status of the channel, cognitive users can use the unutilized radio spectrum for dynamic spectrum access [53]. However, detecting the receiver becomes a difficult task as the leakage power of the oscillator is low, thereby restricting the reliable detection range below 20 m [6]. On the other hand, most of the recent works are focused on primary transmitter detection based on local observations of CR users. Different ways of enabling the sensing techniques based on primary user transmitter signal are reported in [36]. The transmitter based signal detection approach is considered throughout the work. In this method, weak signal from a primary transmitter can be detected through the local observations of CR users. Thus, in transmitter based signal detection (non-cooperative detection), secondary system should have a capability to discriminate the used and unused radio frequency bands [2].

Accordingly, spectrum sensing research has been active, resulting in numerous sensing algorithms, which are well summarized in [3, 58]. However, all the methods can be broadly classified into three main categories: coherence detection, non-coherence (blind) detection, and feature detection. The sensing methods related to the above categories are briefly explained below:

5.1.2.1 Matched Filtering Detection

Matched-filtering is a coherent detection method. It is an optimal detection method that requires perfect synchronization between the licensed user transmitter (PU Tx) and cognitive radio receiver (CR Rx) [57]. This method requires signal characteristics (i.e., carrier frequency, bandwidth, modulation type, etc.) of the frequency band to be scanned. The matched filtering achieves a certain probability of false alarm (P_f)

or probability of misdetection (P_m) accurately as compared to other methods [2]. But, in case of the cognitive radio, the prior information about the entire frequency band is impracticable. It is reported that the required number of samples for matched filtering increases as O(1/SNR) for a target probability of false alarm at low SNRs [49]. Moreover, the implementation complexity of sensing unit (receiver) is large owing to the reception of all modulated signals from different standards. Another disadvantage of matched filtering is the large power consumption by the detector.

5.1.2.2 Energy Detection

Energy detector is a non coherent detector. Hence, it does not require any prior information of channel characteristics of the frequency band to be scanned. It is the commonly used spectrum sensing method due to its low computational and implementation complexities [57]. The PU signal can be detected by comparing the energy of received signal with a pre determined threshold (which depends on the noise variance) [57]. One of the challenges with energy detector based sensing is the discrimination between PU's signal and noise in a low SNR regime. This detection method does not work efficiently for detecting spread spectrum signals.

5.1.2.3 Cyclostationary Feature Detection

Cyclostationary feature detection is a method for detecting primary user transmissions by exploring the cyclostationary features of the received signal. Cyclostationary features are caused by the periodicity in the signal or in its statistics like mean and autocorrelation. These features can also be intentionally induced to assist spectrum sensing. Instead of power spectral density (PSD), cyclic correlation function or spectral coherence function (SCF) is used for detecting PU's signal in a given frequency band. Cyclostationary based detection algorithm can differentiate noise from primary user's signals in a low SNR environment. This is because of randomness in the noisy data, which has no correlation [57], whereas modulated signals are cyclostationary with spectral correlation. Furthermore, cyclostationary properties can be used for classification of the received signals.

5.1.2.4 Other Sensing Methods

Apart from the above sensing algorithms, other methods being used in the literature includes, radio identification based sensing, wavelet transform based sensing, covariance based detection, entropy estimation, detection based on Bayesian criterion, interference-based detection, eigen value-based spectrum sensing, compressive sensing, multi-taper spectral estimation, and time-frequency analysis [3, 57].

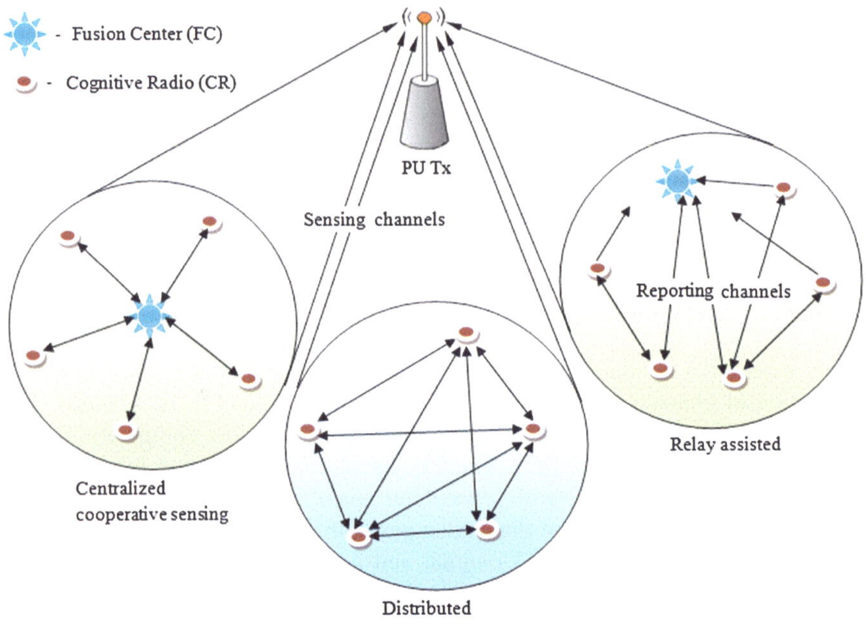

Fig. 5.3 Cooperative spectrum sensing classification [1]

5.1.3 Cooperative Sensing

In cooperation, cognitive users share their sensing information for making a combined decision more accurately than the individual decisions [1]. This helps to improve the detection performance. Hence, cooperative spectrum sensing is an effective method to address noise impediments in the network. The authors [1] presented and classified the cooperative sensing methods into three main categories based on the sensing data shared in the network, as, centralized, distributed, and relay-assisted cooperation methods. The methodologies of cooperative sensing are illustrated in Fig. 5.3.

In the case of the centralized cooperation method, a central node also referred as fusion center (FC) is considered to evaluate the global decision [1]. The fusion center collects the individual decisions of particular channel from all cooperative cognitive users. Based on the combination of individual decisions, the CR makes the global decision that determines the presence of primary user signal and relays the decision back to the all cooperative cognitive users for dynamic spectrum access. The centralized cooperative sensing model is shown in the left circle of Fig. 5.3. For local sensing, all nodes are tuned to the selected licensed channel or frequency band. The channel between the PU transmitter and each CR user (to observe the primary signal) is called as sensing channel. For data reporting, all CR users are tuned to a control/reporting channel. Physical point-to-point link between each cooperating cognitive user and the central node (for sending the sensing results) is termed as reporting channel [1]. In centralized cooperative sensing, cognitive base station (BS) is naturally the fusion center. Unlike centralized cooperative sensing, distributed cooperative sensing does not rely on a central node for making the global decision. In this case, cognitive users communicate among themselves and converge to a unified decision regarding the presence or absence of primary signal by iterations as shown in the middle circle of Fig. 5.3. In this model, each cognitive user sends its own sensing data to other users and combines with the received sensing data based on a distributed algorithm for sensing a frequency band [1].

The cognitive users which are far away from PU Tx may not perform spectrum sensing with great efficiency due to lack of signal strength and fading in the channel. In this condition, spectrum sensing can be improved using relay nodes [5]. Hence, in the case of the relay-assisted cooperative sensing, a CR user observing a weak sensing channel and strong reporting channel or a CR user with strong sensing channel and a weak reporting channel can cooperate by using relays to improve the cooperative sensing accuracy [1]. This is shown in extreme right circle of the Fig. 5.3. In our work, the performance of the sensing is analysed using centralized cooperative sensing.

5.1.4 Challenges in Spectrum Sensing

5.1.4.1 Hardware Implementation Requirements

Spectrum sensing for cognitive radio application requires high sampling rate, high resolution A/D converters with large dynamic range, and high speed signal processors [57]. In cognitive radio network, each node is required to sense the wideband for dynamic spectrum access. Hence, it should be able to capture and analyze a relatively larger band and to increase the throughput of the cognitive users. The large operating bandwidths impose additional components on the radio frequency (RF) front-end such as antennas of wideband receiving capability and power amplifiers. These components should be able to operate over wide operating frequency range. In addition to this, high speed processing units (DSPs or FPGAs) are required for performing computationally complex signal processing tasks with relatively low latency. The authors [57] presented two different sensing architectures. One of them is single radio architecture, where the radio has particular/specific time interval to sense the spectrum. In this case, the transmission of secondary data/signal happens after the sensing is completed. Hence, cognitive users are unable to transmit their data during the sensing period. Owing to this the throughput of the cognitive user decreases although it has less complex architecture and low cost. The other one is dual-radio sensing architecture, where one radio chain is dedicated for data transmission and reception while the other chain is dedicated for spectrum monitoring. The drawback of such an approach is the increased power consumption and hardware cost. Moreover, the researchers or designers should develop energy efficient algorithms/architectures for the system to work with low power.

5.1.4.2 Sensing Time and Operation Frequency

Primary user has a right to use their frequency bands anytime even though the CR is transmitting its data in that channel. Due to this, both the PU and CR will suffer from the noise interference caused by one another. To avoid this problem, CR user has to sense spectrum within a short time (<2 s). In addition, the CR should vacate the channel (or move from one band to another band) once the primary users intend to use their channel [57, 58]. Hence, the sensing time is a crucial parameter to increase the efficiency of spectrum utilization and throughput of the secondary users. One more parameter which influences the spectrum sensing is operating frequency, i.e., how often a CR user will carry out spectrum sensing. If cognitive user wants to perform spectrum sensing very often, it leads to interference to the primary users. So interference tolerance of licensed user is another factor one has to consider during the sensing. Selection of above sensing parameters/arguments such as sensing time, move time, and operating frequency is an important task. According to the IEEE 802.22 WRAN standard, the sensing period must be ≤2 s. In addition to the sensing frequency, the channel detection time, channel move time and some other timing related parameters are also defined in the standard [7]. In summary, sensing should be done as frequently as possible in order to prevent interference to the PU systems.

5.1.4.3 Distinguish Between Different Waveforms in a Channel

This is one of the interesting and desirable challenges to know the signal type in a frequency band to be scanned. During the scanning process the CR may receive the signal either from a licensed user or from a cognitive user or from a malicious/selfish user [9].

5.1.4.4 Detecting Spread Spectrum Signal

The two major spread spectrum technologies available in the literature are frequency hopping spread spectrum (FHSS) and direct sequence spread-spectrum (DSSS). In general, fixed frequency devices operate at a single frequency or channel [57]. In the case of FHSS, the PU signal is rapidly switching by a carrier to multiple narrowband frequency channels based on hopping pattern (a sequence known to both transmitter and receiver). Cognitive radios that receive the spread spectrum signals are difficult to sense the PU status due to power distribution of signal over different frequency bands. Spectrum Sharing Strategy for Frequency Hopping Signal in industrial, scientific and medical (ISM) band is presented in [4]. Detection and estimation of frequency hopping signals using wavelet transform is presented in [38]. Still, development of efficient detection algorithm for the spread spectrum signals in a low SNR environment is an active research domain.

5.1.4.5 Cooperative Sensing Fusion Methods

In the case of the cooperative spectrum sensing (CSS), the global decision making based on local sensing decisions from various cognitive users is a challenging task [57]. In general, the local sensing results/decisions are either based on hard decision (0 or 1) or soft decision (energy, power, entropy measurements, etc.) [1, 29]. In the case of the soft decision fusion methods, multiple cognitive users transmit soft decisions (measurements) to a central node (fusion center). Then, the FC combines the soft decisions and evaluates the global decision. On the other hand, in the case of the hard decision fusion methods, FC makes the final decision based on binary values (either 0 or 1) received from each cognitive radio user in the cooperation. It is reported that the soft decision fusion methods outperform the hard decision fusion method in terms of the probability of detection [47]. On the other hand, the authors [32] reported that the hard-decision fusion methods also perform equally when more number of nodes present in the cooperation. It is reported that the distance between secondary users and the fusion center (FC) poses reliability issue in the cooperative sensing. This is because control channels (from the CR user to FC) are also subject to fading and shadowing effects like the sensing channels [8, 26]. The noise effects in the sensing channel are addressed in cooperative spectrum sensing but the noise effects in the control channels are need to be addressed. Reference [41] proposed a forward error correction (FEC) technique to increase the sensing performance in presence of noise in the control channel. However, there is a need to develop an efficient correction method in noisy control channel to make the global decision more reliable.

5.1.4.6 Security

Cooperative/multinode sensing is being used to overcome the single node sensing issue that arises due to channel impediments. Although cooperation among multiple cognitive users improve the sensing performance, presence of few malicious/suspicious cognitive users may severely degrade the efficiency of the system. The prominent issues in cooperative sensing are the possible emulation attacks and statistical spectrum sensing data falsification (SSDF) attacks of suspicious cognitive radio (SCR) users, where suspicious/malicious cognitive users intentionally report/send the false measurement to other cognitive users, and thereby wrongly influencing the cooperative or global decision. This kind of behaviors or attacks are investigated and presented in [21, 37]. A more challenging problem is to develop an effective counter measures once an attack is identified. Most of the literature works are focused on single suspicious user elimination in the CRN [20]. But, if the network has more than a single suspicious user, detection performance will degrade considerably. Multiple suspicious user elimination methods are proposed to improve the reliability of cooperative sensing [37, 45]. The authors [30] proposed public key encryption based primary user identification method to prevent SUs pretended as PUs in the CRNs.

5.1.4.7 Wideband Spectrum Sensing

Spectrum sensing can be done in a single narrowband or in a wideband spectrum. Narrowband sensing techniques can detect the PU signal in one particular frequency band. These are less complex and are suitable where spectrum utilization is high, without a constraint on throughput. However, when the spectrum utilization is high, wideband spectrum sensing techniques are essential to increase the throughput of the secondary users by exploring (finding) the higher number of spectrum holes for dynamic spectrum access. But, the wideband sensing techniques are more complex compared to the narrowband sensing. The authors [18, 48] discussed and presented different wideband sensing algorithms along with the challenging issues involved in the wideband spectrum sensing.

5.1.5 Standardization Efforts

Wireless standards developed recently or currently under development have started incorporating cognitive features due to the evolution of cognitive radio technology [9, 17]. IEEE 802.22 is the first world wide effort to define a standardized air interface based on cognitive radio techniques for the opportunistic use of TV white spaces (TVWS) [7, 34]. The standard is designed for the secondary usage of TVWS on a non-interfering basis so as to prevent any harmful interference to the incumbent operation (such as digital TV and analog TV broadcasting) and low power licensed devices (such as wireless microphones and medical telemetry devices). The primary application of this standard is fixed broadband access especially for hard-to-reach, low population density areas (typical of rural environments) and thus has a great potential for worldwide applicability [9]. Cognitive functionalities included in the standard are PU detection, geo-location, coexistence with other WRANs, and frequency agility. The implementation of a database is mandatory for PU detection while sensing is optional.

Other initiatives related to cognitive radios standard are IEEE 802.11, dynamic spectrum access networks standards committee (DySPAN-SC), IEEE 802.11af standard, which is currently under development, aims to define modifications to IEEE 802.11 PHY/MAC for TVWS operation [9]. The DySPAN-SC develops standard for radio and spectrum management. It was also formerly known as IEEE Standards Coordinating Committee 41 (SCC41) and IEEE P1900 standards committee. IEEE 802.19 focuses on coexistence between different unlicensed wireless networks in 802.11 group of standards like IEEE 802.11 (WLAN), IEEE 802.15 (WPAN), 802.16 (WMAN), 802.22, etc. IEEE 802.19 task group 1 focuses on wireless coexistence in the TVWS. In addition, many groups/unions such as European Telecommunication Standards Institute (ETSI) and International Telecommunication Union (ITU), etc., are working on standards for radio spectrum allocation.

5.2 Whitespace Detection Using Single Node

5.2.1 Narrowband Detection

One of the most important components of the cognitive radio is the ability to sense and aware of parameters related to the availability of spectrum and its radio channel characteristics [57]. To achieve this, cognitive users should have reliable sensing unit (to sense the spectrum holes) and reconfigurable unit (to alter its transmission parameters) according to the spectrum hole characteristics. Hence, spectrum sensing is the most important component of the cognitive radio. In statistical perspective, spectrum sensing is a problem of detecting the presence or obscene of a signal in a noisy environment. Before going to the signal detection, one should know the critical differences between detection and demodulation. The demodulation is the process of decoding a message based on the received signal. The main problem in demodulation is to decode the transmitted message from the received (possibly corrupted) signal [49]. However, the receiver knows that there is a primary signal present and extracts the information contained in the signal. On the other hand, in signal detection, the detector/sensor/sensing algorithm detects the presence of signal at low signal-to-noise ratio (SNR) environment. The main focus of this chapter is to develop signal processing techniques for whitespace detection. In the case of spectrum sensing, detecting the absence of PU signal is a critical task because there is no way to test the absence of a PU signal. The only possible way to test this is to contradict the existence of signal when PU is present actually in the channel [49]. Moreover, contradicting the existence of a signal is hard when the signal power is very low. Thus, the sensing problem becomes a hypothetical question. In application point of view, the primary objective of spectrum sensing is to provide more spectrum access opportunities to secondary users without causing harmful interference to the legacy networks. Moreover, it has to detect the signal with high probability in a low SNR environment. For instance, cognitive radio system must be capable of detecting digital TV signals at an SNR of -21 dB: when the noise is over one hundred times stronger than the actual signal [23].

The choice of a detection criteria is based on the optimization of the desired objective function involving different performance parameters (P_f, P_d, SNR, sensing time, and the detection range). There are several detection criteria [22, 35]: Bayesian, Neyman-Pearson, minimax, locally optimum, sequential detection, etc. Bayesian formulation can be used to minimize the Bayes risk, which depends on the prior probabilities of two hypotheses, cost assignments, and conditional densities of the observations under the two hypotheses. However, the required prior probabilities of the hypotheses and cost assignments may not be necessarily available to implement the optimal Bayesian decision rule. Neyman-Pearson (NP) formulation maximizes the probability of detection for a given constraint on the false alarm probability. Noise statistics are required for the NP implementation and may be estimated. Yet another criterion for detection is minimax which minimizes the maximum Bayes risk by using the Bayes decision rule corresponding to the least favorable prior probability assignment. Minimax concept results in robust detection where optimum detectors are designed for certain least favorable models like heavy-tailed noise models. Robust detection techniques are used when the observation statistics are not known exactly but only approximately [9]. Accordingly, spectrum sensing research has been active, resulting in numerous sensing algorithms, which are well summarized in [3]. In practice, detection performance can be primarily determined on the basis of false alarm probability (P_f), which denotes the probabilities of CR user declaring that a PU is present when the spectrum is actually free, and probability of detection (P_d), which denotes the probability of a CR user declaring that a PU is present when the spectrum is indeed occupied by the PU. To achieve optimal detection performance, the probability of detection should be maximized subject to the constraint of the probability of false alarm. The goal of this chapter is to study and analyze the energy detection method.

The objective of spectrum sensing is to detect the presence of a signal in the sensing band. Hence, the spectrum sensing problem can be modeled into a binary hypothesis test as follows

$$
\begin{aligned}
H_0 &: r[n] = w[n], \\
H_1 &: r[n] = hs[n] + w[n], \quad n = 0, 1, 2, \ldots, N-1
\end{aligned}
$$

where H_0 is the hypothesis when the primary user signal $s[n]$ is absent, H_1 is the hypothesis when the PU signal is present in the sensing band, $r[n]$ is the received signal sequence by the cognitive user, N is the number of samples in a particular duration, $w[n]$ is the additive white Gaussian noise (AWGN), and 'h' is the channel gain.

In this work, the algorithms are analyzed based on the following assumptions, (1) the noise in each band follows Gaussian, independent and identically distribution ($i.i.d$) with zero mean and variance σ_w^2, (2) the received signal is a stochastic signal, and it follows Gaussian $i.i.d$ with mean μ_s and variance ζ^2, (3) the channel gain (h), transmitted signal, and noise are independent of each other, and (4) the channel is time invariant during the sensing period.

5.2.2 Detection Based on Energy Measurement

The block diagram representation of energy detector is shown in Fig. 5.4. The received signal $r(t)$ is filtered by a band-pass filter to select the channel to be scanned. Detector computes the energy of the received signal by taking discrete time samples from analog signal to discrete (A/D) signal converter and compares it with the predetermined threshold value (λ_e). The objective of spectrum sensing is to discriminate between the above two hypotheses H_0 (PU is absent) and H_1 (PU is in operation).

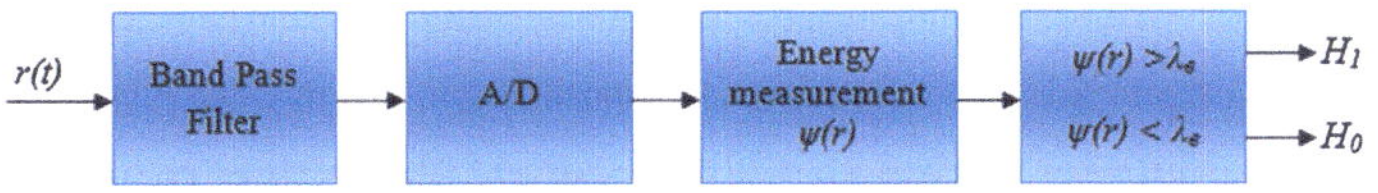

Fig. 5.4 Block diagram of energy detector

In general, Neyman Pearson (NP) criterion is being used in energy detection technique. The optimal NP test is to compare the log-likelihood ratio with a predetermined threshold (λ_e),

$$\log\left(\frac{P(r_0, r_1, \ldots r_{(N-1)})|H_1)}{P(r_0, r_1, \ldots r_{(N-1)})|H_0)}\right) \underset{H_0}{\overset{H_1}{\gtrless}} \lambda_e \tag{5.1}$$

where $P(r|H_1)$, $P(r|H_0)$ are the probability density functions of H_1 and H_0 respectively. Obviously, the log-likelihood ratio depends on the distribution of the signal to be detected.

In practice the detector does not have any known characteristics of a signal to be detected in a particular band. We adopt the signal model as [24], where the channels between the primary and the CR users are corrupted with AWGN. Under such conditions, the above test statistic can be approximated as,

$$\psi(r) = \sum_{n=0}^{N-1} |r[n]|^2 \underset{H_0}{\overset{H_1}{\gtrless}} \lambda_e \tag{5.2}$$

where ψ is the energy measurement and (λ_e) is a detection threshold which depends on the tolerable false alarm probability, expressed as [55]

$$\lambda_e = F^{-1}(1 - P_{f\text{-}des}|U, \sigma_w^2) \tag{5.3}$$

where F^{-1} denotes the inverse of gamma function, $P_{f\text{-}des}$ is the desired false alarm probability, U and σ_w^2 are the shaping and sizing parameters and depends on the sample size (N) and variance of the noise, respectively.

In a non-fading environment, the theoretical detection and false alarm probability for single node are given by [52]

$$P_d = P\{\psi(r) > \lambda_e/H_1\} = Q_u(\sqrt{2\gamma}, \sqrt{\lambda_e}) \tag{5.4}$$

$$P_f = P\{\psi(r) > \lambda_e/H_0\} = \frac{\Gamma(U, \lambda_e/2)}{\Gamma(U)} \tag{5.5}$$

where γ is the signal to noise ratio (SNR), $U = T.W$ is the time bandwidth product, $\Gamma(.)$ and $\Gamma(.,.)$ are complete and incomplete gamma functions and $Qu(.,.)$ is the generalized Marcum Q-function given by [52]

In case of the complex valued phase shift keying (PSK) modulated PU signal and circularly symmetric complex Gaussian (CSCG) noise, the theoretical detection and false alarm probability for a given λ_e with considered observation period (t_s) is given by [27]

$$P_d\left(\lambda_e, t_s\right) = P\left(\psi(r) > \lambda_e / H_1\right) \tag{5.6}$$

$$= Q\left(\left(\frac{\lambda_e}{\sigma_w^2} - \gamma - 1\right)\sqrt{\frac{t_s f_s}{2\gamma + 1}}\right) \tag{5.7}$$

$$P_f\left(\lambda_e, t_s\right) = P\left(\psi(r) > \lambda_e / H_0\right) = Q\left(\left(\frac{\lambda_e}{\sigma_w^2}\right)\sqrt{t_s f_s}\right) \tag{5.8}$$

where t_s is the sensing period, f_s is the sampling rate, and $Q(.)$ is the complementary distribution function formulated as

$$Q(x) = \frac{1}{\sqrt{2\pi}} \int_x^\infty \exp\left(-\frac{t^2}{2}\right) dt.$$

The detection threshold (λ_e) based on central limit theorem can be derived from [27]

$$\left(\frac{\lambda_e}{\sigma_w^2}\right) = Q^{-1}(P_d)\sqrt{\frac{2\gamma + 1}{t_s f_s}} + \gamma + 1 \tag{5.9}$$

where $P_d(\lambda_e, t_s)$, and $P_f(\lambda_e, t_s)$ are the detection probability and false alarm probability for a given threshold and sensing time.

In Rayleigh fading environment, the received signal energy fluctuates due to multipath components received at each CR. It happens when there is no line of sight (LOS) component in the received multipath signals. Owing to this, the signal strength at the receiver varies and it follows a Rayleigh distribution. The probability of detection P_d varies with SNR as given by [11, 16]

$$P_{d\text{-fad}} = \int Q_u(\sqrt{2\gamma}, \sqrt{\lambda}) f_\gamma(r) \mathrm{d}r \tag{5.10}$$

where $f_\gamma(r)$ is the probability density function of SNR under fading. The closed form expression is known for Eq. (5.10) for Rayleigh fading channel. The authors, [11] have also presented the closed form expressions for the detection probability for Nakagami and Rician fading channels.

5.2.3 Wideband Detection

In contrast to narrowband techniques as mentioned above, wideband spectrum sensing techniques aim to sense the wide bandwidth of spectrum. In order to achieve higher opportunistic aggregate throughput in cognitive radio networks, cognitive users must sense the signals in multiple bands. Hence, it is assumed that the total frequency bandwidth (wideband) need to be scanned is divided into 'K' non-overlapping subbands, β subbands ($1 \leq \beta \leq K$) are vacant for particular time duration and these vacant bands are available for opportunistic spectrum access [59].

The fundamental problem of wideband spectrum sensing in CR is to discriminate the following two composite hypotheses, which decide whether the kth sub-band (SB) is occupied or vacant.

$$H_0^k : \mathbf{R}(n) = \mathbf{W}(n), \quad n = 0, 1, 2, \ldots, (N-1)$$
$$H_1^k : \mathbf{R}(n) = h.\mathbf{S}(n) + \mathbf{W}(n)$$

where $\mathbf{R}(n)$, $\mathbf{W}(n)$, and $\mathbf{S}(n)$ are represented as

$$\mathbf{R}(n) = [\mathbf{r}^0, \mathbf{r}^1, \mathbf{r}^2, \ldots, \mathbf{r}^{(K-1)}]$$
$$\mathbf{W}(n) = [\mathbf{w}^0, \mathbf{w}^1, \mathbf{w}^2, \ldots, \mathbf{w}^{(K-1)}]$$
$$\mathbf{S}(n) = [\mathbf{s}^0, \mathbf{s}^1, \mathbf{s}^2, \ldots, \mathbf{s}^{(K-1)}]$$

where $\mathbf{r}^k$, $\mathbf{w}^k$ and $\mathbf{s}^k$ are the received signal, noise and primary user signal of length N in kth subband respectively. Based on Eq. (5.2) multiple narrowband within the considered wideband is evaluated.

5.2.4 Simulation Results and Discussion

5.2.4.1 Simulation Setup

In simulation, we have considered DVB-T (2k mode) signal with sample/frame size (N) of 256 as primary user transmitted signal. The Simulink model of DVB-T signal generation based on EN 300 744 standard for terrestrial transmission of digital television signals is presented in [13, 54]. For signal detection, the signal specification and simulation parameters are tabulated in Table 5.1. The observation time for signal detection is chosen such that the sample size (N) is equal to $2U$, where U is the product of observation time and bandwidth. Detector computes the test statistic from the received signal in the observed time period and compares it with the corresponding pre-computed threshold for a desired value of probability of false alarm. The

Table 5.1 Simulation parameters

PU signal	DVB-T signal
Carrier frequency, observed time duration	4.8 MHz, 10–120 μs
Sample size or frame size (N)	64, 128, 256, 512
Number of frames (ξ)	10,000
Number of bins (L)	15, 25, 35, 45
Channel band width (W)	6 MHz
P_f, P_d values	0.1, 0.9
Detection range	3–10 km
SNR range	30–0 dB
Population size, number of generations	30, 100

shaping and sizing parameter values to compute the threshold are taken as $U = \frac{N}{2}$ and $\sigma_w^2 = 1$, respectively. Simulations are carried out for energy detection. The fading effects are considered in the simulation due to Doppler shifts of the received signal during the transmission. A signal with wavelength λ_s experiences a frequency shift given by $\delta_f = \frac{v}{\lambda_s} = \frac{v}{c}.f_c$, where v is the speed of the transmitter relative to the transmitter, c is the speed of light, and f_c is the carrier frequency. Doppler shift is considered as 1.5 % of the carrier frequency. Theoretically, the closed form expressions are available or derived to obtain the closed form solutions over possible channel impediments. But, for practical (simulation) case, we have to approximate the solution to the theoretical solution by increasing the number of iterations. Therefore, there are no closed form solutions for all detection methods. Owing to this non existence of closed form solutions for P_f and P_d, the performance of the detection is analyzed using Monte-Carlo methods of 10,000 iterations [39]. In summary, we approximated it to 4 decimal points. In this chapter, the SNR_{wall} denotes the minimum SNR that a detector can detect the signal with desired detection and false alarm probabilities ($P_d = 0.9$ and $P_f = 0.1$), below that SNR, the detector cannot achieve the desired performance.

5.2.4.2 Performance Criteria

The receiver operating characteristic (ROC) curves interms of P_d versus P_f (or $P_m = (1 - P_d)$ versus P_f) and SNR versus P_d are carried out to analyze the sensing performance of the algorithms. Performance of spectrum sensing algorithms may differ in different scenarios. It is very important to compare and choose the best algorithm for a given scenario [9]. At the same time, it is necessary to choose proper performance criteria for a fair comparison. In this section, important simulation parameters which can be used to evaluate the sensing algorithms are presented.

False alarm probability: It is defined as the probability that the detector declares the existence/presence of PU, when the PU is actually absent. It is denoted as P_f. False alarm (in signal processing) is also called as Type I error (in statistics). If there are too many false alarms, the spectrum opportunities may be overlooked resulting in an inefficient spectrum reuse. Therefore controlling the false alarm probability is crucial for efficient spectrum usage.

Missed detection probability: It is defined as the probability that the detector declares the absence of PU, when the PU is actually present. It is denoted as P_m. Missed detection (in signal processing) is also called as Type II error (in statistics). Too many missed detections may lead to collisions of the PU and SU transmissions causing interference to the PU. Therefore controlling the missed detection probability is crucial for keeping the interference to the PU under the permissible limits [9].

Sensing time: In practice, it is desirable that the sensing time is shorter and the data transmission duration is longer. If the sensing time is too long, the data transmission duration reduces thereby reducing the throughput of the secondary users. In the case of dual radio system, the device can sense and transmit its data simultaneously [57]. Hence, single radio systems are inefficient in perspective of sensing time.

Signal strength (SNR): The cognitive radios receive the radio signal in each band through PU transmitter to sense the PU activity in the particular channel. The received signal strength (or SNR) at each CR depends on various parameters such as PU transmitted power, fading possibilities, path loss (CR distance from PU Tx), propagation medium, channel gain, etc. The performance of the detector is directly related to the SNR of the received signal at the detector [9]. The error probabilities P_f and P_m are related each other through sensing time, SNR, and detection threshold.

Detection range: It is the maximum distance between the CR user and the PU Tx such that the detector should detect the PU signal reliably. Detection range depends on the signal strength at the receiver, sensing time, and propagation environment. In addition to this, cognitive users should have an ability to detect the PU signal reliably in low SNR regime, though the cognitive radio receiver is far away from the primary user transmitter.

Knowledge of PU parameters and noise distribution: The performance of spectrum sensing is also depends on prior knowledge on statistical properties of primary user signal and noise distribution. The more knowledge on these parameters will give the better detector performance. In practice, the PU signal may be deterministic or random. Whereas, noise may be white Gaussian with known variance or colored with unknown variance. In addition to these properties, knowledge on other activities/properties of PU such as On-Off time (hold time) and geographic location can improve the performance of the detector [9].

Probable noise impediments: Probable noise factors that affect the primary user signal between primary user transmitter (PU Tx) and cognitive radio receiver (CR Rx) are,

(i) Additive white Gaussian noise: It is a general noise presents in the channel. Its spectral density (expressed as watts/hertz of bandwidth) is constant over considered bandwidth that means all frequencies are effected equally due to noise.
(ii) Multipath fading: The signal from PU Tx that travels different paths may or may not add coherently at CR Rx. Due to this the signal strength at the sensor

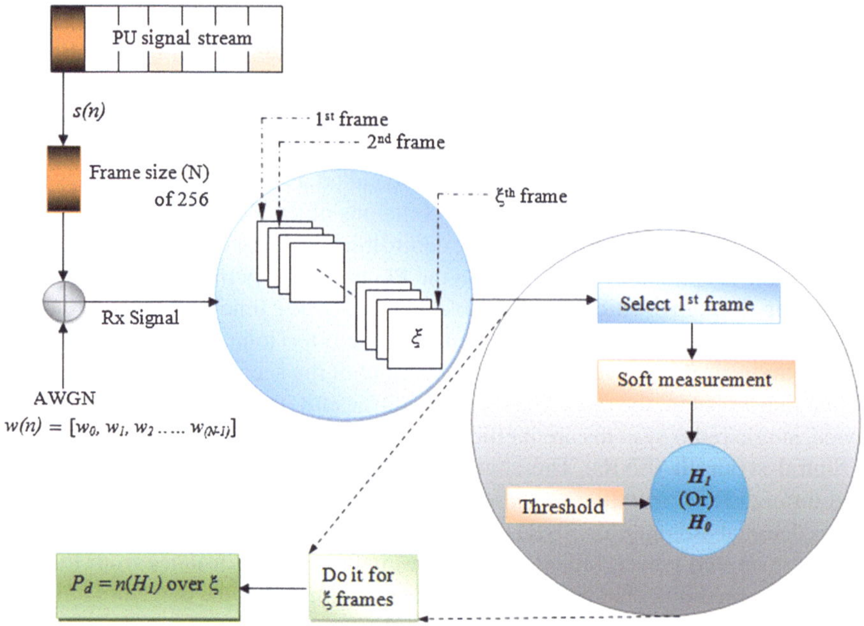

Fig. 5.5 Simulation set up for evaluating detection probability

changes over time, which reduces the detection accuracy of the sensor. There are different kinds of fading effects occur during the sensing process such as Nakagami fading, Log-normal shadow fading, Rayleigh fading, Rician fading, and Weibull fading. In this work, we have considered only Rayleigh fading effect.

(iii) Shadowing: In this case, an individual sensing node may be blocked by an obstacle (or bad channel). Therefore, the CR user may not be able to see the primary user, causing high probability of sensing errors.

5.2.4.3 Computation of Detection Probability

It is defined as the probability that the detector declares the presence of PU, when the PU is actually present. Figure 5.5 explains the evaluation of detection probability for all kind of detection methods. In practice, two types of procedures are being used in detection process, (i) sequential detection method, (ii) snap shot detection method [43, 60]. In the former case, the hypothesis test is performed on ξ contiguous frames in a primary user (PU) signal stream. In the later case, the hypothesis test is performed by selecting a single frame of desired length (or a snap short) from the PU signal stream. Based on the samples in the snap shot and white Gaussian noise, different frames (ξ) of desired length can be generated. This is one of the processes of boot strapping method in Monte Carlo techniques. Then, the detection test has been done

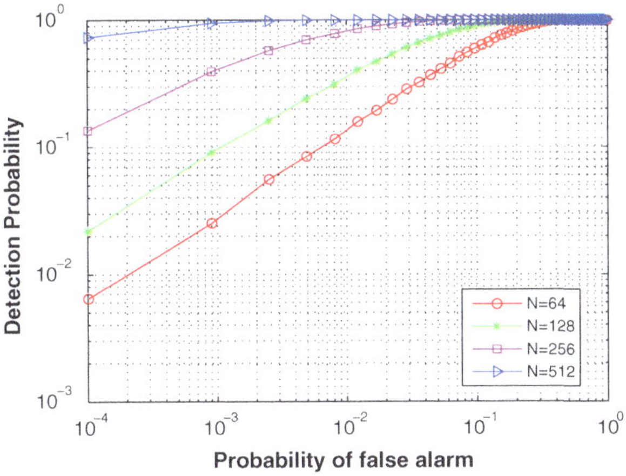

Fig. 5.6 ROC curves of energy detector at SNR = −5 dB

for each frame and accumulated the decisions. Finally, the detection probability can be computed using the basic probability formula which is ratio between number of times the detector supports alternate hypothesis (H_1) over the number of times (ξ) the test is performed.

Figure 5.6 illustrates the Receiver Operating Characteristics (ROC) curves of energy detector with variable sample size or frame length (N = 64, 128, 256, and 512) at a fixed signal strengths of −5 and −20 dB respectively. In the simulation, we fix the SNR of the signal and vary the frame size. From the figures, it is clear that, the performance of energy detector increases with the sample size.

Figure 5.7 illustrates the performance of single node sensing using energy detector at different fixed false alarm probabilities (P_f = 0.1 and 0.01). The simulation parameters are N = 256. From this figure, it is clear that the detection probability increases as the false alarm probability increases. For instance, it is evident that, at required P_f and P_d ($P_f \leq 0.1$ and $P_d \geq 0.9$), the proposed detection algorithm is able to detect noisy DVB-T signals of SNR up to −8 dB. In conclusion, the detection probability is directly proportional to the signal-to noise ratio of the received signals.

Figure 5.8 plots the ROC curves of wideband spectrum sensing with energy detection using sample size of 256. In this simulation, it is considered that, nine subbands (SB1 to SB9) are present within the considered wideband. Each SB have different SNR ranging from −10 to −30 dB. From this figure, it is evident that the subbands under deep noisy environment have low detection probability.

Figure 5.9 illustrates the bar graph of detection probability of each subband with different sample size (N = 64, 128, 256, and 512) using energy detection at P_f = 0.1 and 0.01. From the figure, it is clear that the probability of finding a PU signal is more as number of sample size increases. Moreover, the detection probability reduces as we reduce the target probability of false alarm from 0.1 to 0.01. For instance, the sensing algorithm detects SB1, SB3, SB4, SB5, SB7 and SB9 frequency bands

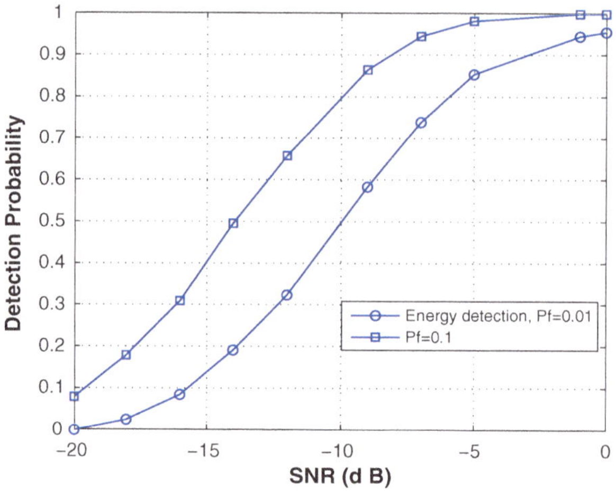

Fig. 5.7 Performance of energy detector at different P_f

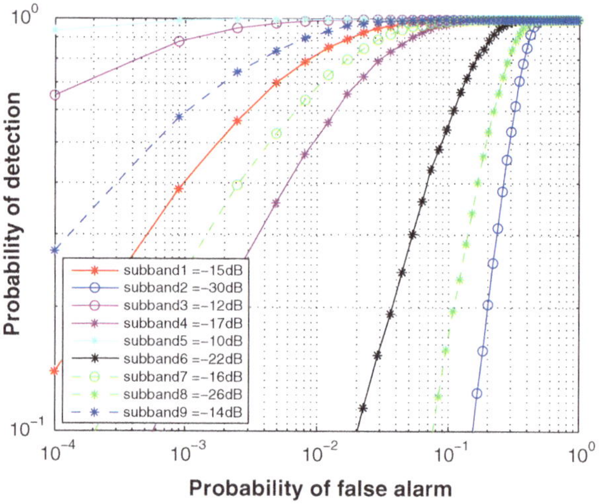

Fig. 5.8 ROC curves of wideband sensing

occupied by PUs signal with sample size of 256, $P_d \geq 0.9$, and $P_f \leq 0.1$ as per IEEE 802.22 WRAN standard. The sensing algorithm detected SB5 (-10dB) as the only band occupied by PU with less number of samples i.e., $N = 64$ at required probabilities. Similarly, the sensing algorithm detected SB3, SB5, and SB9 frequency bands occupied by PUs signal at $P_f = 0.1$. But, at $P_f = 0.01$, SB5 is the only band occupied by the primary user signal.

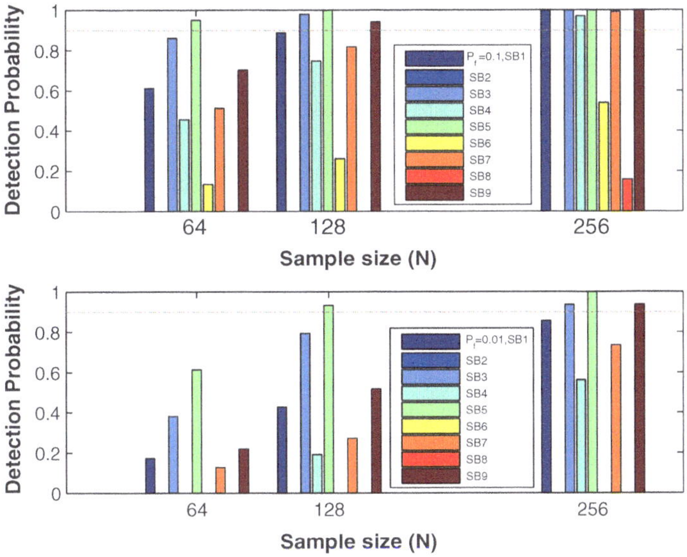

Fig. 5.9 Sample size (N) against (p_d) of a wideband

5.3 Whitespace Detection Using Multinode

5.3.1 Cooperative Narrowband Detection

In CRNs, the noise impediments such as multipath fading, shadowing, and the receiver uncertainty issues degrade the detection performance of single node [1]. Spectrum sensing based on multiple cognitive users can improve the sensing performance using space diversity techniques [12]. Figure 5.10 illustrates the cooperative sensing model with probable noise impediments. In the figure, CR1, CR2, and CR3 are located inside the transmission range of primary user transmitter (PU Tx) while CR4 is outside the range. The nodes outside the PU transmission range are commonly referred as uncertain receiver. Due to the reception of multipath signals from the PU transmitter, CR3 experiences multipath fading effect, and hence the PU's signal may not be correctly detected. CR1 experiences a shadowing effect because of blocking by a hill. However, due to spatial diversity, it is less probable for all spatially distributed CR users in a cognitive radio network (CRN) to experience the fading or receiver uncertainty at the same time. Among all the cognitive users in cooperation, few cognitive users like CR2 (Fig. 5.10) can observe a strong PU signal. In this model, each CR user encounters different channel conditions. The combined/global decision based on the individual decisions from each CR user can overcome the deficiency of single node sensing. Owing to this cooperative/collaborative/multinode spectrum sensing is an impressive and practical way to overcome multipath fading, shadowing,

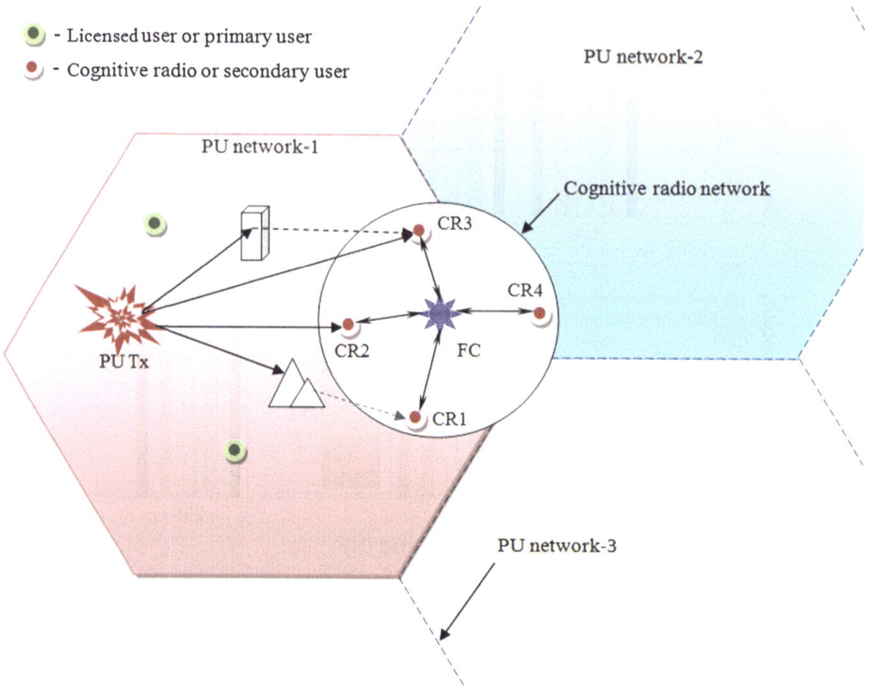

Fig. 5.10 Noise impediments in cooperative spectrum sensing

and the receiver uncertainty. In conclusion, the main objective of cooperative sensing is to improve the sensing performance by exploring spatial diversity [1, 50].

In Fig. 5.10, it is illustrated that the decision statistics from all the cooperative nodes are collected in a central node known as fusion center (FC). Each node uses local spectrum sensing using any of the defined signal processing technique. Fusion center takes the global decision about the presence or absence of a signal in the sensing band using local sensing decisions. This helps to improve the detection probability with an increased overhead traffic, design complexity [28, 56]. Cooperation schemes are classified into soft and hard decision fusion methods depending on the form of decision collected from each node [14]. In soft decision fusion techniques, weighted gain combining (WGC) and equal gain combining (EGC) methods and in hard decision techniques, logical OR, AND, and MOST methods are being widely used at the fusion center. In former case, all users transmit soft decisions to a central node that combines the values and makes the global decision. In later case, each sensor takes its own decision and transmits only a binary value to the fusion center. The weighted gain combining based on differential evolution algorithm is proposed to enhance the sensing performance. It is an evolutionary computation method and has been applied in diverse domains of science and engineering applications [10, 51]. DE finds optimal values for a set of parameters by making repeatedly pseudo-random changes to their values. The number of parameters is referred as dimension of the problem. After making changes, the algorithm evaluates the fitness of the solution. It became a popular evolutionary algorithm because, it is simple to implement, better performance in comparison with other evolutionary algorithms, and it has less number of control parameters and less space complexity.

The problem of whitespace detection using multiple nodes in a CRN can be defined by assuming M nodes in the network. The hypotheses test for multinode detection can be written as

$$\begin{aligned} H_0 &: r_m[n] = w_m[n], \quad n = 0, 1, \ldots, (N-1) \\ H_1 &: r_m[n] = h_m . s_m[n] + w_m[n], \quad m = 1, \ldots, M \end{aligned} \tag{5.11}$$

where h_m is the channel gain. It is assumed that the channel is slowly varying such that the channel frequency response or channel gain remains constant during the sensing duration. Due to free space path loss, the SNR of the received signal varies and depends on the distance that the CR user is located from the primary user transmitter. The free space path loss in decibels (dB) can be expressed as [44]

$$\gamma_{pl} = 20\log_{10}\left(\frac{4\pi}{\nu}d_m\right) \tag{5.12}$$

where d_m is the distance of mth node from PU transmitter and 'ν' s the velocity of radio waves in free space.

The cooperation can be done using hard decision or soft decision fusion methods. In the class of hard decision fusion, logical AND, OR rules are commonly used. In case of OR rule, the FC decides that a signal is present if any of the nodes reports about signal detection. Detection probability of OR is, $C_{d-OR} = 1 - \prod_{m=1}^{M} (1 - P_{d,m})$. Where $P_{d,m}$ is the detection probability of the mth node. In case of the AND rule, a signal is detected if all nodes have detected a signal. Detection probability of AND is $C_{\text{d-AND}} = \prod_{m=1}^{M} P_{d,m}$.

In this chapter, we concentrate mainly on the soft decision fusion techniques due to its reliability and performance improvement compared to hard decision fusion methods. To enhance the cooperative sensing performance, different weights are assigned to the cognitive users according to their received signal strength.

WGC Method Using Log Likelihood Ratio Test. In this method, the weights to each cognitive radio in the network are evaluated using log-likelihood ratio (LLR) test [46], expressed as

$$\log\left[\frac{P(\mathbf{Z}|H_1)}{P(\mathbf{Z}|H_0)}\right] \underset{H_0}{\overset{H_1}{\gtrless}} \lambda_e \tag{5.13}$$

where $\mathbf{Z} = (r_1, r_2, \ldots, r_M)$ is the received soft decision vector at FC from each CR user. Then, Eq. (5.13) can be rewritten as

$$\log\left(\frac{P(\mathbf{Z}|H_1)}{P(\mathbf{Z}|H_0)}\right) = \sum_{m=0}^{M-1} \log\left(\frac{\frac{1}{\pi^N(\sigma_{w-m}^2+\zeta_m^2)^N}\exp^{-\frac{\|r_m\|^2}{\sigma_{w-m}^2+\zeta_m^2}}}{\frac{1}{\pi^N\sigma_{w-m}^{2N}}\exp^{-\frac{\|r_m\|^2}{\sigma_{w-m}^2}}}\right)$$

This can be approximated as

$$= \sum_{m=0}^{M-1} \log\left(\exp\left(\frac{-\|r_m\|^2}{\sigma_{w-m}^2+\zeta_m^2} + \frac{\|r_m\|^2}{\sigma_{w-m}^2}\right)\right)$$

$$= \sum_{m=0}^{M-1} \|r_m\|^2 . \frac{\zeta_m^2}{\sigma_{w-m}^2(\sigma_{w-m}^2+\zeta_m^2)}$$

Thus, the test statistic for cooperative sensing can be written as

$$\sum_{n=0}^{N-1} |r_m[n]|^2 . \Theta_m \underset{H_0}{\overset{H_1}{\gtrless}} \lambda_e \tag{5.14}$$

where the test statistic $\|r_m\|^2 = \psi_m$ is the energy measurement of the mth node, Θ_m is the weight factor for mth node, given as

$$\Theta_m = \frac{\zeta_m^2}{\sigma_{w-m}^2(\sigma_{w-m}^2+\zeta_m^2)} \tag{5.15}$$

and where ζ_m^2 and σ_{w-m}^2 are the variances of signal and noise at mth node. Hence the cooperative detection probability using energy detection can be expressed as

$$C_{d-wgc} = \sum_{m=1}^{M} \psi_m . \Theta_m \underset{H_0}{\overset{H_1}{\gtrless}} \lambda_e \tag{5.16}$$

where λ_e is the threshold as given in Eq. (5.3) which depends on the desired false alarm probability.

In general, the central fusion node does not have prior information about signal strength. Therefore, in case of the EGC, equal weights are given to all nodes and aggregate their measurements to make the global decision.

WGC Method Using Differential Evolution Algorithm. Though the LLR method enhances the performance of the cooperative sensing, the weight evaluation completely depends on the signal characteristics. Hence, the differential evolution algorithm is applied to evaluate the optimal weights for weighted gain cooperative sensing which is independent of signal and noise characteristics [42].

Problem formulation: In this case, the problem is formulated as to find a set of weight values that maximizes the cumulative sum. Mathematically, it can be expressed as

$$\max \sum_{m=1}^{M} \psi_m \Theta_m, \; s.t \sum_{m=1}^{M} \Theta_m = 1, \quad 0 < \Theta_m < 1 \tag{5.17}$$

In DE algorithm, maximization of the sum of products (soft decision and its corresponding weight) of all cooperative nodes can be considered as an objective function. In this algorithm, for weight optimization, population of size 'P' is initialized as

$$\mathbf{\Theta_I} = \left[\mathbf{\Theta_{1,G}}, \mathbf{\Theta_{2,G}}, \ldots, \mathbf{\Theta_{P,G}}\right], \; i = 1, 2, \ldots, P$$

where $\mathbf{\Theta_{i,G}}$ is a vector containing 'M' number of random weights at 'Gth' generation. The best weight set in each generation is the one which gives optimal values for maximizing the cumulative sum. The next generations of vectors are generated as follows. For every vector, $\mathbf{\Theta_{i,G}}$ (target vector), the following three steps are performed.

Mutation: Three mutually distinct random vectors $\mathbf{\Theta}_{r1,\mathbf{G}}, \mathbf{\Theta}_{r2,\mathbf{G}}, \mathbf{\Theta}_{r3,\mathbf{G}}$ are taken such that $i \neq r1 \neq r2 \neq r3$. The mutant vector/donor vector is generated according to the expression, given as,

$$\mathbf{V_{i,G+1}} = \mathbf{\Theta}_{r1,\mathbf{G}} + F.(\mathbf{\Theta}_{r2,\mathbf{G}} - \mathbf{\Theta}_{r3,\mathbf{G}}) \tag{5.18}$$

where $F \in [0,2]$ is a constant which controls the magnitude of the differential variation.

Crossover: The diversity of the vector set is increased by developing a trial vector as

$$\begin{aligned} u_{j,i,G+1} &= V_{j,i,G+1} \quad \text{if}(\text{rand}(j) \leq C_R) \text{ or } j = \text{rnbr}(i) \\ &= \Theta_{j,i,G} \quad \text{if}(\text{rand}(j) > C_R) \text{ or } j \neq \text{rnbr}(i) \end{aligned} \tag{5.19}$$

where $j = 1, 2, \ldots, M$, $rand(j)$ is the random number generator with outcome $\in [0,1]$. C_R is the crossover constant $\in [0,1]$ which has to be chosen by the user, and $rnbr(i)$ is a randomly chosen index from $\{1,2, \ldots, M\}$ which ensures that the trial vector $\mathbf{u_{i,G+1}}$ gets at least one parameter from donor vector $\mathbf{v_{i,G+1}}$.

Selection: In this process, the trial vector $\mathbf{u_{i,G+1}}$ is compared with the target vector $\mathbf{\Theta_{i,G+1}}$ and the one that gives the best values for cumulative sum is passed on to the next generation as $\mathbf{\Theta_{i,G+1}}$. The algorithm is continued till the optimum weight vector ($\mathbf{\Theta_{opt}} = [\Theta_{opt(1)}, \Theta_{opt(2)}, \ldots, \Theta_{opt(M)}]$) is found. The cooperative

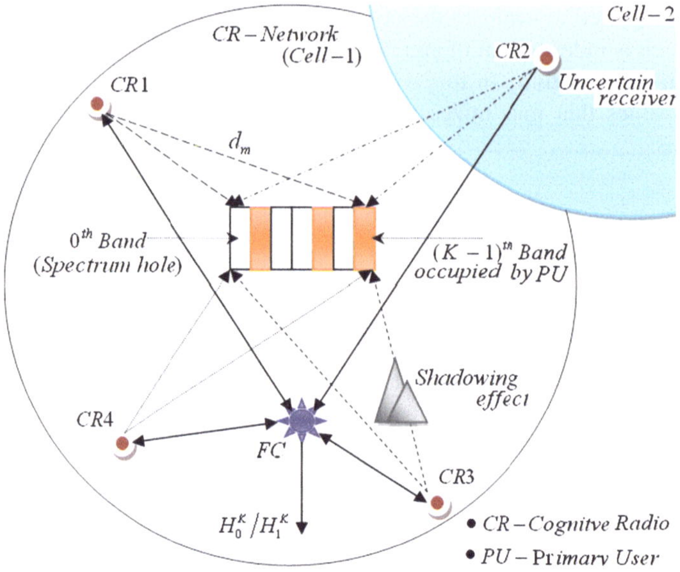

Fig. 5.11 Cooperative wideband sensing model in a CRN [40]

detection probability with the optimal weights can be determined by replacing the weights evaluated using LLR test.

5.3.2 Cooperative Wideband Detection

Figure 5.11 represents an overview of cooperative wideband spectrum sensing (CWSS) model. In this model, all the cooperative CR users are distributed over the CRN at different distance from PU transmitter. Each CR user senses the entire wideband and sends the measurement or decision to the fusion center (FC). Finally, FC makes the global decision by aggregating the received local sensing information in each band and informs the global decision to all cooperative users.

Assuming that there are M nodes in the cooperation and the received signals of all nodes are independent, then the objective of sensing is to discriminate the following two composite hypotheses

$$H_0^k : \mathbf{R}_m^k(n) = \mathbf{W}_m^k(n), \quad m = 0, 1, 2, \ldots, M-1$$
$$H_1^k : \mathbf{R}_m^k(n) = h_m^k * \mathbf{S}_m^k(n) + \mathbf{W}_m^k(n), \quad n = 0, 1, \ldots, N-1$$

where $\mathbf{R}_m^k(n)$, $\mathbf{W}_m^k(n)$, and $\mathbf{S}_m^k(n)$ can be compactly represented as

$$\mathbf{R}_m^k(n) = [\mathbf{r}_m^0, \mathbf{r}_m^1, \mathbf{r}_m^2, \ldots, \mathbf{r}_m^{(K-1)}], \quad m = 1, 2, \ldots, M$$
$$\mathbf{W}_m^k(n) = [\mathbf{w}_m^0, \mathbf{w}_m^1, \mathbf{w}_m^2, \ldots, \mathbf{w}_m^{(K-1)}], \quad m = 1, 2, \ldots, M$$
$$\mathbf{S}_m^k(n) = [\mathbf{s}_m^0, \mathbf{s}_m^1, \mathbf{s}_m^2, \ldots, \mathbf{s}_m^{(K-1)}], \quad m = 1, 2, \ldots, M$$

where $\mathbf{r}_m^k$, $\mathbf{w}_m^k$ and $\mathbf{s}_m^k$ are the received signal, noise and primary user signal samples of mth node in kth SB such that $k \in \{0, \ldots, (K-1)\}$, and N is the total number of samples considered for spectrum sensing.

In this work, the algorithms are analyzed based on the following assumptions, (1) the noise in each SB ($\mathbf{w}^k$) follows Gaussian, independent and identically distribution ($i.i.d$) with zero mean and variance σ_w^2, (2) the received signal in each subband ($\mathbf{s}^k$) is a stochastic signal, and it follows Gaussian $i.i.d$ with mean μ_s and variance ζ^2, (3) the channel gain (h), transmitted signal ($\mathbf{s}^k$), and the additive white Gaussian noise ($\mathbf{w}^k$) are independent of each other, and (4) the channel is time invariant during the sensing period.

Intuitively, for K number of SBs, the test statistic for cooperative wideband sensing can be formulated as [42],

$$C_{d-wgc}^{(k)} = \sum_{m=1}^{M} \psi_{avg-m}^{k}(r).\Theta_m \underset{H_0^k}{\overset{H_1^k}{\gtrless}} \lambda_e^k, \quad k = 0, 1, \ldots, (K-1) \tag{5.20}$$

where $\psi_{avg-m}^k(r) = E(\psi_m^k(r)) = \frac{1}{M}\sum_{m=1}^{M} \psi_m^k(r)$, $\psi_m^k(r)$ represents the energy measurement of the mth CR (or mth node) on kth SB.

5.3.3 Simulation Results and Discussion

5.3.3.1 Simulation Results for Cooperative Narrowband Sensing

In the simulation, cooperative cognitive users are assumed to have configurations as shown in Fig. 5.11. The fading and path loss effects are considered in the simulation. The cooperative nodes are assumed to be randomly distributed over the considered geographic area and located within 3–10 km from the licensed user transmitter.

Figures 5.12, 5.13, 5.14, and 5.15 describe ROC curves with different number of CR nodes in cooperation for logical OR, AND, LLR based WGC and EGC, respectively. Detection performance is examined with variable number of CR nodes (M as 1, 3, 5 and 8) in the cooperation. A close observation of these figures show that WGC fusion rule gives better P_d for same P_f compared to other fusion rules. One more observation is that the detection probability increases as the number of nodes in cooperation increases, except for AND rule as evident from Fig. 5.13. The ROC

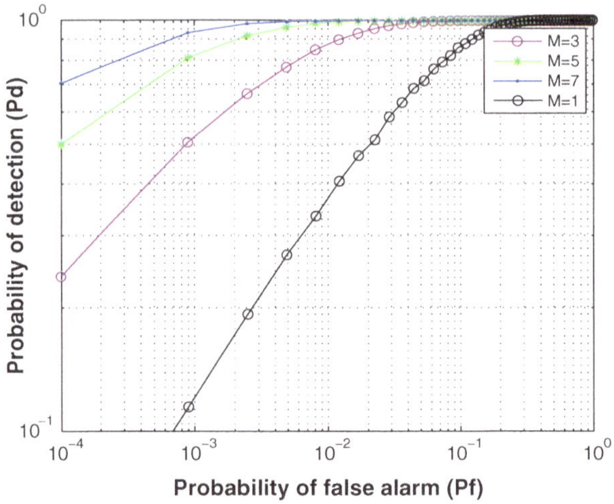

Fig. 5.12 ROC curves of OR fusion

Fig. 5.13 ROC curves of AND fusion

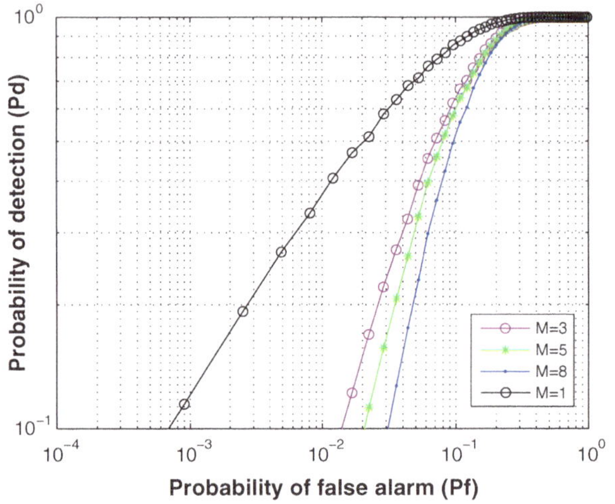

for EGC is also follows the same trend and the detection performance is less than WGC. Figures 5.12, 5.14, and 5.15 reveals that as the SNR increases the probability of successful detection increases. This is because, OR rule decides about the presence of signal if any of the CR users in cooperation have detected the signal. WGC and EGC also claims about the presence of signal based on the combined SNR of all nodes. In case of the AND rule, P_d decreases as the number of nodes in sensing increases as in Fig. 5.13. This is because, this rule decides that a signal is present if all CR's have detected that the signal is present. In conclusion, cooperative sensing is able to enhance the performance of the detection as compared to single node sensing.

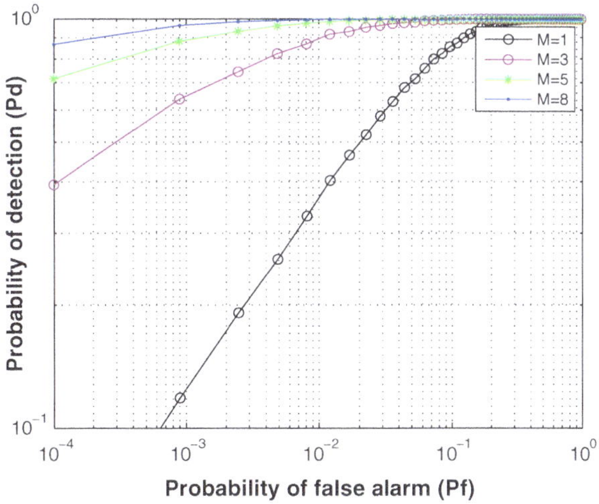

Fig. 5.14 ROC curves of WGC fusion

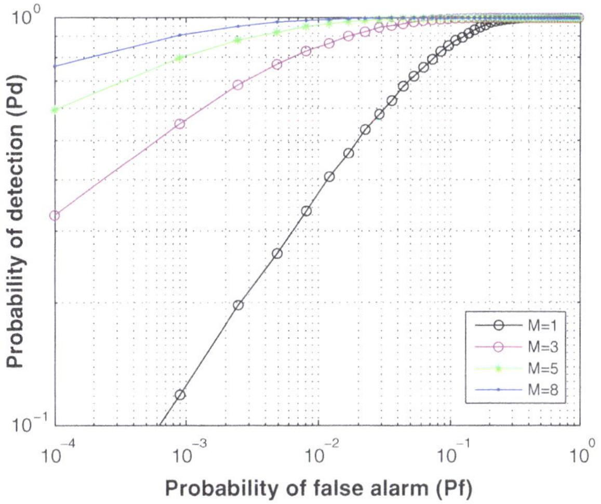

Fig. 5.15 ROC curves of EGC fusion

Figure 5.16 illustrates the performance of the differential evolution algorithm interms of number of iterations against detection probability. In the simulation, we have considered number of nodes (M) as 5, population size of 30, number of generations 100, and the SNR of all the nodes are varied in between -20 and 0 dB. From the figure, it can be seen that the proposed DE solution converges after approximately 25 iterations.

Figure 5.17 shows the performance comparison of the proposed DE based weighted gain combining, LLR-based weighted gain combining, and equal gain combining algorithms under probable noise impediments in the channel. The number of secondary users in collaboration is considered as 5 and the SNR of all the nodes are

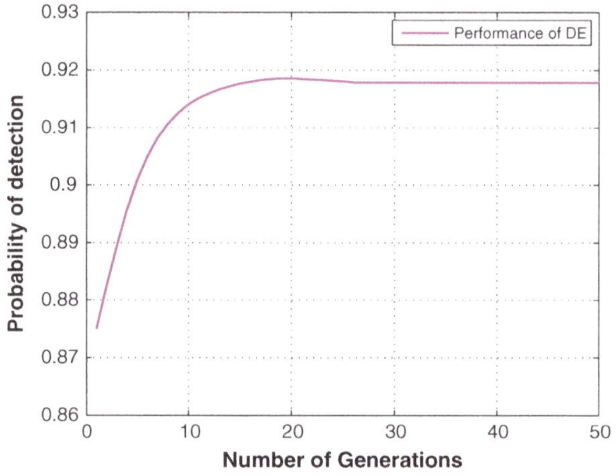

Fig. 5.16 Performance of the differential evolution algorithm

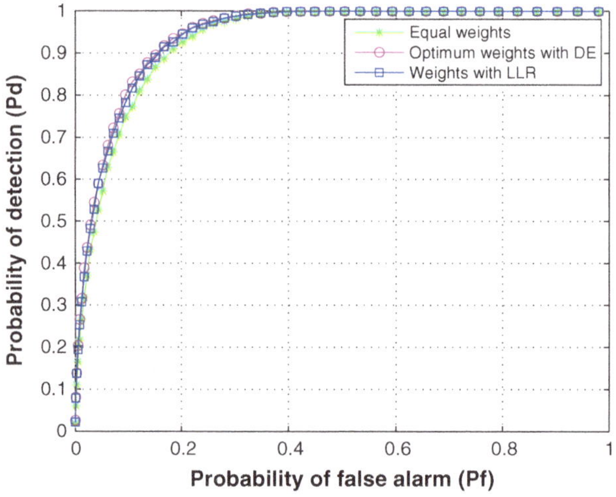

Fig. 5.17 ROC curves of weighed gain cooperation sensing methods

varied in between −20 and 0 dB. In case of the equal gain combining method, the weights are generated equally with a constraint that the sum of the weights equal to one. The detection probability of the EGC method is not an optimal method because of equal importance to each cognitive node in cooperation and there is significant performance stagnation due to improper assignment of weights to each node. In case of the LLR-based weighted gain combining method, the weights are evaluated as given in Eq. (5.15), which depends on variances of noise and signal. This technique is the optimal method for cooperation when the received signal characteristics are known. But, it is not always be the case, is available for all the frequency bands.

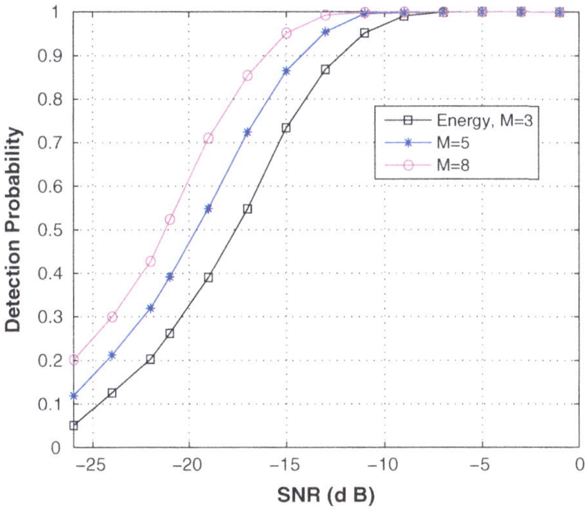

Fig. 5.18 Performance of DE based WGC for energy detection method

In case of the proposed weight generation method, the weights are generated using differential evolution algorithm with cumulative sum as the objective function as given in Eq. (5.17). From the simulation results, it can be seen that the performance of the DE method matches with the LLR method with added advantage of weight generation, which is independent of the received signal characteristics. Hence, the technique can be used for real time cognitive radio networks for cooperative spectrum sensing.

Figure 5.18 shows the multinode sensing performance based on energy detection method using DE based WGC fusion with fixed probability of false alarm ($P_f = 0.1$) and different number of nodes ($M = 3$, 5 and 8) in cooperation. From this figure, it is observed that, cooperative sensing enhances the performance of the system. From the figure, it is observed that for the energy detection, the least SNR required or SNR_{wall} to achieve the desired performance using DE based WGC fusion logic is -12 dB with three nodes, -14 dB with five nodes, and -16 dB using eight nodes.

5.3.3.2 Simulation Results for Cooperative Wideband Sensing

In our simulation, cooperative CRs are assumed to have a configuration as shown in Fig. 5.11. The performance of proposed CWSS is evaluated using different collaborative scenarios (soft decision fusion techniques). In this simulation, two cases of CR network geometry, ideal distance case (IDC) and different/random distance case (DDC) are considered. In case of IDC, it is assumed that all cooperative CR users are located at equal distance from the PU transmitter where the path-loss is negligible. In case of DDC, it is assumed that, all cooperative CRs are distributed randomly over

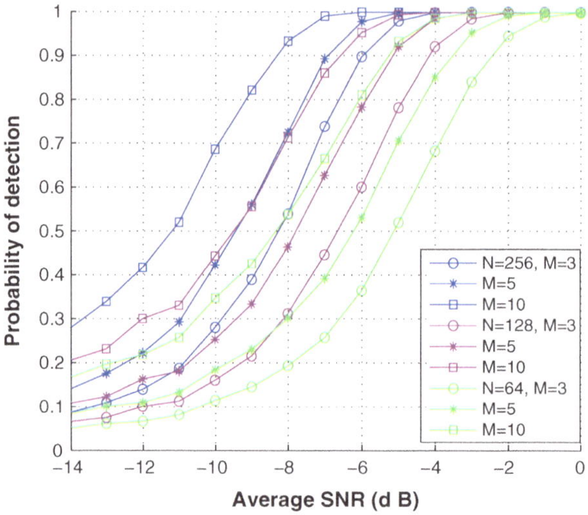

Fig. 5.19 Average SNR versus P_d for variable sample size using EGC fusion

the considered geographic area. We have considered, randomly distributed CR users are located within 3–10 km from PU transmitter.

Figure 5.19 shows the sensitivity of EGC fusion logic under AWGN and Rayleigh fading channel environment with variable number of samples ($N = 64, 128$, and 256) at fixed false alarm $P_f = 0.01$. In this simulation, variable number ($M = 3, 5$ and 10) of cognitive users are considered in cooperation. From this figure, it is clear that the P_d increases with increase in sample size and number of CR users that are participated in the cooperation.

5.4 Conclusions

From the study, we ascertain that the main requirements of whitespace detection are the prediction of signal status in frequency bands in a low signal-to-noise ratio (SNR) environment. Hence, in this chapter, energy detection method is used to predict the status of multiple frequency bands. The presented method is able to detect the PU signals of SNR up to -8 dB using single node at desired performance ($P_f \leq 0.1$ and $P_d \geq 0.9$) with $N = 256$. Henceforth, the single node detection method is extended to the multinode detection, where multiple cognitive users collaborate to improve the sensing performance using spatial diversity. The weighted gain combiniation method based on differential evolution is proposed and applied for cooperative sensing. It can detect -16 dB signals with eight users in cooperation. The advantage of cooperative sensing with differential evolution algorithm is that, it does not require any prior information of signal strength for weight generation where as log-likelihood ratio based weighted combing method requires a prior information. In conclusion, the researchers and engineers can develop efficient whitespace detection algorithms and hardware implementation prototypes for cognitive radio networks.

Acknowledgments The authors are thankful to the University Grants Commission (UGC), Government of India and University of Cape Town, South Africa for providing necessary support to carry out this work.

References

1. Akyildiz, I.F., Brandon, F.L., Ravikumar, B.: Cooperative spectrum sensing in cognitive radio networks: a survey. Phys. Commun. **4**(1), 40–62 (2011)
2. Akyildiz, I.F., Lee, W.Y., Vuran, M.C., Mohanty, S.: Next generation/dynamic spectrum access/cognitive radio wireless networks: a survey. Comput. Netw. **50**(13), 2127–2159 (2006)
3. Axell, E., Leus, G., Larsson, E., Poor, H.V.: Spectrum sensing for cognitive radio: state-of-the-art and recent advances. IEEE Signal Process. Mag. **29**(3), 101–116 (2012)
4. Bao, N., Shen, L.: Quick sensing and spectrum sharing strategy for frequency hopping signal in ISM band. In: International Conference on Wireless Communications Signal Processing (WCSP), pp. 1–5 (2009)
5. Bhowmick, A., Das, M., Biswas, J., Roy, S., Kundu, S.: Relay based cooperative spectrum sensing in cognitive radio network. In: IEEE International Advance Computing Conference (IACC), pp. 333–337 (2014)
6. Brown, T.X.: An analysis of unlicensed device operation in licensed broadcast service bands. In: First IEEE International Symposium on New Frontiers in Dynamic Spectrum Access Networks (DySPAN), pp. 11–29 (2005)
7. Carlos, K.C., Birru, D.: IEEE 802.22: an introduction to the first wireless standard based on cognitive radios. IEEE J. Commun. **1**(1), 38–47 (2006)
8. Chaudhari, S., Lunden, J., Koivunen, V., Poor, H.V.: Cooperative sensing with imperfect reporting channels: hard decisions or Soft decisions? IEEE Trans. Signal Process. **60**(1), 18–28 (2012)
9. Chaudhari, S.: Spectrum sensing for cognitive radios: algorithms, performance, and limitations. Technical report, Ph.D thesis submitted at Aalto University (2012)
10. Das, S., Suganthan, P.: Differential evolution: a survey of the state-of-the-art. IEEE Trans. Evol. Comput. **15**(1), 4–31 (2011)
11. Digham, F., Alouini, M., Simon, M.K.: On the energy detection of unknown signals over fading channels. In: IEEE International Conference on Communications (ICC), pp. 3575–3579 (2003)
12. Duan, D., Yang, L., Jose, C.: Cooperative diversity of spectrum sensing for cognitive radio systems. IEEE Trans. Signal Process. **58**(6), 3218–3227 (2010)
13. ETSI: ETSI standard EN 300 744: Digital video broadcasting (DVB); framing structure, channel coding and modulation for digital terrestrial television. Technical report, European Telecommunications Standards Institute (1997)
14. Fazeli-Dehkordy, S., Plataniotis, K., Pasupathy, S.: Wide-band collaborative spectrum search strategy for cognitive radio networks. IEEE Trans. Signal Process. **59**(8), 3903–3914 (2011)
15. FCC: Spectrum policy task force report (ET docket no.02-135). Technical report, Federal Communications Commission (2002)
16. Ghasemi, A., Sousa, E.: Collaborative spectrum sensing for opportunistic access in fading environments. In: First International Symposium on New Frontiers in Dynamic Spectrum Access Networks, pp. 131–136 (2005)
17. Ghosh, C., Roy, S., Cavalcanti, D.: Coexistence challenges for heterogeneous cognitive wireless networks in tv white spaces. IEEE Wirel. Commun. **18**(4), 22–31 (2011)
18. Hattab, G., Ibnkahla, M.: Multiband spectrum access: great promises for future cognitive radio networks. Proc. IEEE **102**(3), 282–306 (2014)

19. Hoyhtya, M., Hekkala, A., Katz, M., Mammela, A.: Spectrum awareness: techniques and challenges for active spectrum sensing. Cogn. Wirel. Netw. **18**(7), 353–372 (2007)
20. Kalamkar, S., Banerjee, A., Roychowdhury, A.: Malicious user suppression for cooperative spectrum sensing in cognitive radio networks using dixon's outlier detection method. In: National Conference on Communications (NCC), pp. 1–5 (2012)
21. Kaligineedi, P., Khabbazian, M., Bhargava, V.: Malicious user detection in a cognitive radio cooperative sensing system. IEEE Trans. wirel. Commun. **9**(8), 2488–2497 (2010)
22. Kay, S.M.: Fundamentals of Statistical Signal Processing, Volume 2: Detection theory. Prentice Hall, Upper Saddle River (1998)
23. Kim, J., Andrews, J.: Sensitive white space detection with spectral covariance sensing. IEEE Trans. wirel. Commun. **9**(9), 2945–2955 (2012)
24. Kimtho, P., Jun-Ichi, T.: Signal detection for analog and digital TV signals for cognitive radio. Technical report, IEICE Technical Report (Institute of Electronics, Information and Communication Engineers) (2006)
25. Lee, W.Y.: Spectrum management in cognitive radio wireless networks. Technical report, Ph.D thesis submitted at Georgia Institute of Technology (2009)
26. Letaief, K., Zhang, W.: Cooperative communications for cognitive radio networks. Proc. IEEE **97**(5), 878–893 (2009)
27. Liang, Y.C., Zeng, Y., Peh, E., Hoang, A.T.: Sensing-throughput tradeoff for cognitive radio networks. IEEE Trans. wirel. Commun. **7**(4), 1326–1337 (2008)
28. Lin, M., Vinod, A.: A low complexity high resolution cooperative spectrum sensing scheme for cognitive radios. Circuits Syst. Signal Process. **31**(3), 1127–1145 (2012)
29. Ma, J., Zhao, G., Li, Y.: Soft combination and detection for cooperative spectrum sensing in cognitive radio networks. IEEE Trans. wirel. Commun. **7**(11), 4502–4507 (2008)
30. Mathur, C., Subbalakshmi, K.P.: Digital signatures for centralized DSA networks. In: 4th IEEE Consumer Communications and Networking Conference (CCNC), pp. 1037–1041 (2007)
31. McHenry, M.A.: NSF spectrum occupancy measurements projects summary. Technical report, Shared Spectrum Company Report (2005)
32. Mishra, S., Sahai, A., Brodersen, R.W.: Cooperative sensing among cognitive radios. In: International Conference on Communications (ICC), vol. 4, pp. 1658–1663 (2006)
33. Mitola, J., Maguire, G.Q.: Cognitive radio: making software radios more personal. IEEE Pers. Commun. **6**(4), 13–18 (1999)
34. P802.22/D3.0, I.: IEEE draft standard for information technology-telecommunications and information exchange between systems—WRANs—specific requirements. Technical report, IEEE Communication society (2011)
35. Poor, H.V.: An Introduction to Signal Detection and Estimation, 2nd edn. Springer, New York (1998)
36. Raza Umar, A.U.H.S.: A comparative study of spectrum awareness techniques for cognitive radio oriented wireless networks. Phys. Commun. (2012)
37. Sanket S. Kalamkar, P.K.S., Banerjee, A.: Block outlier methods for malicious user detection in cooperative spectrum sensing. In: 79th IEEE Vehicular Technology Conference-Spring (VTC-Spring), pp. 1–5 (2014)
38. Sirotiya, M., Banerjee, A.: Detection and estimation of frequency hopping signals using wavelet transform. In: Second UK-India-IDRC International Workshop on Cognitive Wireless Systems (UKIWCWS), pp. 1–5 (2010)
39. Srinu, S., Sabat, S.L.: Cooperative wideband sensing based on entropy and cyclic features under noise uncertainty. IET Signal Proc. **7**(8), 655–663 (2013)
40. Srinu, S., Sabat, S.L.: Cooperative wideband spectrum sensing in suspicious cognitive radio network. IET wirel. Sens. Syst. **3**(2), 153–161 (2013)
41. Srinu, S., Sabat, S.L.: Multinode sensing with forward error correction and differential evolution algorithms for noisy cognitive radio networks. Comput. Electr. Eng. **40**(4), 1090–1100 (2014)
42. Srinu, S., Sabat, S.: Optimal multinode sensing in a malicious cognitive radio network. IEEE Syst. J. **PP**(99), 1–10 (2013)

43. Srinu, S.: Entropy based reliable cooperative spectrum sensing for cognitive radio networks. Technical report, Ph.D thesis submitted at the University of Hyderabad (2013)
44. Srinu, S., Sabat, S.L.: FPGA implementation and performance study of spectrum sensing based on entropy estimation using cyclic features. Comput. Electr. Eng. **38**(6), 1658–1669 (2012)
45. Srinu, S., Sabat, S.L.: Cooperative wideband sensing based on cyclostationary features with multiple malicious user elimination. AEU Int. J. Electron. Commun. **67**(8), 702–707 (2013)
46. Srinu, S., Sabat, S.L.: Effective cooperative wideband sensing using energy detection under suspicious cognitive radio network. Comput. Electr. Eng. **39**(4), 1153–1163 (2013)
47. Srinu, S., Sabat, S.L., Udgata, S.K.: FPGA implementation of cooperative spectrum sensing for cognitive radio networks. In: Second UK-India-IDRC International Workshop on Cognitive Wireless Systems (UKIWCWS), pp. 1–5 (2010)
48. Sun, H., Nallanathan, A.: Wideband spectrum sensing for cognitive radio networks: a survey. IEEE Wirel. Commun. **20**(2), 74–81 (2013)
49. Tandra, R., Sahai, A.: SNR walls for signal detection. IEEE J. Sel. Top. Sign. Proces. **2**(1), 4–17 (2008)
50. Unnikrishnan, J., Veeravalli, V.V.: Cooperative sensing for primary detection in cognitive radio. IEEE J. Sel. Top. Sign. Proces. **2**(1), 18–27 (2008)
51. Vesterstrom, J., Thomsen, R.: A comparative study of differential evolution, particle swarm optimization, and evolutionary algorithms on numerical benchmark problems. In: Congress on Evolutionary Computation, pp. 1980–1987 (2004)
52. Visser, F., Janssen, G., Paweczak, P.: Multinode spectrm sensing based on energy detection for dynamic spectrum access. In: IEEE Vehicular Technology Conference (VTC), pp. 1394–1398 (2008)
53. Wild, B., Ramchandran, K.: Detecting primary receivers for cognitive radio applications. In: First IEEE International Symposium on New Frontiers in Dynamic Spectrum Access Networks (DySPAN), pp. 124–130 (2005)
54. http://www.mathworks.com.: Digital Video Broadcasting—Terrestrial—MATLAB & Simulink Example. Mathworks (2004)
55. Xia, W., Cheng, W.: Correlation-based spectrum sensing in cognitive radio. In: IEEE Proceedings, pp. 67–72 (2009)
56. Xing, C., Zhisong, B., Weiling, W.: Detection efficiency of cooperative spectrum sensing in cognitive radio network. J. Chin. Univ. Posts Telecommun. **15**(3), 1–7 (2008)
57. Yucek, T., Arslan, H.: A survey of spectrum sensing algorithms for cognitive radio applications. IEEE Telecommun. Surv. Tutorials **11**(1), 116–130 (2009)
58. Zeng, Y., Liang, Y.C., Hoang, A., Zhang, R.: A review on spectrum sensing for cognitive radio: challenges and solutions. EURASIP J. Adv. Signal Process. **2010**(1), 381465 (2010)
59. Zhi, Q., Shuguang, C., Sayed, A.H., Poor, H.V.: Optimal multiband joint detection for spectrum sensing in cognitive radio networks. IEEE Trans. Signal Process. **57**(3), 1128–1140 (2009)
60. Zou, Q., Zheng, S., Sayed, A.: Cooperative sensing via sequential detection. IEEE Trans. Signal Process. **58**(12), 6266–6283 (2010)

Chapter 6
Geo-location White Space Spectrum Databases: Review of Models and Design of a Dynamic Spectrum Access Coexistence Planner and Manager

Luzango Mfupe, Fisseha Mekuria, Litsietsi Montsi and Mjumo Mzyece

Abstract Geo-location white space spectrum databases (GL-WSDB) are currently the most preferred technique for enabling spectrum sharing between primary users and secondary users or white space devices (WSDs) in the very-high frequency (VHF) and ultra-high frequency (UHF) bands. This is true because the technologies for making low-cost WSDs that are capable of autonomous sensing and detection of available white space (WS) spectrum are not yet feasible. The primary roles of such GL-WSDB systems are three-fold: first, to translate the protective WS spectrum usage regulations provided by the national spectrum regulatory authorities. The second role is the allocation of locally available spectrum channels in such a way that any harmful interference to the primary user networks and secondary transceiver stations is kept below acceptable thresholds. The third important role played by the GL-WSDB is to provide technical mechanisms for enabling WSDs to access locally available WS spectrum channels. This chapter reviewed the necessary enabling technical conditions to allow coexistence and compatibility of primary and secondary systems in the VHF and UHF spectrum through a GL-WSDB approach. The practical implementation of real-life GL-WSDB was performed. Results of WS channels available from five cities in South Africa calculated from the implemented GL-WSDB was compared with a commercially variant GL-WSDB and was found to be 68 % similar. Additionally results from the implemented GL-WSDB were compared with the measurements obtained from the field spectrum scanning campaign at

L. Mfupe (✉) · F. Mekuria · L. Montsi
Meraka Institute, Council for Scientific and Industrial Research (CSIR),
PO Box 395, Pretoria 0001, South Africa
e-mail: Lmfupe@csir.co.za

F. Mekuria
e-mail: Fmekuria@csir.co.za

L. Montsi
e-mail: Lmontsi@csir.co.za

L. Mfupe · M. Mzyece
Department of Electrical Engineering, FSATI, Tshwane University of Technology (TUT),
Private Bag X680, Pretoria 0001, South Africa
e-mail: Mjumo.Mzyece@gmail.com

A.K. Mishra and D.L. Johnson (eds.), *White Space Communication*,
Signals and Communication Technology, DOI 10.1007/978-3-319-08747-4_6

a location in Cape Town South Africa and was found to be 64 % similar. The chapter recommended some important enabling TVWS policy for national regulators.

Keywords Database · Primary users · Spectrum · GL-WSDB · WSDs · TVWS · DSA

6.1 Introduction

This chapter discusses the concept of geo-location white space spectrum databases (GL-WSDBs) as an enabling technology for spectrum sharing. GL-WSDB is presented as compatibility—coexistence planner and manager of white space devices (WSDs), white space broadband networks (WSBNs) and licensed spectrum networks. The chapter discusses step-by-step methodological approaches for designing national white space spectrum database systems. The focus is placed on the very high frequency (VHF) and ultra-high frequency (UHF) bands, with the possibility to extend it to other useful radio frequency bands of interest.

The main contributions of the chapter are twofold. Firstly, to highlight the exact functional position and role of GL-WSDBs within terrestrial broadcasting TV networks architecture, examining terrestrial broadcasting TV network planning models, methodologies and parameters. Secondly, the chapter focuses on the development of GL-WSDBs, explaining the computational technical parameters and policy aspects required for enabling secondary users to utilize white space spectrum without causing any harmful interference to the incumbent users.

Furthermore, experimental results of a GL-WSDB system developed by the CSIR-Meraka Institute are presented. The results are evaluated using two methods. Firstly, compared with selected commercially available GL-WSDB and secondly compared with the spectrum occupancy scanning measurements that were performed in the field. The remainder of the chapter is arranged as follows: Sect. 6.2 provides motivation of the study. In Sect. 6.3, national terrestrial broadcast TV panning models are introduced and discussed. Section 6.4 introduces GL-WSDB planning parameters, while Sect. 6.5 discusses GL-WSDB implementation methodologies. Section 6.6 discusses the practical implementation. Section 6.7 states the problem. Section 6.8 evaluates the results. Section 6.9 discusses the enabling policy and Sect. 6.10 concludes the chapter.

6.1.1 TVWhite Space Networks

A typical TVWS network is described in Fig. 6.1 as an overlay on a broader terrestrial broadcast TV network. Fixed WSDs query the GL-WSDB to access locally available TV channels in order to provide wireless broadband Internet connectivity to unserved or underserved areas.

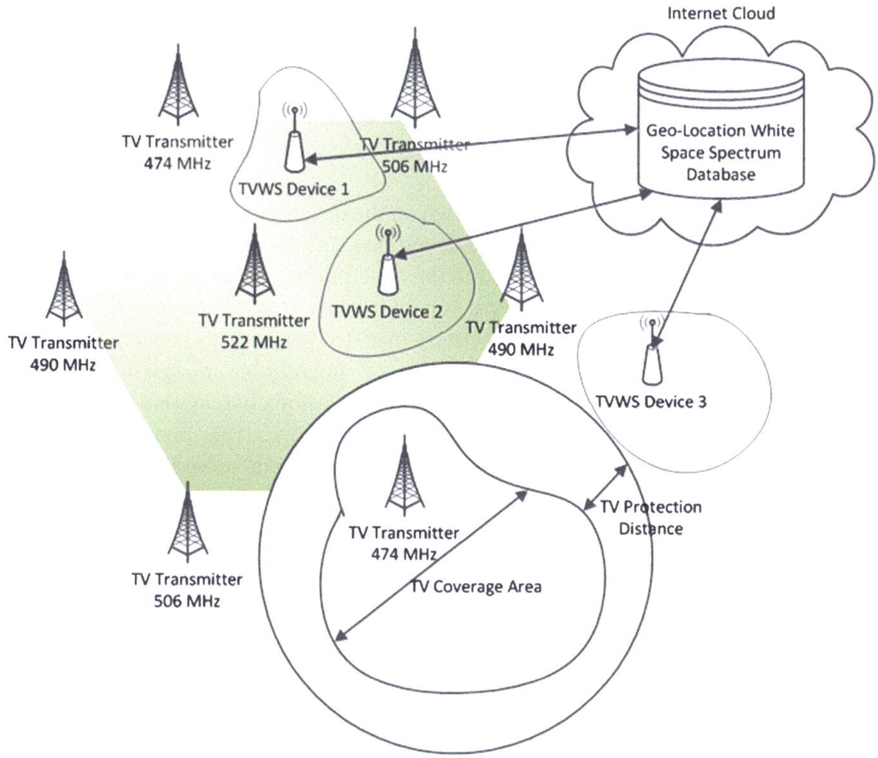

Fig. 6.1 Geo-location white space spectrum database in a UHF band terrestrial broadcast TV network structure

6.2 Motivation

Radio frequency (RF) spectrum is a precious non-depleting but limited wireless national information and communication technology (ICT) infrastructure resource. RF spectrum enables wireless devices to communicate. Therefore spectral utilisation efficiency is a major pillar towards the realisation of a successful national broadband policy. The forecasted demand for RF spectrum needed for mobile and other wireless services is projected to increase rapidly in the next five years [1]. This demand creates an urgent need for more RF spectrum. However, spectrum audit measurement reports in many parts of the world have shown that there is a great under-utilisation of this resource. For example, spectrum audit measurements conducted in South Africa five major cities by the Independent Communications Authority of South Africa (ICASA) reveals gross under-utilisation of the allocated UHF spectrum by up to 99 % [2].

Wireless access technologies are seen as an alternative and cost effective means of communication over fixed-line or wired infrastructure in developing regions, where traditionally there is little or no wired infrastructure. To achieve this, national spectrum regulators particularly in developing countries need to start putting in place enabling policy measures necessary for the adoption of efficient dynamic spectrum allocation and utilisation. In so doing, regulators will accommodate the emerging dynamic spectrum access (DSA) technologies necessary to keep up with growing demand of spectrum needed for mobile and wireless services and bridging the so-called digital divide.

It is noteworthy that a lot is already known on the developments of DSA and white space (WS) spectrum technologies in the developed countries. However, very little is known and often is not directly relevant to the developing regions context. Currently GL-WSDB technologies are the preferred enabler for WSDs in establishing WSBNs for providing broadband Internet connectivity over TV bands. This is because WSDs that are capable of autonomous spectrum sensing are not reliable enough in protecting primary users. TV band spectrum has superior propagation characteristics; penetrating man-made and natural obstacles and has longer range. This makes WSBNs a suitable choice for connecting the rural under-served and un-served areas with low population densities where mainstream mobile and other wireless services operators are reluctant to serve.

6.3 National Terrestrial Broadcast TV Planning Models

Constructing a national geo-location whitespace spectrum database requires one to have knowledge of an incumbent national terrestrial television-planning model well in advance. Broadly speaking, the national planning model outlines worst-case technical conditions sufficient to ensure quality coverage at the edge of a reference network (RN) as well as protecting incumbent transmitters from any harmful interference. Such technical conditions can be extracted from empirical measurements (i.e., field strength curves) [3] or derived from calculated statistical approximations [4].

6.3.1 Reference Planning Configurations

Strictly speaking, the models discussed above assist a GL-WSDB constructor to understand the typical reference planning configurations (RPCs) that have been utilised for each scenario in a particular terrestrial TV broadcast geographical coverage area. RPCs are results of deliberations stemming from the radio regulations (RR) of the International Telecommunications Union (ITU), the ITU regional radio-communication conferences (RRCs) [5], ITU multilateral agreements such as [6, 7], as well as unilateral (country-specific) decisions [3]. The key approaches defined by the RPCs generally relate to the configuration geometries of the terrestrial broadcast TV stations transmitter-receiver pairs; the most commonly used such geometries are:

- *Fixed outdoor reception geometry*. This configuration assumes a rooftop TV receiving antenna height of 10 m above ground level (AGL).
- *Portable/or mobile outdoor geometry*. This configuration assumes an outdoor receiving TV antenna height of 1.5 m AGL.
- *Portable indoor geometry*. This configuration assumes an indoor TV receiving antenna height of 1.5 m AGL.

It is important to note that RPC geometries are classified according to the TV transmission technology to be used. The two dominant transmission technologies are analogue and digital; in most cases analogue technology is only used for rooftop reception scenarios while digital technology is used for fixed reception and portable/mobile reception scenarios. Furthermore, the RPCs also specify parameters such as the reference frequency, the criteria for TV receiving locations probability, the tolerable interference and the minimum median field strengths in the VHF and UHF bands for each scenario.

6.3.2 Terrestrial Television Networks Structures

A typical terrestrial television network consists of a plurality of transmitter sites located in the geographical areas to be covered. These transmitters have varying heights and powers ranging from high-power transmitters to very low-power transmitters. Usually the high-power sites are used as the main transmitters to the respective targeted geographical market. Main transmitters are sometimes supplemented with the low-power relay sites that are used for the purpose of gap-filing the planned coverage areas within which the TV transmission from the high-power sites is received with low quality below the desirable minimum threshold. The low-power sites are categorised into two main types: (1) Sites that are owned by the broadcaster (2) Sites that are owned by a particular community for the purpose of enhancing the reception quality from a particular TV broadcasting network. Furthermore, terrestrial TV networks are usually configured into two distinct topologies:

- *The Single Frequency Network (SFN)*. In this topology all transmitters in a particular network are configured to utilise the same frequency in a delay-synchronized manner; that is the time taken for a broadcast signal from transmitters in a SFN to a receiver must falls within the pre-defined guard interval. Therefore, the signal with shorter delay time is treated as wanted while that with longer delay time is treated as unwanted. All transmitters in the SFN broadcast identical content nationally or in different geographical areas. That is, they are using the same channel for providing a continuous coverage. Existing technologies makes it unfeasible to implement regional or local-specific programming, as this will require a break-up from the nation-wide SFN channel. However, several SFNs can re-use the same frequencies at a separation distance that is safe not to cause harmful interference to each other. SFN topology is mostly suited for use in the digital TV transmission technologies. Figure 6.2 illustrates this further.

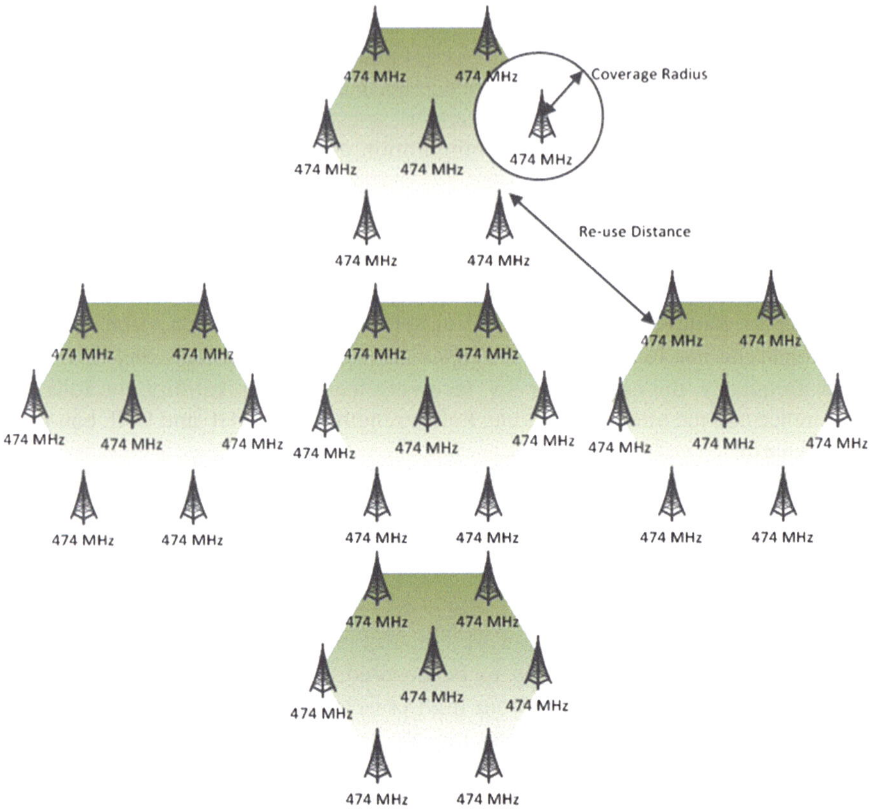

Fig. 6.2 Multiple single frequency networks structure

- *The multiple frequency network (MFN).* In this topology transmitters in a particular network are configured to utilise different frequencies in different geographical areas. That is they are using different channels for providing coverage when broadcasting same or different programmes. Regional-specific broadcasting is possible under this topology. MFNs are structured in such a way that channels assignments plans are performed in uniform lattice systems to avoid interference in the adjacent channels and to provide safe re-use separation distances for co-channels. MFNs topology is most prevalent in the analogue TV transmission technology and during the analogue to digital migration period. A structure of a MFN is illustrated in Fig. 6.3.

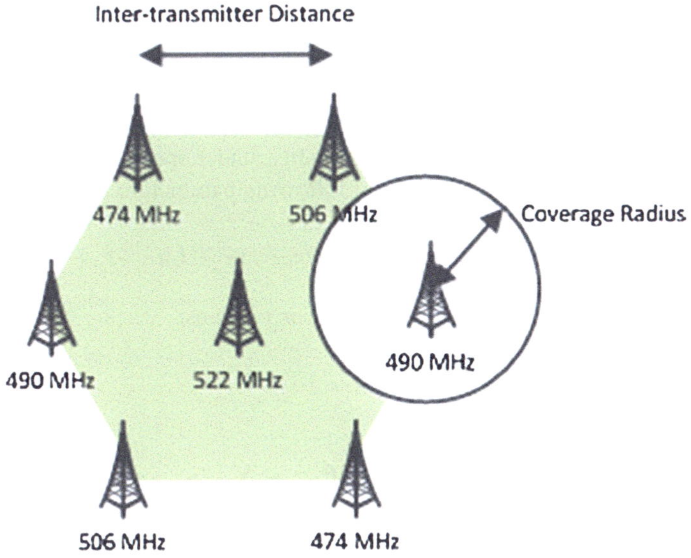

Fig. 6.3 Structure of multiple frequency networks

6.3.3 Terrestrial Broadcast Television Networks Coverage Modes

A coverage area in terrestrial broadcast TV is broadly defined as the geographical area within which the wanted field strength measured at a receiving location is greater than or equal to the minimum median field strength (this is the minimum value of the field strength necessary to permit a desired reception quality, under specified reception conditions). The quality of received TV signal at a particular height of the viewer's receiving antenna may be influenced by the natural obstacles such as trees, valleys and adverse weather conditions along its path. Additionally, man-made obstacles such as buildings as well as other transmitters using the same frequencies may also influence the signal. There are two approaches used to characterize the mode of coverage areas as a function of the quality of service received by the TV viewers located within the particular coverage area:

- *Noise limited contour (NLC) approach*. This approach defines an area (contour) for a TV transmitter where TV receivers can receive a TV signal and where incumbent TV receivers need to be protected from harmful interfering noise from other TV transmitters or other devices using the same radio frequency. The generation of noise from transmitters using same frequency can be described as network self-interference. The coverage of the NLC is an area within which the pre-defined minimum carrier-to-noise-ratio (CNR) threshold is satisfied by the difference between the minimum wanted signal power and the receiver noise floor as follows:

$$N_t = KTB \tag{6.1}$$
$$N = N_t + N_f \tag{6.2}$$

The quantity median received power is the minimum value of the received signal necessary to permit a desired reception quality, under specified reception conditions. This is expressed as a function of following parameters:

$$P_{med} = f(P_{Tx}, g_{Tx\rightarrow Rcv}, g_{Rcv\rightarrow Tx}, PL_{Rcv\rightarrow Tx}) \tag{6.3}$$

Hence the minimum wanted received power is given as:

$$P_{min} = P_{med} - (\sigma \times \varphi) \tag{6.4}$$

Therefore:

$$CNR_{[\mathrm{dB}]} \geq P_{min} - N \tag{6.5}$$

where:

N_t	Johnson-Nyquist noise in dBm
N_f	Receiver noise figure in dB
K	Boltzmann's constant in J/K
N	Receiver noise floor in dBm
T	Noise temperature in K
B	Receiver bandwidth in Hz
P_{min}	Minimum wanted power/sensitivity in dBm
P_{Tx}	Transmitter power in dBm
$g_{Tx\rightarrow Rcv}$	Gain of the transmitter to receiver path in dBi
$g_{Rcv\rightarrow Tx}$	Gain of the receiver to transmitter path in dBi
$PL_{Rcv\rightarrow Tx}$	Path loss of the receiver to transmitter path in dB
P_{med}	Median received power in dBm
σ	Standard deviation of signal in dB
φ	Log-normal distribution factor of signal in dB

Moreover, it is important to note that in the analogue terrestrial TV transmission technology; the signal-to-noise-ratio (SNR) is 8.047 dB less than that of the CNR [8]. For brevity, the difference between the two ratios is due to the fact that the SNR is a result of measurements taken at the output of the demodulator and is expressed as a ratio of the peak-to-peak video baseband signal voltage (typical value for the PAL-I standard is 0.7 and 0.714 V for the NTSC standard) to the noise bandwidth of the video signal [8]. CNR on the other hand is the ratio of the RF carrier power to the noise at the transport path (i.e., measured at the input of the receiver). The difference between CNR and SNR is expressed as follows:

$$CNR_{[\mathrm{dB}]} = SNR + 8.047 - CF \tag{6.6}$$

where:

CF Demodulator correlation factor
(typical value = 0.6 dB)

- *Interference limited contour (ILC) approach.* This is an approach that allows the new assignments of multiple radio frequencies to multiple TV broadcast transmitters based on a principle of inter-transmitter separation distance with respect to the existing transmitters. Such new assignments can cause tolerable interference to the existing TV viewers or vice versa. The coverage of an ILC is therefore defined as an area within which the pre-defined minimum carrier-to-noise-plus-interference-ratio ($CNIR$) threshold is satisfied by the difference between the minimum wanted signal power and the interfering signal power. In case of a single interfering signal power; the interfering transmitter power is expressed as a function of the following parameters:

$$P_i = oob(B_{Txi \to Rcv}), g_{Txi \to Rcv}, g_{Rcv \to Txi}, PL_{Rcv \to Txi} \tag{6.7}$$

Therefore:

$$CNIR \geq P_{min} - P_i \tag{6.8}$$

where:

P_i Interfering signal power in dBm
oob Out-of-band emission of interfering transmitter relative to wanted receiver in dBc
$B_{Txi \to Rcv}$ Frequency offset between the interfering transmitter and wanted receiver in Hz
$g_{Txi \to Rcv}$ Gain of the interfering transmitter to wanted receiver path dBi
$g_{Rcv \to Txi}$ Gain of the receiver to interfering transmitter path dBi
$PL_{Rcv \to Txi}$ Path loss of the wanted receiver to the interfering transmitter path in dB

Moreover, the term nuisance field strength is defined as the field strength E_n of the wanted signal required for achieving a desired signal quality. With the aid of (Eq. 6.8) this is obtained by the summation of a single interfering signal and the required protection ratio.

$$E_n = CNIR + P_i \tag{6.9}$$

Likewise, in case of the multiple interfering signal powers, the term usable field strength E_u is defined as the field strength of wanted signal required for achieving a desired signal quality. This is obtained by the power sum of the multiple interfering signals and the required protection ratios.

$$E_n = CNIR + sum[P_i] \tag{6.10}$$

Moreover, the relationship between the received signal power and the electric field strength is derived as follows; the received signal power is given by: $P_r = \frac{E^2}{120\pi} \times A_e \times G_{ant}$ in watts. However, since the effective aperture of a receiving antenna is given by $A_e = \frac{\lambda^2}{4\pi}$ in square meters while the signal wavelength is expressed as $\lambda = \left(\frac{300^2}{f^2 MHZ}\right)$ in meters, therefore this becomes: $P_{r[Watt]} = \frac{E^2}{120\pi} \times G_{ant} \times \frac{7160}{f^2}$, and by converting the received signal power into milliwatts and the electric field strength into microvolt per meter the expression for received signal power becomes:

$$P_{r[milliwatts]} = 1.9 \times 10^{-8} \frac{E\left(\frac{\mu v}{m^2}\right)}{f^2(MHz)} \times G_{ant} \tag{6.11}$$

Equation 6.11 can be expressed in the logarithmic form as follows:

$$P_{r[dBm]} = -772 + 20log\left(E, \frac{\mu v}{m^2}\right) - 20\log(f, MHz) + 10\log(G_{ant}),$$

from which we can finally re-arrange as:

$$P_{r[dBm]} = \left(E, \frac{dB\mu v}{m}\right) - 77.2 - 20\log(f, MHz) + G_{ant(dB)}, \tag{6.12}$$

where:

A_e Antenna effective aperture in m^2
λ Signal wavelength,
G_{ant} Gain of receiving antenna in dB
E Electric field strength in dB μ V/m

6.3.4 Terrestrial Broadcast TV Protection Ratios

In general terms, protection ratios are the minimum required signal threshold levels necessary to protect terrestrial TV from any harmful interference in order to provide the desired signal quality. As discussed in Sect. 6.3.3, the ratios: $(CNR, CNIR, CIR, SNR)$, as well as the noise to interference plus noise ratio $(NINR)$, and the interference to noise ratio $(INR)$are hereby referred to as protection ratios. It is worth noting that these ratios are standard and technology-specific that is, various analogue and digital TV standards have different values of protection ratios [9–11]. The relationship between various protection ratios is expressed as follows:

Consider that noise INR is introduced to the CIR of the system. The system should be desensitised by:

$$NINR_{[dB]} = 1 + INR \tag{6.13}$$

The following relationships are based on standard definitions:

$$CIR_{[dB]} = CNR - INR \tag{6.14}$$

And subsequently:

$$CNIR_{[dB]} = CNR - NINR + INR \tag{6.15}$$

6.3.5 Radio Propagation Models

Radio propagation models are the crucial tools in a national terrestrial broadcast TV planning model. The models are used to predict the worst-case scenarios for path losses, the received signal power or field strength at the receiver. Generally, radio propagation models are categorised according to their specific usage and a suitable model must be selected for each planning scenario. Furthermore some propagation models take into account the physical characteristics of the terrain profile and clutter between the transmitter and the receiver while other models do not. Table 6.1 lists popular radio propagation models and their specific usage. Table 6.2 compares performance of selected propagation models.

- *Deterministic propagation models*. These models predict the transmitted power or field strength from the radiating centre of the transmitter to the height of the receiver taking into account the physical path terrain elevation profile. The models also take into account the free space losses and diffraction losses along the path. Therefore, these models are suitable for site-specific planning scenarios where high precision prediction is required.
- *Empirical propagation models*. These models also referred to as statistical; predicts the transmitted power or field strength from the radiating centre of the transmitter to the height of the receiver independent of the physical path terrain elevation profile. Largely, the calculations in these models rely on the physical data obtained from extensive measurement campaigns from different geographical locations and environmental conditions. The models have minimum reliance on terrain elevation profile data. Therefore, these models are suitable for general planning and coordination scenarios where high precision prediction is not required.
- *Mixed propagation models*. These models are considered to possess both the deterministic and empirical characteristics.

Table 6.1 Illustration of selected propagation models [12]

Model	Frequency range	Distance	Category	Terrain depen-dency	Typical application
ITU-R P.452-14	Above 0.1 GHz	Up to 10,000 km	Empirical	Low	Point-to-point interference prediction procedure between stations on surface. Uses multi-edge diffraction, troposcatter and ducting.
ITU-R P.525	0.1–100 GHz	Un-specified	Deterministic	No	Point-to-point terrestrial services
ITU-R P.526-12	0.01–100 GHz	Up to radio horizon	Deterministic	High	Point-to-point propagation by diffraction. Uses path terrain profile between transmitter and receiver
ITU-R P.1812-2	0.03–3 GHz	Up to and beyond radio horizon	Deterministic	High	Point-to-multipoint site-specific and coordination planning for terrestrial broadcast
Extended hata	0.03–3 GHz	Up to 40 km	Empirical	Low	Point-to-point short –to- medium range planning of terrestrial broadcast station with short-to-medium height antennas. Uses measured terrain data in the form of curves
Longley-Rice [3]	0.02–40 GHz	1–2000 km	Mixed: Empirical/Deterministic	High	Point-to-point and point-to-multipoint planning and generic coordination- planning of terrestrial broadcast stations. Uses terrain profile elevation and measured data
ITU-R P.1546-6	0.03–3 GHz	1–1000 km	Mixed: Empirical/Deterministic	Low	Point-to-multipoint generic coordination –planning of Terrestrial broadcast stations. Uses measured terrain data in the form of curves and terrain profile elevation 3 – 15 km from the transmitter
ITWOM ver. 3 [14]	0.02–20 GHz	1–2000 km	Mixed: Empirical/Deterministic	High	Point-to-point and point-to-multipoint planning and generic coordination- planning of terrestrial broadcast stations. Uses terrain profile elevation and measured data
TM-91-1 [13]	0.04–1 GHz	Less than 16 km	Empirical	Low	Point-to-point planning for short distances

Table 6.2 Performance comparison of selected propagation models simulated on flat Earth using Monte-Carlo (MC) method

Scenario 1: Frequency = 650 MHz, Transmitter Height = 90 m, Receiver Height = 10 m, Distance = 1 m = 100 km, Number of Samples = 10,000			
Propagation model	Mean path loss (dB)	Median path loss (dB)	Standard deviation path loss (dB)
ITU-R P.452-14	140.49	143.34	23.04
ITU-R P.1546-4	157.55	162.61	20.34
Extended Hata	126.47	127.55	16.29
Longley-Rice	145.03	150.54	24.08
Scenario 2: Frequency = 650 MHz, Transmitter Height = 5 m, Receiver Height = 1.5 m, Distance = 1 m–1.5 km, Number of Samples = 10,000			
Propagation model	Mean path loss (dB)	Median path loss (dB)	Standard deviation path loss (dB)
ITU-R P.452-14	89.13	91.14	15.08
ITU-R P.1546-4	100.79	104.30	13.95
Extended Hata	98.15	101.10	9.95
Longley-Rice	83.68	86.28	8.54

Table 6.3 Comparison of antenna gain (dBd) of most commonly used sub-band specific and wideband DTT receiving antennas in South Africa [16]

Frequency	Gain (dBd)	
	Sub-band specific antenna	Wideband antenna
470–600	10.179	6.25
600–730	10.356	8.759
730–862	10.136	9.210

6.3.6 Terrestrial Broadcast TV Antennas: Patterns, Pointing and Polarisation

Two types of antennas are used in terrestrial broadcast TV: (1) transmitting antennas and, (2) receiving antennas. Furthermore, antennas can be characterised according to their frequency band of operation (band-specific) or non-band specific ones (wideband). As well as their directivity: (1) directional antennas and, (2) non-directional antennas (Omni directional). The gains and directivity discrimination of terrestrial broadcast TV receiving antennas are described in [15]. Table 6.3 compares typical gains of the most commonly used band-specific and wideband DTT receiving antennas found in South Africa [16].

Note that, radiation patterns of a terrestrial broadcast antenna define the maximum power gain of the antenna in any specific direction. Antenna alignment describes the angular displacement of the transmitter-receiver pair's central axis. Such angles can either lie between (0 and 360 °)in the azimuth direction or lie between (-90 and 90 °) in the elevation direction.

Polarization of an antenna describes the orientation of the electric field of the radio wave with respect to the Earth's surface and is determined by the physical structure of the antenna and its orientation. Strictly speaking in terrestrial broadcast planning, transmitter-receiver antenna pair's central axis should be pointing to each other and should have equivalent polarisation (e.g., both vertical and horizontal). Accordingly, using the correct antenna gain patterns, antenna direction (azimuth and elevation) and antenna polarization is critical to planning to ensure that the maximum amount of signal power is transmitted to where the intended viewership population is located. Figure 6.4 illustrates a broadcast transmitting antenna.

6.3.7 Terrestrial Broadcast TV Coverage Determination: Methodologies

The targeted large geographical area of interest is sub-divided into many smaller size geographical areas of typical resolutions (100×100 m, 200×200 m, up to1, 000 m); it is worth noting, however, that significant variations of field strength begin to be noticeable only at a distance of about [20]. Using a suitable propagation model, the quality of received coverage would vary from one location to another within

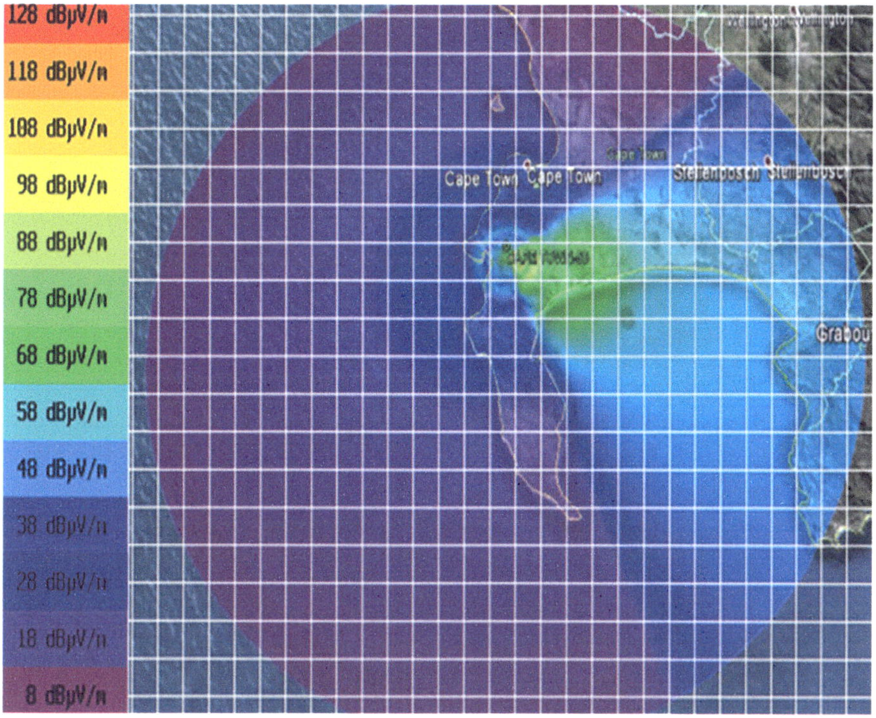

Fig. 6.4 Horizontally polarised DTT transmitting antenna, azimuth 105°, frequency 706 MHz, the figure depicts an overlay of signal coverage prediction in Cape Town Central Business District, South Africa

these small areas due to the fading phenomenon caused by the local terrain and clutter. As a result of this received signal variation, a geographical area is considered covered if the median minimum field strength threshold has been exceeded at a large percentage of locations within a small area [17]. Suppose that the received signal is characterised by a Gaussian distribution function. The cumulative distribution function of the coverage location probability on a small area A can be expressed as follows:

$$q_{E_u}(A) = \frac{1}{2} + \frac{1}{2} erf \left\{ \frac{E_u - \overline{E_{med}}}{\sigma\sqrt{2}} \right\} \tag{6.16}$$

where:

- $q_{E_u}(A) \sim q$ Coverage location probability for usable field strength on a given small area in %
- A Small area in m^2
- E_u Usable field strength in dBu
- E_{med} Minimum median field strength in dBu
- σ Location standard deviation

Table 6.4 Location correction factor for various percentages of location probabilities

$q\%$	σ_w (dB)	φ_w (dB)	$CL(q)$ (dB)
1	5.5	−2.33	−12.8
5	5.5	−1.64	−9.0
10	5.5	−1.26	−7.0
50	5.5	0	0
70	5.5	0.524	2.88
90	5.5	1.28	7.0
95	5.5	1.64	9.0
99	5.5	2.33	12.8

This signal spatial variability is corrected by a quantity known as location correction factor (CL). The CL at any wanted percentage of location probability is calculated as follows:

$$CL(q) = \sigma_w \times \varphi_w \tag{6.17}$$

where:

$CL(q)$ Location correction factor at a wanted percentage of location probability in dB
σ_w Location standard deviation of wanted signal in dB (typical value for fixed-out door reception = 5.5 dB)
φ_w Location log-normal distribution factor for wanted percentage in dB

Tables 6.4 and 6.5 depict the location correction factors and the wanted minimum median field strengths respectively. The FCC states that the minimum median field strength for DTT must be exceeded at 90 % of locations and 50 % of locations for ATT.

Median minimum field strengths E_{med} for DTT frequencies are calculated as follows:

$$E_{med} = E_{ref} + 20log_{10}(f/f_{ref}) \tag{6.18}$$

where:

E_{ref} Reference field strength in dBu
f Frequency in MHz
f_{ref} Reference frequency in MHz

The remainder of this section discusses dominant methods for predicting coverage.

- *Minimum Coupling Loss (MCL)* [20]. The MCL methodology takes a simplistic approach in predicting the required signal isolation necessary to establish coverage by the wanted transmitter in the presence a single interfering transmitter. The methodology does not take into consideration the effect of signal fading. Additionally, MCL assumes that the wanted signal power is above the wanted receiver

Table 6.5 Band specific median minimum field strength values [18, 19]

Analogue Terrestrial TV (ATT)		
Frequency band *(MHz)*	Minimum median field strength (dBu)	
	Grade B contour	Grade C contour
VHF (Band III) 174–238/246–254	55	49
UHF (Band IV) 470–578	65	60
UHF (Band V) 582–862	70	60
Digital Terrestrial TV (DTT)		
Frequency band (MHz)	Location probability (q %)	Minimum median field strength (dBu)
470–502	95	52
558–630	95	54
710–790	95	56

Median minimum field strength E_med for ATT frequency is exceeded at 50 % of locations

Table 6.6 An illustration of the minimum coupling loss (MCL) methodology using a link budget approach

Parameter	Unit
Maximum transmit power of interferer	dBm
Gain of interferer antenna	dB
Margin to account multi-carrier transmissions of interferer	dB
Out-of-band emission of interfering transmitter a frequency offset relative to wanted receiver carrier	dBc
Bandwidth conversion factor between wanted receiver and interferer antennas	dB
Gain of wanted receiver antenna	dBi
Sensitivity of wanted receiver antenna	dBm
Signal isolation = Summation of the above parameters	dB

Please note that variables are not depicted because they are scenario-specific

sensitivity by a margin of 3 dB. The calculation of signal isolation is implemented in a simplified link budget format by adding up the wanted and interferer parameters. Table 6.6 further illustrates this.

The resulting signal isolation in Table 6.6 can be converted into the worst-case scenario distance or frequency separations. This can be achieved by utilising an empirical wave propagation model of interest [12]:

$$PL_{fs} = P_{rop}(f, d, G) \tag{6.19}$$

where:

PL_{fs} Free space path-loss in dB
d Distance from the interferer to the wanted receiver in m
f Frequency in MHz
G Gain in dBi
P_{rop} Propagation function

- *Enhanced Minimum Coupling Loss (EMCL)* [20]. The EMCL methodology predicts the required signal isolation between the wanted transmitter and multiple uniformly distributed interfering transmitters by taking into consideration the effect of signal fading. Moreover, EMCL does not assume a fixed margin of the wanted signal power above the wanted receiver sensitivity. Therefore the resulting signal isolation can be converted into distance or frequency separations as well as the probability of interference. Moreover, the methodology utilises a family of empirical propagation and interference curves [21]. Such curves define TV transmitting antenna height in meters; the field strength in dB above 1 μV/m for 1 kW transmitter effective radiated power (ERP), and distance in kilometres. That is a terrestrial broadcast TV channel should be sufficiently received if the field strength has been exceeded by the predefined percentages of time and location within a predefined small area [21]. Figure 6.5 depicts empirical curves; the wanted field strength can be directly extracted from the curves if the provided transmitter height (in meters) is set to one of the 8 heights values in the curves.

Alternatively, the wanted field strength value can be log-linearly interpolated or extrapolated from Fig. 6.5 as follows:

$$E_w = E_{inf} + (E_{sup} + E_{inf}) \log_{10}(h_1/h_{inf})/\log_{10}(h_{sup}/h_{inf}) \tag{6.20}$$

where:

E_w Wanted field strength in dBu
E_{inf} Field strength h_{inf} at in dBu
E_{sup} Field strength h_{sup} at in dBu
h_1 Effective height of transmitter in m
h_{inf} Nearest nominal effective height of transmitter below h_1 in m
h_{sup} Nearest nominal effective height of transmitter above $h_1 m$

Statistical-based Methodologies. Statistical methods are the best-suited approach to determine coverage in the presence of the signal fading phenomenon as well as the in the presence of multiple sources of interference. There exist several statistical methodologies, these include but are not limited to [22] and [23]:

i. *Schwartz and Yeh.* This is an approximation-based approach, which assumes that the sum of the moment of two interfering field strengths have lognormal distribution with location.

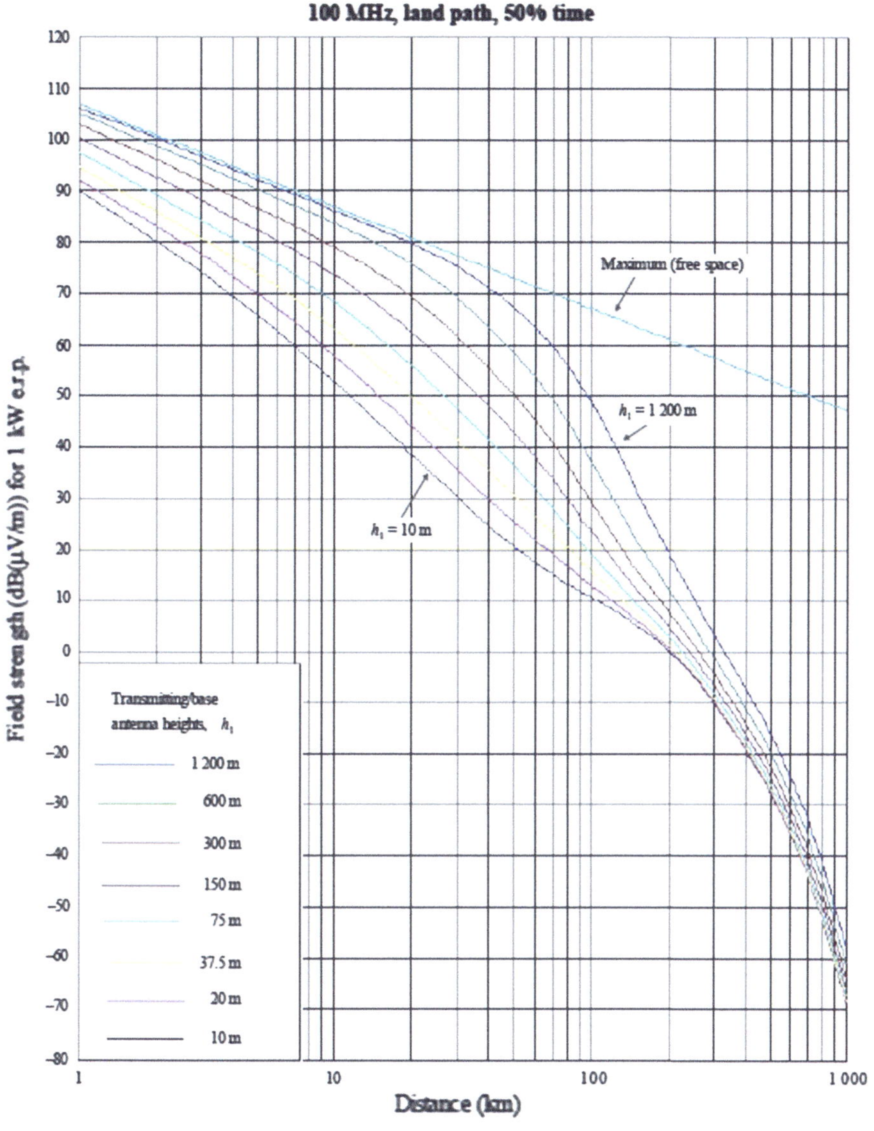

Fig. 6.5 Predicted median field strength exceeded at 50 % of potential receiver locations and exceeded at 50 % of times plotted against distance 1–1000 km for ITU zone I. Transmitter heights that can be directly extracted from the curves are 10, 20, 37.5, 75, 150, 300, 600, and 1200 m. The curves are for a receiving antenna height of 10 m; correction factor is applied for different values of receiving antenna height [21]

ii. *Power sum*. This is an approximation-based approach that calculates the sum of the signal by using a non-statistical treatment of individual summands of signal powers.

iii. *Log normal*. These are approximation-based approaches, which, statistically calculates the sum distribution of several individual summands of lognormal distributed variables.
iv. *Simplified multiplication*. This is an approximation-based approach that statistically multiplies log normally distributed interfering signals in different locations. This approach does not consider the effect of noise.
v *Monte-Carlo (MC) simulations*. This is the most accurate/efficient approach to evaluate the coverage in a small area/pixel. This is achieved by simulating a combination of a large number of victim links and interfering links pairs randomly distributed in multiple reception locations within a pixel. The median, mean and standard deviation of each event are evaluated against the set reception threshold and criteria to analyse the probability of interference. Suppose the probability of interference Pr_I is given by:

$$Pr_I = 1 - Pr_{NI} \tag{6.21}$$

Using (Eq. 6.9) for in the situation where the probability of $\frac{P_{med}}{P_i} > CNIR$ is conditional on the probability of $P_{med} > P_{min}$ this becomes:

$$Pr_{NI} = Pr\left\{\frac{P_{med}}{P_i} > CNIR | P_{med} > P_{min}\right\} \tag{6.22}$$

Applying the intersection property of probability (Eq. 6.22) becomes:

$$Pr_{NI} = \frac{Pr\left\{\frac{P_{med}}{P_i} > CNIR \bigcap P_{med} > P_{min}\right\}}{P_r\left\{P_{med} > P_{min}\right\}} \tag{6.23}$$

Since Monte-Carlo simulation considers many independent events, we rewrite (Eq. 6.23) as follows:

$$Pr_{NI} = \frac{\frac{1}{k}\sum_{j=1}^{k} 1\left\{\frac{P_{med(j)}}{P_i(j)} > CNIR \bigcap P_{med(j)} > P_{min}\right\}}{\frac{1}{k}\sum_{j=1}^{k} 1\left\{P_{med(j)} > P_{min}\right\}} \tag{6.24}$$

For:

$$1_{(criteria)} = \begin{cases} 1, & satisfied \\ 0, & otherwise \end{cases} \tag{6.25}$$

where:

Pr_I Probability of interference in %
Pr_{NI} Probability of no interference in %
K Number of Monte Carlo events

Hence, the relationship between the probability of interference and the location probability in a small coverage area is established as follows:

$$q = 100\,\% - Pr_I \tag{6.26}$$

It is important to note however, that the aforementioned relationship can be true only if both probability quantities were obtained from the same smaller coverage area.

6.3.8 Analysis of Terrestrial Broadcast TV Coverage

There exist two distinct approaches used analyse coverage, interference and population covered in small areas (the small areas are sometimes referred to as "pixels"). Both approaches are based on counting [25].

i. *Proportional counting*. This approach is commonly used for planning terrestrial broadcasting networks specifically when assessing the impact of interference from non-broadcasting networks. The approach labels the location probability of pixels not covered as "zeroes" leaving the location probability of covered pixels untouched. The total coverage area is determined by averaging the sum of all location probabilities.
ii. *Black and White (B&W) counting*. This approach is commonly used for planning terrestrial broadcasting networks specifically when assessing the impact interference from other similar broadcasting networks. The approach either labels the location probability of covered pixels as "ones" and the location probability of uncovered pixels as "zeroes" or vice-versa. Similarly, the total coverage area is determined by averaging the sum of location probabilities. Figure 6.6 depicts the concept of small coverage areas.

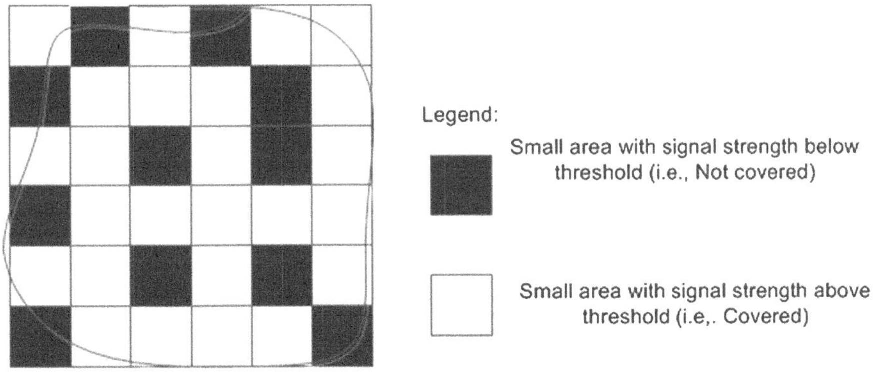

Fig. 6.6 An illustration of terrestrial broadcast TV coverage areas

6.4 Geo-location White Space Spectrum Database Planning Parameters

This section discusses key planning parameters for the implementation of GL-WSDBs with a brief highlight on regulatory-domain specific requirements.

6.4.1 Radio Systems Compatibility-Coexistence Planning Parameters

It is important to consider that a good radio planning practise would allow for a single terrestrial broadcast TV transmitter to provide fixed, portable outdoor and portable indoor reception in respective overlapping areas. Furthermore, it is important to consider that many countries particularly in the emerging and developing regions are still in the dual illumination phase (i.e., simulcasting of analogue and digital terrestrial broadcast TV). Table 6.7 discusses generic parameters applicable for planning the compatibility and co-existence of terrestrial broadcast TV stations and WSDs.

6.4.2 Protection Ratios

Protection ratios are the ratios of wanted signal power over the unwanted signal power at the point of failure of the receiving antenna. In light of protecting terrestrial broadcast TV receivers from potential harmful interference that might be generated by WSDs. Regulators have derived protection ratios for co-channel and adjacent channels specifically for various channel bandwidths. Table 6.8 illustrates this.

6.4.3 White Space Device Spectral Emission Masks

This parameter depends on the regulatory jurisdiction. The spectral mask is used to define the maximum permitted out-of-band (OOB) emissions for operation of WSDs in the VHF and UHF TV bands. Figure 6.7 depicts a typical mask.

6.4.4 Radio Propagation Models

Antenna heights of different classes of WSD range from less than 1.5 m HAAT up to 250 m HAAT. Additionally, since separation distances between WSDs and the terrestrial broadcast TV contours can range from less than few meters up to tens of kilometres. Regulators utilise appropriate propagation models for each scenario in order to accurately calculate the impact of WSDs on TV reception(various propagation models are discussed in Table 6.1).

Table 6.7 Planning Parameters of terrestrial broadcast TV transmitter and receiver

Parameter	Unit	Description
Transmitter		
Transmitter geo-location (Latitude, Longitude)	DMS	Geo-location
Maximum transmitting power	kW	Effective radiated power (ERP)
Transmitter gain	dBi, dBd	Gain of the transmitter
Transmit antenna height	m	Mast height above ground level
Frequency	MHz	Frequency of operation
Channel	–	Number in the VHF or UHF raster
Channel bandwidth	kHz	This parameter depends on the regulatory jurisdiction
Transmit antenna radiation patterns	Deg, dB	Azimuth or elevation angles and power
Transmit antenna polarisation	–	Vertical, Horizontal
Technology type	–	DTT (fixed, mobile), Analogue fixed
Station name/call sign	–	Station identifier
Status	–	Operational or non-operational
Network structure	–	MFN or SFN (see Sect. 6.3.2)
Network coverage mode	–	NLC or ILC (see Sect. 6.3.3)
Receiver		
Receiver height	m	Fixed roof top outdoor antenna or mobile portable outdoor antenna
Receive antenna front-to-back-ratio	dB	Antenna directivity discrimination
Receive antenna gain	dBi, dBd	Gain of the receiving antenna
Receive antenna feeder loses	dB	Losses due to antenna feed
Receiver noise figure	dB	Inherit receiver noise figure
Receive system noise temperature	k	Johnson-Nyquist noise
White Space Device		
Transmitter geo-location (Latitude, Longitude)	DMS	Geo-location
Minimum and maximum allowed transmitting power	W	Effective isotropic radiated power ($EIRP = ERP + 2.15$). This parameter depends on the regulatory jurisdiction
Minimum and maximum allowed and transmitter antenna height	m	Antenna's height above average terrain (HAAT). This parameter depends on the regulatory jurisdiction
Channel bandwidth	kHz	This parameter depends on the regulatory jurisdiction

Table 6.8 Terrestrial broadcast TV receiving antenna protection ratios against WSDs as per Ofcom of UK and FCC of USA

Regulatory body: Ofcom, UK (Class 1 WSD) [24]	
Channel bandwidth	8 MHz
Channel type to be protected	Protection ratio (dB)
Co-channel ($\Delta F = 0$)	17
Adjacent channel ($\Delta F = \pm 1$)	−36
Regulatory body: FCC, USA [25]	
Channel bandwidth	6 MHz
Channel type to be protected	Protection ratio (dB)
Co-channel ($\Delta F = 0$)	23
Adjacent channel ($\Delta F = \pm 1$)	−33

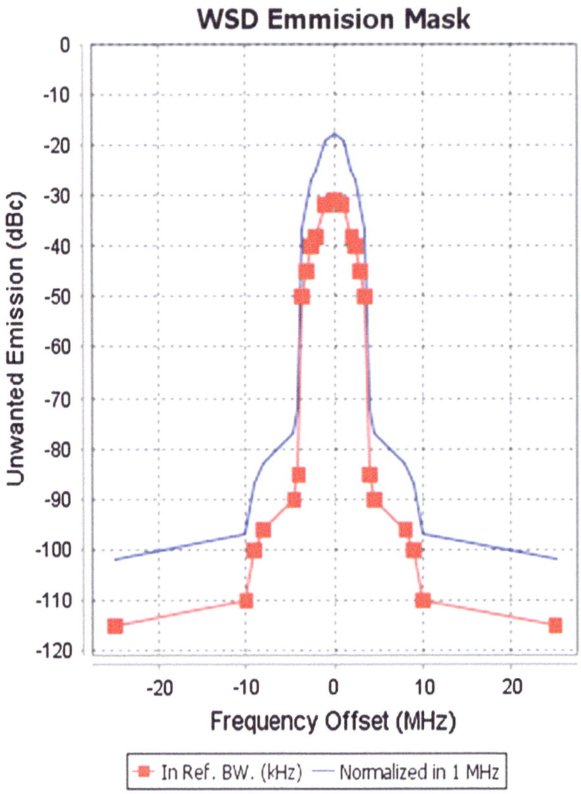

Fig. 6.7 Typical WSD spectral emission mask, the reference bandwidth is 50 kHz

6.5 Geo-location Database Implementation Methodologies

National spectrum regulators are expected to be technical agnostic in methods used in the construction of GL-WSDBs by various would be service providers. In reality, however, it is challenging to separate radio regulations from technical implementation aspects of RF stations that wish to operate in the TV bands on a non-assignment basis. Leading national regulators have shown their preferences on different technical methodological implementation of the GL-WSDB [26–28]. This Section highlights dominant GL-WSDB implementation methodologies.

6.5.1 Vectorised Approach

This is a MCL-based methodology that largely relies on rigid rules of separation distance vectors to each grade of terrestrial broadcast TV contour in order to determine the availability of TVWS channels. The separation distance vector calculations are performed using the field strength propagation and interference curves and appropriate propagation models. This is a preferred approach by both the federal communications commission (FCC) of the USA [25], and the Industry Canada [26].

The algorithm consists of three main parts:

1. *For all technology types of incumbent stations:*
   1.1. *Using appropriate propagation model perform calculation of signal strength coverage for each contour grade in the geographical area of interest.*
2. *For each class of WSD {Transmit power, Antenna HAAT}:*
   2.1 *Using appropriate propagation model or curves perform calculation of separation distances from the edge of each incumbent contour grade.*
3. *For each test point of interest {WSD's geo-location, HAAT & all VHF/UHF channels}:*
   3.1 *Calculate and analyse available TV white space channels based on the contour separation distances and protection ratios.*
   3.2 *Hence, a channel is available if a WSD is outside of the separation distance*

6.5.2 CINR Threshold Approach

This is basically an EMCL-based methodology; the approach compares the received signal power level within a small geographical area of interest (pixel) in each channel against the minimum CINR threshold to determine if a channel is occupied or not occupied [27]. Process is repeated across all pixels.

The algorithm consists of three main parts:

1. *For all technology types of incumbent stations:*

    1.1. *Using appropriate propagation model perform calculation of received signal power levels in all channels within a small geographical area (pixel) of interest.*

2. *For all channels in a small geographical area of interest (pixel):*

    2.1 *Compare the received signal strength against the minimum CINR threshold.*
    2.2 *Hence, a channel is occupied if the received signal strength power level is above the minimum CINR threshold. Otherwise a channel is available*

3. *For all available channels in a small geographical area of interest (pixel) and WSD's HAAT:*

    3.1. *Calculate and analyse maximum allowed WSD transmitting power levels based on the protection ratios and adjacent channel selectivity threshold.*

6.5.3 Degradation of Location Probability Approach

This is a statistical approach that utilise Monte Carlo (MC) simulations methodology to determine the degradation inlocation probability of a DTT receiver in small geographical coverage areas (pixels). Any presence of WSD interfering signal within a pixel reduces the location probability of an incumbent receiver. This degradation is subsequently used to calculate the availability of TV whitespace channels. This is a preferred methodology by the Ofcom of UK [24] and the European Conference on Postal and Telecommunications Administrations (CEPT) [28].

The algorithm consists of four main parts:

1. *For all technology types of incumbent stations:*

    1.1. *Using appropriate propagation model and MC simulations perform calculation of location probability in each geographical small area (pixel) "before" the introduction of an interference signal (i.e., in the presence of system noise only).*

2. *For all technology types of incumbent stations and for each class of WSD {Transmit power, Antenna HAAT}:*

    2.1. *Using appropriate propagation model and MC simulations perform calculation of location probability in each geographical small area (pixel) "after" the introduction of the interference signal.*

3. *For each small area (pixel):*

    3.1. *Using MC simulations perform calculation of the maximum permitted degradation (change) in location probability that is:* $\{\Delta_q = q_{before} - q_{after}\}$.

3.2. *Therefore the permitted change in location probability is set as the threshold* $\Delta_q = E_u$

4. *For each test point of interest {WSD's geo-location & all VHF/UHF channels}:*

4.1. *Calculate and analyse maximum allowed WSD transmitting power levels based on the protection ratios and adjacent channel selectivity threshold.*

6.6 Practical Implementation

In this section a real-life implementation of a Geo-location white space spectrum database (GL-WSDB) covering all regions of South Africa is described. According to the current South Africa terrestrial broadcast frequency allocation table [29], the following entities will require immediate protection from the operation of WSDs in the TVWS:

i. TV stations in the 174–234 and 474–854 MHz bands
ii. Public trunking, maritime radio- navigation, short range devices (SRDs), and wireless microphones in the bands adjacent to 174–234 MHz band
iii. Mobile/fixed services in bands adjacent to the 474–790 MHz band
iv. Radio-astronomy in the 608–614 MHz band

Of the above–listed entities to be protected, the most readily available information from the Independent Communications Authority of South Africa (ICASA), are the parameters describing protection of TV stations. The minimum usable field strength values used to calculate service coverage for ATT and DTT services in bands III, IV and V are listed in Table 6.5.

6.7 Problem Statement

Construct a GL-WSDB for South Africa that is capable of calculating and availing WS channels for the secondary usage in the VHF and UHFTV bands at any given location and time without causing harmful interference to the primary users. The available WS channels should be available to the WSDs with different antenna heights and different transmitting powers levels.

6.7.1 Design Motivation

An MCL methodology, taking into account the 8 MHz TV channel width used in ITU region I, has been implemented. MCL is preferred because it provides robust protection to the primary users in the both grade B and grade C coverage contours. For example, users in marginal areas who cannot receive grade B quality can still be protected under grade C contours. Many rural users in Africa are located in these fringe reception areas and make do with fairly poor reception quality or do their best to improve their reception by raising their mast height beyond the norm. The degradation of location probability methodology was not preferred because of lack of a reliable national dwelling address system which corresponds with the location TV license holders. In South Africa it is a norm to find several TV license holders living in one house; this could inhibit the process of identifying the location of TV receivers to be protected. Likewise, the CNIR threshold approach was not preferred because it is not reliable enough to protect marginal primary users at the edge of the coverage contours. Due to its heavy reliance of comparing the received signal power levels with the threshold; the approach can easily expose primary users to an interference situation similar to the hidden node problem common to spectrum sensing devices.

6.7.2 The Step-wise Implementation

The implementation utilised three wave propagation models for coverage contours and distance calculations: (1) The FCC-Curves [25] modified to accept ITU region I grade B and grade C contours (2) The Irregular terrain with obstruction model (ITWOM) [14] taking into account ITU region I grade B and grade C contours and, (3) The FCC/OET TM91-1 [13].
In principle, the GL-WSDB performs the following operations:

6.7.2.1 Calculation of incumbents contour distances

Step 1: Identify the TV stations of interest
Given the fact that South Africa is still in dual illumination period (this is a transition period from analogue TV transmission technology to digital TV transmission technology of which a particular country utilises both TV transmission technologies simultaneously). There exist over 1,000 entries of incumbent stations in the terrestrial broadcast frequency plan; (operational and planned high power and gap-filler DTT multiplex sites and analogue sites). The data is obtained from the regulator (ICASA). We are considering an interference limited contour (ILC) mode in a multi-frequency network (MFN) structure with single transmitting tower at each site. We therefore distinguish each entry by a combination of frequency assignment in use(in VHF and UHF bands,) and site name. We initially prepare and load this data together with other relevant technical information of each TV station into a relational database. For each snapshot of calculation, the focus is however placed only on a particular area of interest (i.e., within and around the area relative to the WSD's present location at that particular time.) That is, the GL-WSDB must identify all incumbent stations present in a radius of interest within which a WSD might potentially cause harmful interference and cull the incumbent stations outside this area of interest. The

significance of culling procedure for each snapshot is the reduction of computational time as opposed to when the entire database of transmitters is used each time. The GL-WSDB implements Haversine point-to-point geodesic great-circle distance formula to find the minima and maxima of the bounding box (the formula is accurate for distances up to 475 km):

$$Hv\left(\frac{d}{R}\right) = Hv(lat_1 - lat_2) + \cos(lat_2)Hv(\Delta lon) \quad (6.27)$$

where:

Hv	Haversine function
d	Spherical distance between two points in km
R	Radius of sphere in km
lat	Latitude in radians
Δlon	Change in longitude in radians

Figure 6.8 depicts the how TV transmitters of interest are being identified.

Step 2: Calculate antenna HAATs of incumbent stations
The GL-WSDB utilises the geo-location of the identified TV stations of interest in Fig. 6.8. Calculates HAAT of each identified TV station taking into consideration the

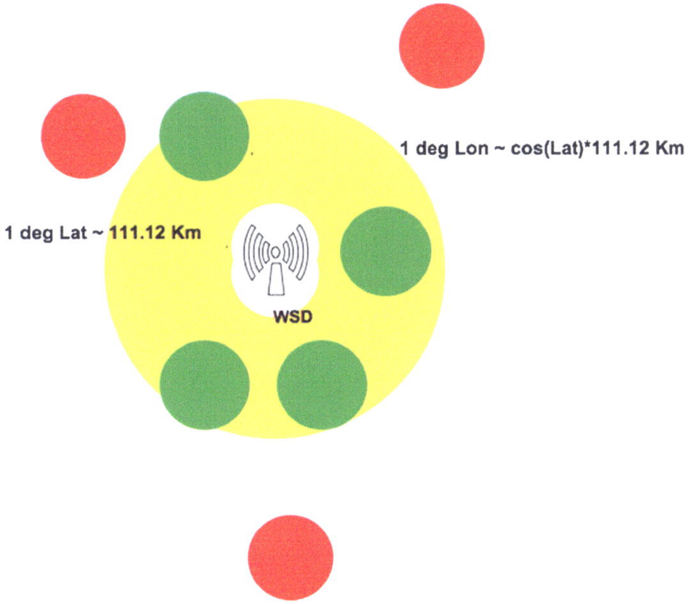

Fig. 6.8 An illustration of TV stations culling:green circles represent a snapshot of TV stations within a given boundary of interest relative to the WSD location. Red circles are station not to be considered in the snapshot

surrounding terrain profile within a distance of 3.2 to 15 km. The resulting HAAT pattern is a plot obtained from the evenly spaced increment of 2.5° starting from (0° as true North) to 360°. Corresponding numerical values for each degree increment is also obtained. The antenna's radiation centre above mean sea level (RACMSL) is also obtained. The GL-WSDB utilise 3-arc-second digital terrain data to perform this operation [30].

Step 3: Derive the relative field value(s)
The typical normalised relative field value of an antenna is approximated to be 0.707 at Half Power Bandwidth (HPBW). Alternatively, The GL-WSDB calculates the relative field value by using the TV station transmitting antenna depression angle:

$$\alpha = 0.0277\sqrt{h} \tag{6.28}$$

where:

α Depression angle of antenna in degrees
h Radiation centre above mean sea level of an antenna (RCAMSL)of antenna in

The depression angle is linearly interpolated in the VHF/UHF antenna vertical patterns graph to extrapolate the corresponding normalised relative field value [31].

Step 4: Calculate the radial ERP
The radial ERP (ERP_r) in kilowatts is a product of maximum ERP of the TV station and the square of the normalised relative field value:

$$ERP_r = maxERP \times E^2_{\varphi,\theta} \tag{6.29}$$

where:

$E_{\varphi,\theta}$ Normalise relative field values.

Step 5: Convert the obtained in (Eq. 6.29) into radial power in decibel referenced to 1 kW (dBk_r). The GL-WSDB performs the following calculation:

$$dBK_r = 10Log_{10}ERP_r \tag{6.30}$$

Step 6: Calculate the radial field strength E_r
For each frequency band TV of operation. The GL-WSDB subtracts the dBk_r obtained in (Eq. 6.30) from the appropriate protected contour field strength values specified in Table 6.5:

$$E_r = E_{med} - dBK_r \tag{6.31}$$

Step 7: Interpolate the contour coverage distances (contour grade B and contour grade C)of each TV station of interest as a function of field strength. The GL-WSDB log-linearly interpolates the contour distance using ITU-R P.1546-5propagation model [21] at 95 % locations and 50 % time:

$$D_c = D_{inf}(\frac{D_{sup}}{D_{inf}})\Delta E \tag{6.32}$$

Given that:

$$\Delta E = (E_r - Einf) \times (E_{sup} - E_{inf}) \tag{6.33}$$

where:

D_c Contour coverage distance in km
E_{inf} Nearest field strength value below E_r in dBu
E_{sup} Nearest field strength values above E_r in dBu
D_{inf} Distance value for E_{inf} in km
D_{sup} Distance value for E_{sup} in km

6.7.2.2 Calculation of incumbents contour separation distances

Step 8:
Utilise TM-91-1 [13] propagation model and protection ratios for calculation of shorter separation distances at shorter WSD antenna heights. TM-91-1 is preferred because it is more accurate for wave propagation predictions in shorter distances and for devices with shorter antenna heights:

$$E_{WSD} = E_{med} - CIR \tag{6.34}$$

Hence:

$$E_{WSD} = 141.4 + 20\text{Log}_{10}h_1h_2 - 40\log_{10} D_{sep} + 10\log_{10} P_{WSD} + B \tag{6.35}$$

where:

E_{WSD} Field strength of WSD in dBu
CIR Incumbent's protection ratio for co-channel and adjacent channel in dB
h_1, h_2 RCAMSL of antennas in m
P_{WSD} Effective radiated power of WSD in W
B Building penetration loss in dB
D_{Sep} Contour separation distance in m

Likewise, utilise ITWOM (ITU-R P.1546-5and L-R) interference model at 50 % locations and 10 % times and protection ratios for interpolation of longer separation distances at longer WSD antenna heights. Table 6.9 illustrates the calculated separation distance of various classes of WSDs.

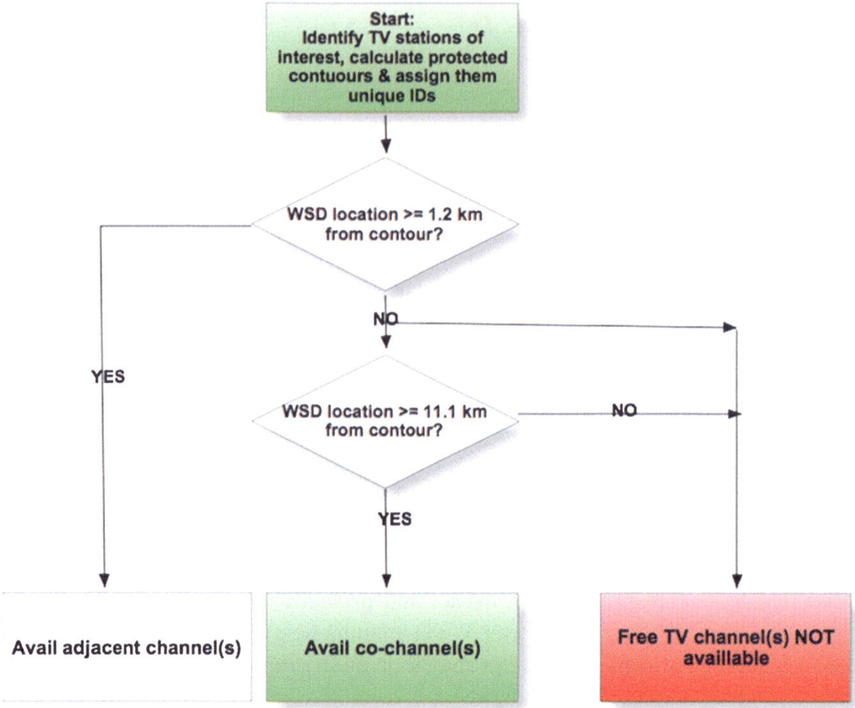

Fig. 6.9 High-level illustration of channel availability flow chart. Distance values used are for illustration purpose only

6.7.2.3 Calculate the available TVWS channel(s)

A high-level flow-chart in Fig. 6.9 illustrates a sample implementation utilizing WSD transmit power and incumbent contour separation distance constraints parameters shown in Table 6.9. The GL-WSDB should avail TVWS channels under the following conditions: A WSD must be operating within 1.2 km or beyond the protected contour for adjacent channels. Similarly, free co-channels will be made available if the WSD is operating within 11.1 km or beyond. These distance values used are for illustration purpose only.

6.7.3 Results and Discussions

We took a rather conservative approach during the calculation of the distance to protected contours for ATT stations. We chose to include the minimum median field strength values prescribed for both grade B and grade C service areas. This is attributable to the fact that some of the TV viewers in the rural communities particularly those located further beyond the edges of the protected grade B service

Table 6.9 WSD antenna HAATs, incumbent contour separation distances and transmit power constraints

WSD antenna HAAT (m)	WSD transmit EIRP (dBm)	Incumbent contour separation distances (km)	
		Adjacent channel	Co-channel
< 3	36	0.260651739	4.635116213
3 < 10	36	0.532988726	8.462525688
10 < 30	36	0.923163553	10.08
30 < 50	36	1.598966178	13.44
50 < 75	36	1.921718181	15.68
75 < 100	36	2.197568814	19.04
100 < 150	36	2.66494363	22.4
150 < 200	36	3.153203826	28
200–250	36	3.575397067	33.6
3–30	30	0.652775209	10.36443494
3–30	27	0.548916333	8.715416183
3–30	24	0.461581777	7.328762226
< 3 m	20	0.116081671	1.843086124
< 3 m	16	0.092316355	1.465752445

The calculation of separation distances is described in step 8, (Eqs. 6.34 and 6.35)

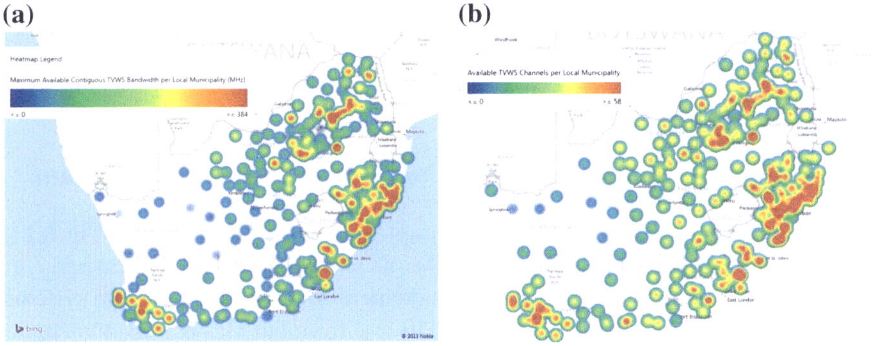

Fig. 6.10 **a** Heat map depicting aggregated available TVWS channels in South African local municipalities. **b** Heat map depicting aggregated available maximum contiguous TVWS channel bandwidth in South African local municipalities. Please note that the map aggregates the amount of available TVWS channels/contiguous bandwidth found in municipalities that are close to each other

coverage areas would still wish to receive TV broadcast using extended outdoors antennas (longer than 10 m). Such conservatism approach to some extent can lead into much larger protected TV broadcast service areas. Table 6.10 and Fig. 6.10 depict snapshots of the results in various South African local municipalities.

Table 6.10 Snapshot depicting available TVWS channels and contiguous TVWS bandwidth per local South Africa municipality

Province	Local municipality	Number of TVWS channels	Maximum available contiguous bandwidth (MHz)	Population	Population density (km/sq.)
Western cape	Beaufort west	25	104	49,586	2
Eastern cape	Blue crane route	40	64	36,002	3
Gauteng	Midvaal	38	112	95,301	55
Gauteng	Mogale city	39	112	362,422	270
Mpumalanga	Albert luthuli	21	104	186,010	33
Mpumalanga	Bushbuckridge	28	104	541,248	53
Limpopo	Ba-Phalaborwa	45	128	150,637	20
KwaZulu - Natal	Ezingoleni	45	256	52,540	81

Population data is obtained from Statistics South Africa [32]

6.8 Results Evaluation

This Section evaluates results of our GL-WSDB implementation using two methods: Firstly by performing a comparison against a selected commercial variant and secondly by performing a comparison against spectrum occupancy scan measurements in the field.

6.8.1 Results Evaluation against Commercial GL-WSDB Variant

This sub-section compares results of our implementation [33] shown in Fig. 6.11, with a commercial variant [34], shown in Fig. 6.13. We introduced the term *percentage of similarity* as the sum of all available identical channels found in both systems (i.e., from the commercial variant GL-WSDB and from our GL-WSDB implementation), over the sum of all channels found in both systems. For the purpose of this comparison, our implementation uses FCC-Curves [25] model modified with ITU region I grade B contour in Table 6.5. To the best of our knowledge the commercial variant GL-WSDB utilises FCC-Curves [25] model with FCC contours (i.e., 36 dBu for the VHF band and 41 dBu for the UHF band) used in the North America. The channels are to be available for a 4 Watt fixed WSD with an antenna height AGL of 30 m. Moreover, the comparison has been done for five cities in South Africa: (Johannesburg, East London, Polokwane, Durban, and Bloemfontein). The overall average percentage of similarity is 68 %. Results are presented in Table 6.11. The difference in the type of contours used by the two systems is confirmed by the results obtained. The commercial variant GL-WSDB produced fewer channels in most occasions due to the larger protected FCC contours as compared to the smaller ITU region I grade

Table 6.11 Comparison between CSIR-Meraka GL-WSDB [33] and a commercial variant GL-WSDB [34] for the available TVWS channels in five South African cities for a 4 Watt fixed WSD at 30 meters transmitting antenna height

City	System	Available TVWS channel number per system	% of Similarity
Johannesburg	CSIR-Meraka GL-WSDB	41, 62, 66	100
	Commercial GL-WSDB	41, 62, 66	
East london	CSIR-Meraka GL-WSDB	11, 21, 22, 24, 25, 26, 27, 28, 29, 30, 32, 33, 34, 38, 39, 40, 41, 42, 43, 44, 47, 48, 49, 56, 57, 59, 60, 61, 67, 68	47
	Commercial GL-WSDB	11, 21, 22, 23, 25, 26, 27, 28, 29, 30, 34, 38, 39, 40, 41, 47, 51, 52, 56, 60, 64, 66	
Durban	CSIR-Meraka GL-WSDB	36, 38, 33, 40, 44, 46, 48, 52, 53, 54, 55, 56, 57, 58, 59, 60, 61, 62, 63, 64, 65, 66, 67, 68	67
	Commercial GL-WSDB	53, 54, 55, 56, 57, 58, 59, 60, 61, 62, 63, 64, 65, 66, 67, 68	
Polokwane	CSIR-Meraka GL-WSDB	21, 22, 23, 30, 31, 32, 36, 37, 38, 39, 40, 41, 42, 46, 47, 48, 49, 50, 51, 52, 53, 54, 58, 62, 66	57
	Commercial GL-WSDB	21, 25, 29, 30, 31, 32, 36, 37, 38, 39, 40, 41, 42, 46, 50, 54, 66	
Bloemfontein	CSIR-Meraka GL-WSDB	4, 11, 21, 22, 23, 24, 25, 26, 27, 29, 30, 31, 36, 37, 38, 42, 46, 50, 55, 56, 57, 59, 60, 61, 65, 66, 67, 68	71
	Commercial GL-WSDB	4, 11, 21, 22, 23, 24, 25, 26, 27, 28, 29, 30, 31, 35, 36, 37, 38, 42, 50, 57, 58, 59, 60, 61, 62, 63, 64, 65, 66, 67	
Average overall percentage of similarity			68

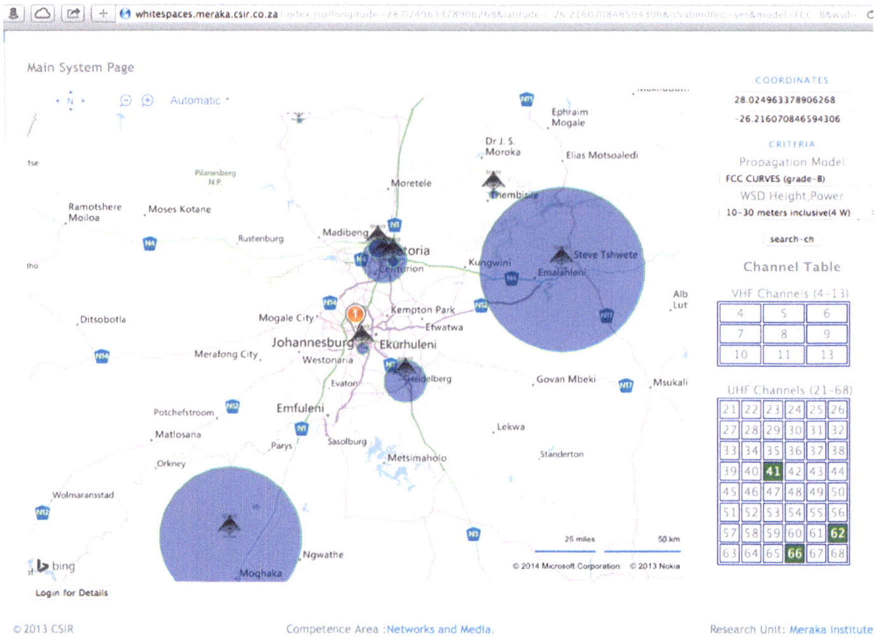

Fig. 6.11 CSIR-Meraka WS-GLSDB using FCC-Curves wave propagation model showing 3 available channels (UHF channels 41, 62 and 66) for a 4 W fixed WSD with antenna HAAT of 10 to 30 m in Johannesburg area, South Africa [33]

B contour (see Table 6.5). The results underscore an important aspect for national regulators particularly in developing countries to enforce jurisdiction-based rules for the operation of GL-WSDBs.

6.8.2 Results Evaluation against Field Spectrum Measurements

This sub-section compares results of our implementation [33] with spectrum occupancy field measurement scans [35]. Figure 6.12 shows 37 available TVWS channels in the VHF and UHF bands (channels 4, 6, 7, 9, 10, 13, 24, 29, 31, 32, 36, 37, 39, 40, 41, 43, 44, 45, 47, 48, 52 to 68), these were calculated by our GL-WSDB. The channels are available at Settlers primary school, Cape Town, South Africa. The channels are available for a 1 Watt fixed WSD with an antenna HAAT of between 3 to 30 m. Figure 6.14 shows results of the spectrum occupancy field measurements scan performed at Settlers primary school, Cape Town, South Africa [35]. 3+2 scans with each scan lasting for about 12 min were performed on the UHF channels 21 – 68. The type of antenna used was Antenna Craft ST4 (Omnidirectional discone outdoor TV antenna). The antenna was installed at 13 m AGL. Rhode & Schwartz PR100

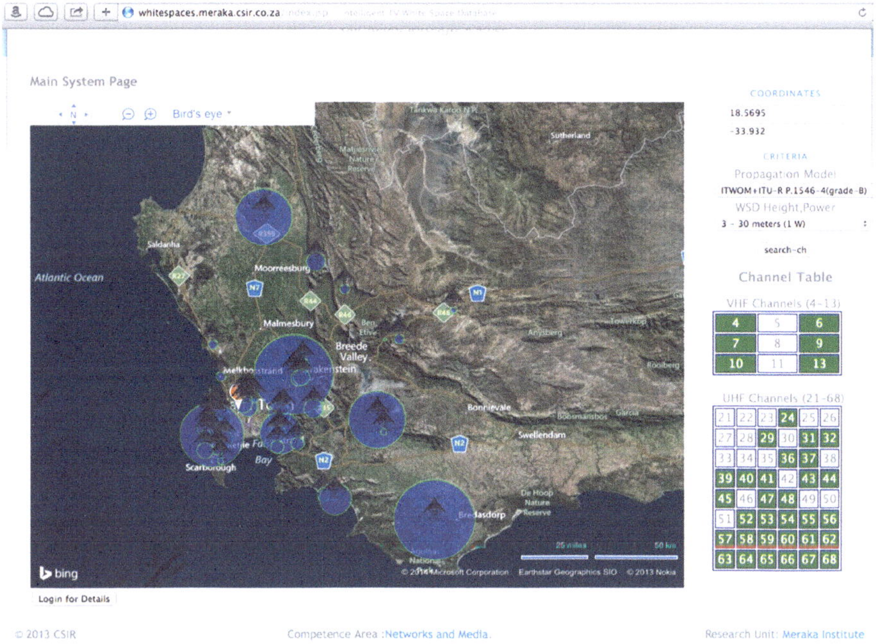

Fig. 6.12 CSIR-Meraka WS-GLSDB using ITWOM wave propagation model showing 37 available TVWS channels (VHF and UHF channels 4, 6, 7, 9, 10, 13, 24, 29, 31, 32, 36, 37, 39, 40, 41, 43, 44, 45, 47, 48, 52 to 68) for a 1 watt fixed WSD with antenna HAAT of 3 to 30 m at Settlers primary school, Cape Town, South Africa [33]

spectrum analyser was used to capture the measurements. The scans suggested that the following 25 UHF channels might be available (channels 21, 23, 25, 27, 29, 31–32, part of 33, 35, 36, part of 37, 39–41, part of 43, part of 44, 49, 51–55, 57, 61, and 63). The comparison shows that 16 out of the 25 UHF channels obtained in the field measurement are also found in the WS-GLSD UHF results. This result confirms what is known about the limitations of sensing at a single point where a WSD would want to operate. Some of the channels that appear available on the spectrum scan would prove to be unavailable if spectrum analysis was done in other nearby sample points in the area. Likewise, due to the presence of low-power gap-filler stations that reuse channels at distant inter-transmitter distances but within the GL-WSDB culling distance, the GL-WSDB is able to avail TVWS channels that are not locally detected by the spectrum measurement scans

6.9 Enabling Policy

Broadband access to all communities in South Africa by 2020 is a government strategy. To achieve this enabling spectrum policy frameworks play an important role. ICASA is currently trialling TVWS networks in South Africa to help determine

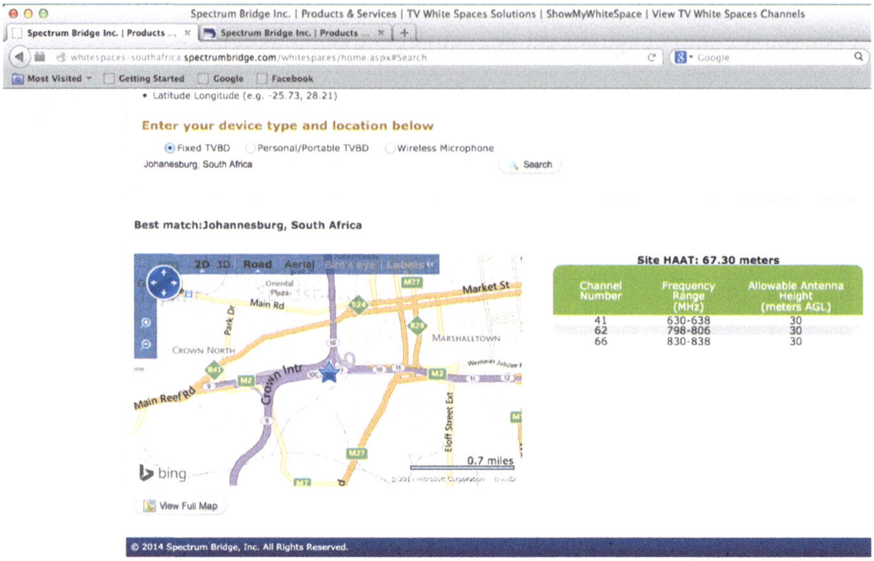

Fig. 6.13 A commercial WS-GLSDB using FCC-Curve wave propagation model showing 3 available channels (UHF channels 41, 62 and 66) for a 4 Watt fixed WSD with antenna height of 30 m AGL in Johannesburg area, South Africa [34]

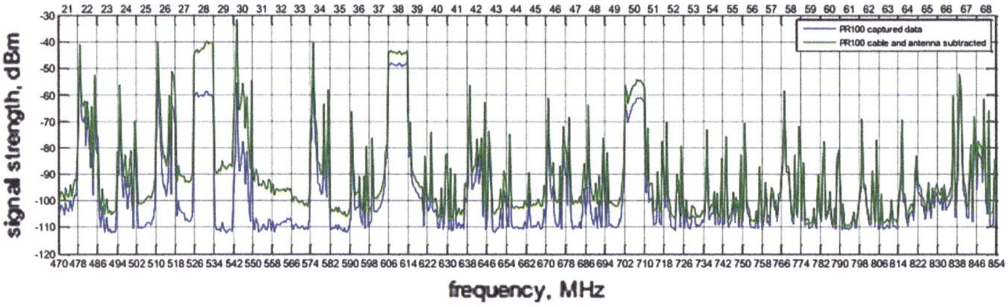

Fig. 6.14 Channel occupancy measurement data obtained at Settlers primary school, Cape Town, South Africa. 3+2 channel scan measurements were performed on the UHF channels 21 – 68, each scan lasted for about 12 min. The type of antenna used was Antenna Craft ST4 (Omnidirectional discone outdoor TV antenna) captured with Rhode & SchwartzPR100 spectrum analyser. The RCAMSL of antenna was at 13 m AGL [35]

the best policy for the country. The first of such trials took place in Cape Town [36]. Other African countries such as Kenya, Malawi and Tanzania are also running TV white space trials.

These trials are likely to guide the adoption of Dynamic Spectrum Access technologies for the rest of the continent. Certainly, policy makers will need to address crucial TVWS spectrum technical and regulatory issues such as maximum allowed in-band transmitting power levels of WSDs and their out-of-band (OOB) spectral leakage thresholds, minimum separation distances from the protected contours, incumbent protection ratios, and dynamic spectrum trading options. Even though some will argue that there is no need to re-invent the wheel since most of the aforementioned issues have already been addressed elsewhere in the world. However most of these policies have been formulated in more spectrum-scarce environments in the USA and Europe whereas Africa should take into considerations the fact that majority of Africans reside in under-serviced or un-serviced rural areas, where there is ample spectrum and yet a lack reliable grid power, poor communication infrastructure and challenging terrain.

6.10 Conclusions

In this chapter the functional position and role of the geo-location spectrum database (GL-WSDB)as a coexistence-compatibility planner and manager between primary and secondary users within the terrestrial broadcast TV networks has been explained. Furthermore, the practical design process of a GL-WSDB for South Africa was described. The results produced by the CSIR-Meraka GL-WSDB were evaluated using two methods: First results were compared against a commercial GL-WSDB variant using FCC-Curves model; where an overall 68 % of similarity for the available TVWS channels was found. The results were for a 4 Watt fixed WSD at a 30 m antenna height located atfive cities in South Africa: Johannesburg, East London, Polokwane, Durban, and Bloemfontein. Furthermore, the difference in the type of contours used by the two systems is confirmed by the results obtained. The commercial variant GL-WSDB produced fewer channels in most occasions due to the larger FCC contours (36 dBu and 41 dBu) used in the North America as compared to the smaller ITU region I grade B contour (55 dBu, 65 dBu and 70 dBu). The results underscore an important aspect for national regulators particularly in developing countries to enforce jurisdiction-based policy for the operation of GL-WSDBs. Secondly, the results of the GL-WSDB for a 1 Watt fixed WSD were evaluated against field spectrum occupancy measurements. The results showed that 64 % of channels in the UHF band agree with those of a spectrum scan at a specific sample point in Cape Town. It is important to note that GL-WSDB conservatively avails TVWS channels in a given geo-location and time by using an appropriate wave propagation model while applying the following constraints: (1) WSD antenna height above average terrain (HAAT), (2) WSD transmitting power, (3) incumbent protection ratios and, (4) contour separation distance. The constraints are outlined in Table 6.9. The results confirmed that field measurements are the best way to validate spectrum occupancy since wave models are sometimes susceptible to under-estimation or over-estimation of spectrum occupancy. However, it is challenging and very expensive to perform real-time measurements for large geographical areas. Proper application of models coupled with correction factors, terrain profile modelling, and snapshots of static field measurements can enhance the design and accuracy of GL-WSDBs.

This chapter did not consider the following issues: (1) cross-border dynamic frequency sharing coordination with respect to the five neighboring countries (Lesotho, Swaziland, Botswana, Mozambique, and Zimbabwe) that are bordering South Africa.

(2) Protection of wireless microphones was not considered. Future and ongoing work includes but is not limited to: expansion of a GL-WSDB to cover the entire Southern African Development Community (SADC) region and beyond. CSIR-Meraka as a stakeholder is constantly contributing innovative ideas to the regulator (ICASA) on how best to develop a sound national dynamic spectrum access (DSA) policy framework [36].

Acknowledgments The authors wish to thankfully acknowledge the CSIR and TUT for their continuous financial and material support.

References

1. Cisco, Visual networking index: global mobile data traffic forecast update, 2013–2018, [online], Available http://www.cisco.com/c/en/us/solutions/collateral/service-provider/visual-networking-index-vni/white_paper_c11-520862.html
2. ICASA, 790–862 MHz: Consolidated audit report, [online], Available https://www.icasa.org.za/Portals/0/Regulations/Spectrum%20Usage%20and%20Spare%20Frequencies/790MHz-862MHz%20Consolidated%20Spectrum%20Audit%20Report.pdf
3. FCC, Oet Bulletin No. 69, Longley-rice methodology for evaluating TV coverage and interference, [online] Available http://transition.fcc.gov/Bureaus/Engineering_Technology/Documents/bulletins/oet69/oet69.pdf
4. Brown P.G., Tsioumparakis K., Jordan M., Chong A.: UK Planning model for digital terrestrial television coverage, R&D white paper, WHP 048, BBC, September 22, 2002
5. GE-06: Final acts of the regional radiocommunication conference for planning of the digital terrestrial broadcasting service in parts of regions 1 and 3, in the frequency bands 174–230 and 470–862 MHz, ITU, RRC-06
6. Final acts of europe VHF/UHF broadcasting conference, Stockholm 1961, ITU (1961)
7. The chester 1997 multilateral coordination agreement relating to technical criteria, Coordination principles and procedures for the introduction of terrestrial digital video broadcasting (DVB-T), CEPT, Chester, July 25, 1997
8. Salter J.: Digital TV switchover: An aerial test that uses teletext test page, R&D white paper, WHP 160, BBC, January, 2009
9. Recommendation ITU-R BS.707-4 Transmission of multisound in terrestrial television systems PAL B, D1, G, H and I, and SECAM D, K, K1 and L, 1998
10. Recommendation ITU-R BT.1895, Protection criteria for terrestrial broadcasting systems (2011)
11. Recommendation ITU-R BT.2265, Guidelines for the assessment of interference into broadcasting services (2012)
12. Recommendation ITU-R, Guide to the application of propagation methods of Radiocommunication, pp. 1144–1146. (2012)
13. Daniel W., Wong H.: Propagation in suburban areas at distances less than ten miles, FCC/OET TM91-1, January 25, 1991
14. Shumate S.: Longley-Rice and ITU-R P.1546 combined: A new international terrain-specific propagation model, in the proceedings of 72ND vehicular technology conference, IEEE, pp. 1–5, 6–9 Sept 2010
15. Recommendation BT.419-3, Directivity and polarization discrimination of antennas in the reception of television broadcasting (1990)

16. Southern African Digital Broadcasting Association (SADIBA), Antenna impact study-gauteng, [online], Available http://www.sadiba.org/images/PDFfiles/sadiba_update_antenna_study_doc_20120307_final.pdf
17. The European Conference of Postal and Telecommunications Administrations (CEPT), ECC report 185, Complimentary report to ECC report 159: further definition of technical and operational requirements for the possible operation of white space devices the band 470–790, [online], Available http://www.erodocdb.dk/docs/doc98/official/Pdf/ECCRep185.pdf
18. Recommendation ITU-R BT.417-5, Minimum field strengths for which protection may be sought in planning an analogue terrestrial television service (2002)
19. Recommendation BT.2033-0, Planning criteria, including protection ratios, for second generation of digital terrestrial television broadcasting systems in the VHF/UHF bands (2013)
20. The european conference of postal and telecommunications administrations (CEPT), ERC report 101: a comparison of the minimum coupling loss method, enhanced minimum coupling loss method and the Monte-Carlo simulation, [online], Available http://www.erodocdb.dk/docs/doc98/official/pdf/Rep101.pdf
21. ITU-R, Method for point-to-area predictions for terrestrial services in the frequency range 30–300 MHz, International telecommunication union, Geneva, Recommendation. pp. 1546–1545, September 2013
22. European communications commission (ECC), Technical annex: criteria for planning DVB-T, [online], Available http://www.erodocdb.dk/docs/doc98/official/pdf/ECCRep004_Technical_annex.pdf
23. Itu-R, SM. 2018, Report on Monte carlo simulation methodology for the use in sharing and compatibility studies between different radio services or systems, [online], Available http://www.itu.int/dms_pub/itu-r/opb/rep/R-REP-SM.2028-2001-PDF-E.pdf
24. Ofcom, UK, TV white spaces: approach to co-existence, Technical report, September 4, 2013, [online], Available http://stakeholders.ofcom.org.uk/binaries/consultations/white-space-coexistence/annexes/technical-report.pdf
25. FCC, Third memorandum opinion and order, In the matter of unlicensed operation in the TV broadcast bands, additional spectrum for unlicensed devices below 900 MHz and in the 3 GHz bands, [online], Available http://www.fcc.gov/document/tv-white-spaces-rule-changes
26. Industry canada, Spectrum management and telecommunications, Database requirements: white space (WS) database specifications, DBS-01, Issue 1, May 2013
27. Barbiroli M., Carciofi C., Guiducci D., Missiroli M., Petrini V.: White spaces potentially available in Italian scenarios based on the geo-location database approach, New frontiers in dynamic spectrum access networks (DySPAN), IEEE, pp. 416–421,16–19 October 2012
28. The european conference of postal and telecommunications administrations (CEPT), ECC report 186, Technical and operational requirements for the possible operation of white space devices under geo-location approach, [online], Available http://www.erodocdb.dk/docs/doc98/official/Pdf/ECCRep186.pdf
29. The independent communications commission of south africa (ICASA), Draft frequency migration regulation and frequency migration plan, [online], Available https://www.icasa.org.za/Portals/0/Regulations/Regulations/National%20Frequency%20Plan/DraftFrequencyMigrationRegulation_FrequencyMigrationPlan_GG35598.pdf
30. National space agency (NASA), SRTM3v2, [online], Available ftp://e0srp01u.ecs.nasa.gov/srtm/version2/
31. Tell R.: Reference data for radiofrequency emission hazard analysis, US environmental protection agency, office of radiation programs (Washington D.C. June 1972)
32. Statistics south africa, population statistics, [online], Available http://www.statssa.gov.za/publications/populationstats.asp
33. Council for scientific and industrial research (CSIR) - Meraka Institute, Geo-location spectrum database, [online], Available http://whitespaces.meraka.csir.co.za/
34. Spectrum Bridge Inc., Show my white space, [online], Available http://whitespaces-southafrica.spectrumbridge.com/whitespaces/home.aspx

35. Lysko A., Johnson, D. L.: Report on measurements of spectrum availability performed in cape town area from August 29–31, 2012, CSIR-Meraka Institute, Document referencenumber 221992, November 10, 2012
36. Carlson J., Ntlatlapa N., King J., Sibanda F.M., Hart A., Geerdts C., Song S.: Studies on the use of television white spaces in south africa: recommendations and learnings from the CAPE town trial, [online], Available http://www.tenet.ac.za/tvws/recommendations-and-learnings-from-the-cape-town-tv-white-spaces-trial

Chapter 7
White RHINO—A New SDR Platform for Whitespace Communication

Ojonav Hazarika and Amit Kumar Mishra

Abstract This chapter discusses a new SDR platform called the White RHINO. The White RHINO has been designed to meet the requirements of the 802.22 white space specification but can also be used as a general SDR platform. The SDR was developed at the University Of Cape Town, South Africa and provides powerful functionality in a fully self-contained compact unit that can be produced at a reasonable cost compared to other SDR platforms with similar specifications. This board has a small form factor (121×121 cm), is packed with dual 1 GHz ARM-A9 embedded processors, a fairly large amount of FPGA logic (with 220 DSP slices) and an RF operational bandwidth of 0.3–3 GHz with a maximum instantaneous bandwidth of 28 MHz.

In this chapter, we will discuss the development of the White RHINO SDR platform specifically designed to cater for the requirements of the 802.22 white space radio standard. The compact and inexpensive board was designed at the University of Cape Town. Figure 7.1 shows the White RHINO PC board that can operate as a white space radio or a general SDR platform. This board is designed with a small form factor of 121×121 cm and is packed with dual 1 GHz ARM-A9 embedded processors, a fairly large amount of FPGA logic (with 220 DSP slices) and an RF operational bandwidth of 0.3–3 GHz with a maximum instantaneous bandwidth of 28 MHz. In the rest of the chapter we will discuss various features and design aspects of this hardware platform.

O. Hazarika · A.K. Mishra (✉)
University of Cape Town, Cape Town, South Africa
e-mail: akmishra@ieee.org

A.K. Mishra and D.L. Johnson (eds.), *White Space Communication*,
Signals and Communication Technology, DOI 10.1007/978-3-319-08747-4_7

Fig. 7.1 White RHINO: assembled

7.1 Hardware Architecture

The hardware architecture of the powerful White RHINO as shown in Fig. 7.2, has been logically split up into various subsystems. These specifications of these subsystems have been derived from the WRAN IEEE802.22 requirements [1]. These subsystems are as follows:

7.1.1 Power Supply and Clock Blocks

These blocks are responsible for providing power and all major clock signals to the various devices onboard.

7.1.2 Processing and Memory Blocks

These blocks handle all the configuration and management related tasks of the platform. Adding to that, they shall perform high-speed computational operations for the cognitive plane defined in IEEE802.22 [1].

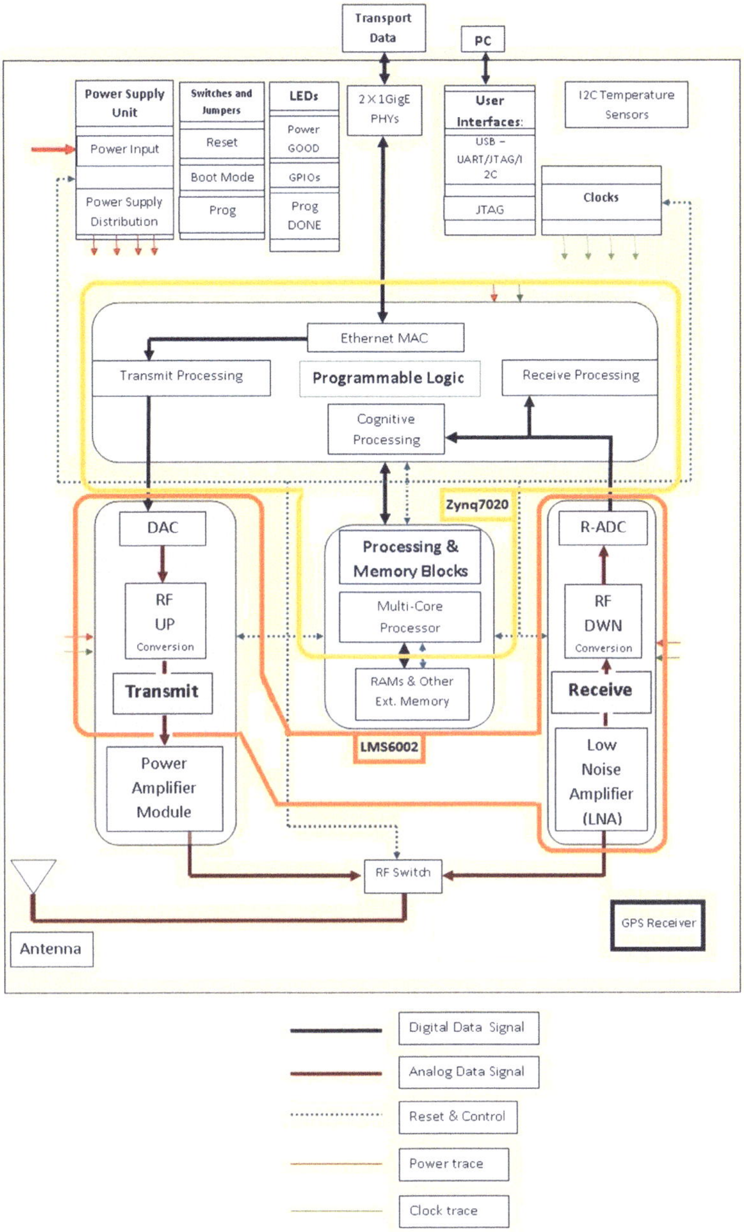

Fig. 7.2 White RHINO hardware: high level block diagram

7.1.3 Data-Path Blocks

White RHINO is a communication system and hence, we require data-path elements on the board. Following are the required Data-Path Blocks in the system:

- **Ethernet PHY:** At the digital back-end, this is the first block in the Data Path. This block is responsible for converting the Ethernet Data from the Media Dependent Interface (like Base-T, Copper or Base-X, Optical fiber) to a Media Independent Interface (MII, GMII, SGMII, RGMII etc.) and vice-versa.
- **Programmable Logic Block:** The Programmable logic then, shall handle the bulk of digital transmit and receive data-path tasks e.g, scheduling/de-scheduling, coding/de-coding, modulation/de-modulation, OFDM/OFDMA etc.
- **RF Blocks:** Finally, we have the RF blocks which shall be responsible for digital to analog/analog to digital conversion, up/down conversion, power amplification, filtering, and TDD switching. After the TDD switch the signal is transmitted or received through the TV Antenna.

7.1.4 Cognitive Blocks

As we know now, the IEEE802.22 defines a cognitive plane. Hence, we require specific blocks for the implementation of this plane. At the hardware level, the only extra block that needs to be implemented is the GPS system for Geolocation. The cognitive processing shall be facilitated through the Data-Path blocks and the Processing and Memory Blocks but its discussion is beyond the scope of the document since they are at the algorithm level.

7.1.5 User Blocks

There shall be various user blocks on the board. These blocks shall facilitate the user to set operating conditions and allow the user to efficiently debug. These include the various switches, jumpers, LED's, USB-UART/JTAG/I2C and JTAG interfaces. Together, they will allow the user to have absolute control over the system and hence, facilitate implementation of efficient applications.

7.1.6 Control and Monitoring Block

This block ensures that the system functions in a reliable way. Its functions shall be to estimate the system performance as well as provide system shutdown in conditions when the system cannot function in a reliable way. There shall be temperature sensosrs at various critical locations on the board. These sensors shall provide the board temperature at those locations through the I2C interface which can be accessed by the processing block. The processing block then will take necessary actions.

7.2 Major Components

In the previous section we discussed the various logical subsystems of the White RHINO board. After splitting up the design into the various logical subsystems, suitable components were selected. The selection of these components were done keeping the cost and size of the platform in view. In order to reduce the size, complexity and hence, the cost of the board, following critical devices have been selected:

7.2.1 Zynq7020 (Xilinx) for Processing and Programmable Logic

The Zynq7020 has embedded dual core 800 MHz ARM Cortex-A9 processors with NEON single precision floating point ALUs, 85 K logic cells and 220 DSP slices [2]. This component has been specified as a user requirement. However, through our feasibility study in Appendix B, we have found that the Zynq7020 is very suited for our purpose. The IEEE802.22 defines some highly resource consuming DSP blocks. The major share of the cognitive processing shall be done in one of the two cores of the embedded ARM Cortex A9 floating point processors. In our study, we use a single complex multiplier for the generation of the covariance matrix. This process is required for the Eigen Value method of Spectrum Sensing which has been proposed as a preferred algorithm in the IEEE802.22 standard documents. Since, the matrix data shall be stored in an external memory, the amount of resources consumed (for LUTs, FFs and DSP slices) shall be less than 3 % of the available respurces as presented in Appendix B. The remaining resources in the FPGA shall be used for other data-path and management blocks. Hence, Zynq7020 is a compact and powerful SoC which provides us with enough computational resources in the form of its programmable logic and processing system blocks.

7.2.2 LMS6002 (Lime Micro) as RF Transceiver

LMS6002 from Lime Micro has been chosen for the platform because of its wide-band performance of 3 GHz and a maximum instantaneous bandwidth of 28 MHz [3]. It has embedded ADCs, DACs, synthesizers, Up/Down Convertors, variable gain amplifiers, Time Division Duplex (TDD) support and hence, is apt for a compact system. However, it has a maximum transmit output power of 6 dBm and hence, an additional amplifier section shall be used to boost it to the specified power levels. Some of its general specifictions have been shown in Fig. 7.3.

7.2.3 RFPA3800(RFMD) as Final Stage Amplifier

There is a dearth of wide-band, low power amplifiers operating in the VHF and UHF frequencies in the open market. However, we aim to build a low cost platform. Hence, this device has been chosen for the first prototype, which has a maximum wide-band output power of around 25 dBm. Its still has to be matched to achieve a wider bandwidth of operation in order to meet the specifications [4]. Some of its specifications at 460 MHz have been shown in Fig. 7.4.

7.2.4 MT41K128M16JT-125 (Micron) as DDR3 SDRAM

The User Requirements define that the processing system architecture of the White RHINO board has to be similar to that of the Zedboard. So, the DDR3 SDRAM interface design has been done with the Zedboard design as the reference. However, the 2 × 128 MB DDR3 part (Micron MT41J128M16HA-15ED) that the Zedboard uses is an out of stock component. Hence, we had to search for an alternate component. A similar but newer part (Micron MT41K128M16JT-125) which has even the same pin mapping and the footprint as the previous device has been chosen as the DDR3 SDRAM part. Two of these 128 MB devices have been used to achieve 512 MB of DRAM memory. Some general specifications have been tabulated in Table 7.1 [5].

General Specifications						
Symbol	**Parameter**	**Condition**	**Min**	**Typ**	**Max**	**Unit**
REFCLK	Reference Clock	No gaps in Freq resolution.	26	-	41	MHz
PN	Phase Noise	1MHz Offset		-130		dBc/Hz
Transmit						
WEVM	Transmitter EVM	UMTS test model 1		2		%
WACLR1	1st Adjacent channel Leakage				-50	dBc
GTx	Transmit Gain control range	TXVGA1, TXVGA2, 1dB steps		60		dB
Receive						
RX NF	Noise figure			5.5		dB
SENS GSM	GSM Sensibility (Listen Mode)	GSM (2dB FE loss)		-105		dBm
WSens	UMTS Sensitivity (Full duplex)	12.2kBs Single Code (2dB FE loss)			-115	dBm

Fig. 7.3 LMS6002 (Lime Micro): general specifications. Photo Courtesy-http://www.limemicro.com/download/Lime_ProductBrief.pdf

Table 7.1 MT41K128M16JT-125 (Micron): general specifications

Package	Density	Clock rate	Width	Depth	Pin count
FBGA	2 Gb	800 MHz	×16	128 MB	96-ball

Parameter	Specification			Unit	Condition
	Min.	Typ.	Max.		
460 MHz					V_{CC}=7.0V, V_{BIAS}=7.0V, I_{CQ}=650mA
Frequency	450	460	470	MHz	EVB tuned for linear operation
Input Power (P_{IN})			23	dBm	V_{CC}<7.5V, load VSWR<2:1
Gain (S21)		18		dB	
OIP3		48		dBm	20dBm/tone, tone spacing=1MHz
P1dB		36.7		dBm	EVB tuned for linear operation
Efficiency at P3dB		50		%	At P3dB, EVB tuned for linear operation
Input Return Loss (S11)		15		dB	
Output Return Loss (S22)		9		dB	
Noise Figure		5		dB	
WCDMA Ch Power at -65dBc ACPR		19.5		dBm	3GPP 3.5, Test Model 1, 64 DPCH
WCDMA Ch Power at -55dBc ACPR		24.5		dBm	3GPP 3.5, Test Model 1, 64 DPCH

Fig. 7.4 RFPA3800: general specifications. Photo Courtesy-www.rfmd.com/CS/Documents/RFPA3800DS.pdf

7.2.5 DP83865 (National Semiconductor) as Ethernet PHY

The White RHINO user requirements specify 2 × 1 Gbps Ethernet connection through copper. In order to facilitate the 1 Gbps ethernet communication, we require Ethernet PHY devices. These devices convert the signal from the MDI interface to the MDIO interface and vice versa. The Ethernet PHY device that has been chosen for the purpose is the DP83865 from National Semiconductors which supports tri-speed Ethernet functionality of 10/100/1,000 Mbps. It also consumes very low power of about 1.1 W and has the auto MDIX capability. The auto MDIX capability implies that, the PHY can identify the Ethernet cable type and correct the packets going to the RGMII interface. The RGMII is the chosen interface for the media independent interface as the Zynq7020 processing system supports only RGMII. The features of the device have been noted below:

- Supports 10/100/1,000 Mbps modes.
- Capable of autonegotiation.
- Capable of auto-MDIX for cable type identification as well as differential pair correction.
- Low power consumption of about 1.1 W.

7.2.6 FT4232 (FTDI) as USB-UART/JTAG/I2C Bridge

The FT4232 device from FTDI is a four channel USB to Serial Protocol bridge [6]. This device is particularly useful. The four channels of the device can be used for various protocols and hence, provide a very space and cost optimized single debug interface. All the four serials get automtically enumerated when connected to a PC through a USB cable. However, to read or write to the respective serial protocol devices (UART/JTAG/I2C), respective drivers need to be present. Hence, there shall be two UARTs connected to the Zynq PS and PL respectively, one JTAG and one I2C (for temperature monitoring). Hence, we elliminate the need for a bulky connector like the DB9.

7.2.7 TDD Switches

The chosen TDD switches for the deisgn are PE42440 (Peregrine) and HMC 849LP4CE (Hittite). As we have seen in Chap. 2, the IEEE802.22 specifies a time division duplex (TDD) system. This means we require switches which can alternate between the transmit and the receive bursts. To accomplish this purpose, we have selected two components which together provide transmit to receive isolation of 90 dB. The PE42440 is the final TR switch which has an output 1 dB compression point of 41 dBm which is apt for our purpose. These switches shall be digitally controlled from the Zynq7020 PL fabric. Together with the RFMD3800, they shall form the RF frontend of the system. Both the devices offer return losses of better than 15 dB, when matched well.

7.3 Onboard Digital Interfaces

The White RHINO has many digital as well as Analog data interfaces. Some of them are high-speed which require special care in order to counteract transmission line effects.

7.3.1 High-Speed Digital Interfaces

7.3.1.1 Medium Dependent Interface (MDI)

This is the interface through which the Ethernet packets are transmitted over a long distance through cables (optical or copper). A signal (especially a high-speed one) transmitted over long distances is prone transmission line effects, like insertion and return losses. Hence, this interface is very medium dependent. This property requires us to take special measures to ensure that the signal quality is maintained even over long distances. The MDI interface is a standardized interface under IEEE802.3 [7] and there are specialized devices known as Ethernet PHYs which handle all the medium dependencies including conversion to and from the media independent interface. In this interface, the data is transmitted or received using four pairs of 100 Ω differential lines which connect to an external cable through an RJ-45 connector.

7.3.1.2 Reduced Gigabit Media Independant Interface (RGMII)

The Ethernet PHY device converts the signal from MDI to a media independent interface. However, currently there are many existing standards for the media independent interface. They are the media independent interface(MII), gigabit media independent interface (GMII), serial gigabit media independent interface (SGMII), reduced gigabit media independent interface(RGMII) and so on. The Zynq7020 supports only the RGMII interface. The GMII uses 8 bit transmit as well as receive data buses. However, the RGMII uses only 4 bit transmit and receive data buses which is advantageous for our purpose. Lesser bus width implies lesser high-speed routing and hence, space is more effectively utilized. The RGMII uses a 125 MHz clock to achieve the required data rate.

7.3.1.3 Zynq-LMS6002 Digital IO Interface

The Zynq-LMS6002 Digital IO Interface is a critical 12-bit data bus which is responsible for carrying baseband data between the Zynq7020 and the LMS6002 ADC/DACs. The sampling clock can reach a maximum frequency of 40 MHz. The interface is serial IQ interleaved. On the Zynq7020 side, the IO lines shall be connected to the PL fabric banks. The bus includes 12-bit data for each transmit and receive, TX/RX clocks and IQ selects.

7.3.1.4 DDR3 Interface

The DDR3 is a high-speed memory interface. The DDR3 IO lines will be connected to the DDR controller present in the Zynq7020 PS fabric. The DDR3 shall be running with a clock rate of 533 MHz. The data bus width with two 16 bit DDR3 SDRAMs form an effective data bus width of 32 bits. The data bus width along with the clock rate provides us with an achievable throughput shall be more than 16 Gbps. Other high speed lines that constitute the interface are 16 bit address bus, bank select, DQS differential strobes and a differential clock. The signalling levels are compatible with the high-speed Stub-Series Terminated Logic (SSTL) 1.5 V family.

7.3.2 Other Digital Interfaces

- **UART:** The Universal Asynchronous Transceiver Receiver (UART) is a slow speed serial communication interface which is mostly used for printing on to the console. This interface is particularly useful during the early bringup stages when hardware bugs are yet to be identified. It works at various datarates (termed as baudrates) like 110, 300, 9,600, 19,200, 115,200 and so on. The two UARTs shall have transmit and receive data lines connecting the FT4232 bridge and the Zynq7200.

- **JTAG:** The Joint Test Action Group (JTAG) is a very popular and powerful debug interface which provides access to the registers of the devices connected in the JTAG boundary chain. In the White RHINO, to keep the JTAG chain simple, only the FT4232 device and the Zynq7020 shall be connected. The JTAG shall be available to the user through the 7-pin JTAG hearder as well as through the USB-JTAG bridge in the form of the FT4232 device. The JTAG interface consists of TCK, TDI, TDO and TMS (Test-clock, data input, data output, mode state respectively). And, it can operate at frequencies like 250 kHz, 500 kHz, 1 MHz, 10 MHz, 40 MHz and so on.
- **QSPI Flash:** QSPI stands for Quad Serial peripheral interface. This interface is particularly used to control a boot flash device from the Zynq7020. The interface has four data lines, a clock and a chip select.
- **Secure Digital (SD) IO interface:** The SD card is an alternative option present on the board to boot the operating system. This interface has been kept because the SD card is portable and the memory is expandable. This adds the flexibility to the booting options on the White RHINO. The SD interface has four data line, a clock, a command signal and a couple of other extra functionlities like write protect and card detect. This interface is connected to the Zynq PS.
- **SPI Interface:** The White RHINO has a Serial Peripheral Interface connected to the Zynq PS. This interface shall be used to configure the LMS6002 device. The interface signals are clock, Master in slave out (MISO), Master out slave in (MOSI), and the Slave enable (SEN).
- **I2C Interface:** There shall be five temperature sensors at various locations on the White RHINO board. These temperature sensors shall provide the board temperature at those locations through the I2C interface. The I2C interface has two interface signals, the serial data (SDA) and the serial clock (SCL). The I2C signals are connected to the PL fabric of the Zynq7020 and the FT4232 device so that the temperature can be monitored on the board as well as through the extermal debug interface.
- **GPS information interface:** As we know that global positioning system (GPS) is required by the IEE802.22 specifications. The GPS data is transferred by the GPS receiver in the form of magnitude and sign. And, the magnitude and the signs are sent out along with a clockout from the GPS receiver. Along with the three above mentioned signals, there is an extra signal which is the data enable signal. The clockout frequency is about 20 MHz. The GPS information interface signals shall be connected to the PL fabric.
- **RF Control:** In this list of digital signals, the RF control interface is the last but an important one. It is an LVCMOS 3.3 V interface which is used to control the TDD switching. The interface is controlled through the PL fabric of the Zynq7020 and is a three bit digital control.

A more refined hardware block diagram has been presented in Fig. 7.5. This diagram contains the major interfaces, devices and specifications.

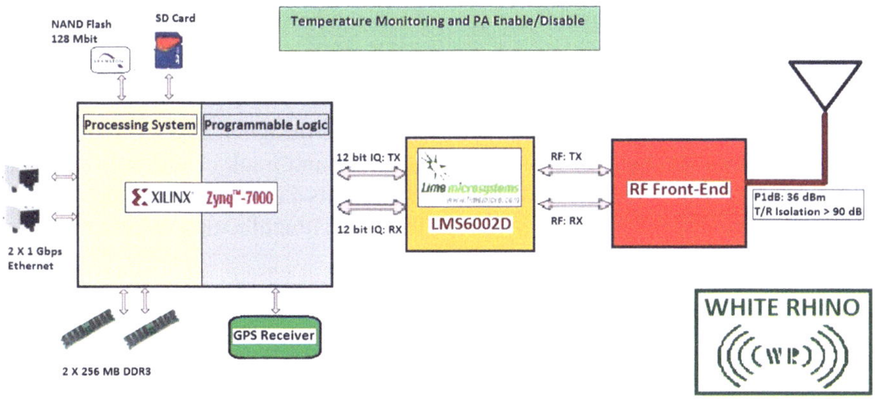

Fig. 7.5 White RHINO hardware: high level block diagram

7.4 Some Critical Design Decisions

The LMS6002 is a 300 MHz–3 GHz transceiver with a maximum instantaneous bandwidth of 28 MHz. However, an RF frontend section comprising of a power amplifier transmitting about 1 W of power is usually band limited. Even if the amplifier may support an instantaneous bandwidth of 28 MHz, it surely cannot have an operational bandwidth of about 3 GHz. So, if we plug such an amplifier in front of the LMS6002 on the same board, we restrict the capability of the LMS6002 device. And then, various reliabiility issues could arise as well due to the high power dissipation of the power amplifiers. Even though, the White RHINO was meant to be a TV Whitespace platform, however, there was no noticeable compromise in choosing to split the White RHINO into two separate boards. These boards were interfaced with RF and digital connectors. The design decisions of splitting the White RHINO into White RHINO core baseband low power RF board, the White RHINO RF frontend board and the PoE+ board provided us with the following advantages:

- We shall be able to use the White RHINO as a generic SDR platform. This SDR platform can have an operational bandwidth of 300–3 GHz with minor lumped component changes.
- The White RHINO RF frontend section remains isolated from the core and hence, we have better control over the RF design which is very critical in terms of performance.
- The thermal performance of the White RHINO with sensitive onboard digital components will be better as the power amplifier contributes to a major share of power dissipation.
- The White RF frontend being a simple board with lesser traces shall not require with more than two PCB layers. Hence, a heat sink can be mounted on the grounded bottom layer. This in turn, shall contribute towards better thermal performance and hence, better linearity performace of the amplifiers.

- The PoE+ stays as a pluggable module which can be used as and when required. The module can be interfaced with on board headers.

We shall, now, discuss the schematic designs of the White RHINO core and the RF boards. The schematics have been drawn in Altium designer. The schematic entry flow has been hierarchical so that is easy to debug and track changes. In order to track the components used for the design, a components database was prepared which contains part numbers, schematic symbols, footprint, manufacturer information and so on.

7.5 Subsystem Design

The White RHINO hardware schematic design has been partitioned into various sub-systems. The various subsystems are the Power Supply distribution, the Zynq-7020, the DDR3, the Peripherals, the Ethernets and the RF sections.

7.5.1 Power Supply Distribution

The first subsystem that we are going to discuss here is the Power supply distribution subsystem. As the name suggests, this subsystem is responsible for generating all the required power rails for the White RHINO. The power supply distribution is depicted in the form of a block diagram in Fig. 7.6. The various power rails required are as follows:

1. Zynq7020—1.0 V Core, 1.5 V DDR, 1.8 V Auxilliary and 3.3 V IO.
2. DDR3—1.5 V.
3. Ethernet PHY(DP83865)—1.8 V Core, 2.5 V Analog and 3.3 V IO.
4. LMS6002 1.8 V internal, 3.3 V IO, 1.8 V TXLO and 1.8 V RXLO.
5. FT4232 1.8 V core and 3.3 V IO.

Many of these power supplies requiring same voltage level demand different supplies and hence, the power supply split has been done accordingly. The Power supply chain starts with the 48 V supplied from the PoE+. This votage is then down-converted to a 12 V supply. Since, the PoE+ board is a different board hence, there is a provision for a 12 V auxilliary supply. The single pull double throw (SPDT) switch shall be used to select the 12 V power source. The 12 V is then downconverted to 5 V which in turn generates all the digital supplies. The RF supplies are generated from a 6 V supply which is generated from the 12 V source. The selection of this supply voltage level has been done on the basis on DC to DC switching regulators readily available in the market. The power sequencing for the digital supplies is done through a simple four channel comparator which takes the power good signals from the rest of the digital supplies and generates the power good all signal. This power good all signal in turn generates the power-on-reset signal for the Zynq7020 device. The various power good signals (in their order of occurance) are as follows:

- **PGOOD_VCC5V0**—5.0 V.
- **PGOOD_VCC1V0**—1.0 V.
- **PGOOD_1V8**—1.8 V Aux
- **PGOOD_DDR**—DDR 1.5 V.

Apart from the power-on-reset signal, there is one more interface signal with the Zynq device. And, that is the PA_EN signal. This is the power amplifier enable signal controlled by the Zynq. This signal shall turn off the power supply to the power amplifier and hence, lessen system power dissipation. Finally, the power consumed by the various devices have been noted in Fig. 7.6. As we can observe, the total amount of power consumed by the major devices on the board is 19.5 W and the power provided by the PoE+ standard is 25 W. Please note that this consumtion chart includes all the major devices present on both the boards. Hence, the White RHINO is capable of getting power up through the PoE+, mentioned as a requirement.

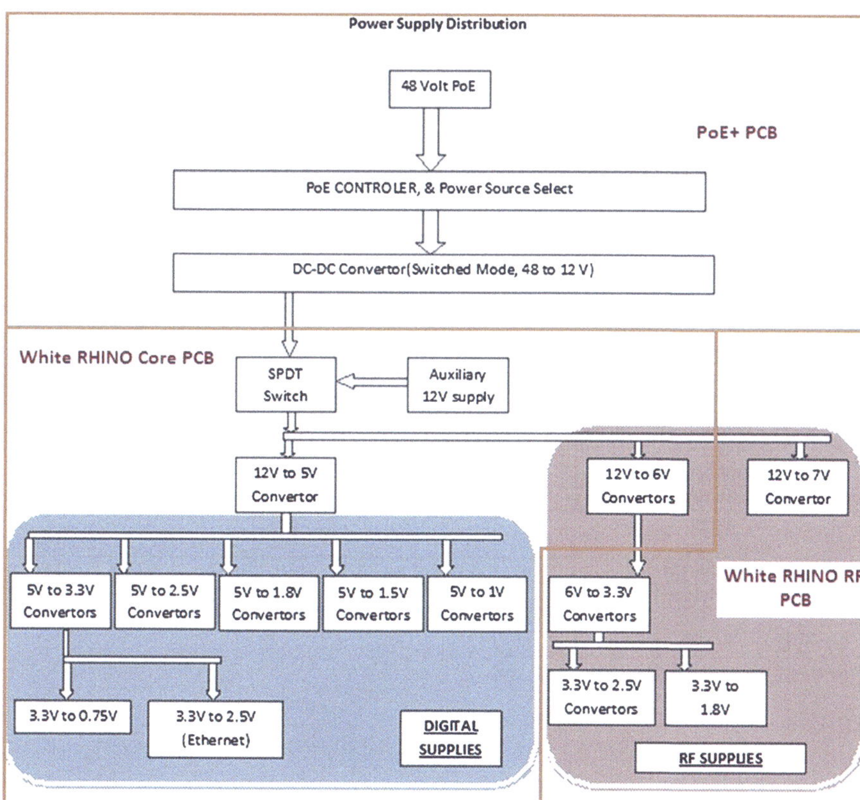

Fig. 7.6 White RHINO: power distribution

7.5.2 Zynq7020 Subsystem

All of the subsystems are connected to the Zynq device. This makes the Zynq subsystem a very important one. The Zynq device has seven banks which includes 2 processing system, 1 DDR and 4 programmable logic banks. The interfaces connected to the various banks are as follows:

- **Config**— JTAG.
- **PL Bank 13**—UART and I2C.
- **PL Bank 33**—GPS and PMOD(Auxilliary peripheral).
- **PL Bank 34**—LMS6002 ADC interface(13 bit LVCMOS + 3bit clock and control) and a user LED.
- **PL Bank 35**—LMS6002 DAC interface(13 bit LVCMOS + 3bit clock and control), LMS6002 Reset, RF frontend control and a user LED.
- **PS Bank 500**—QSPI Flash, LMS6002 SPI, power-on-reset, 3 user LEDs and bootmode select.
- **PS Bank 501**—UART, 2 RGMII Ethernet interfaces and SDIO.
- **PS Bank 502**—DDR interface.

All of these interfaces have been discussed in the previous chapter. However, it is worth mentioning at this point that all the peripheral controllers on the processing system are memory mapped. So, apart from the non-standard interfaces like the LMS6002 digital IO and the GPS, usage of the rest of the interfaces is to a great extent, simplified. Here in, lies the utility of the Zynq device which makes it a very powerful for systems requiring heterogeneous architecures.

	Current Consumption in the Voltage Domains(A)						
Major Power Consuming Devices	1V	1.5V	1.8V	2.5V	3.3V	6V	7V
Zynq(PS)	0.6	0.25			0.3		
Zynq(PL), Power ON	1.2		0.3		0.05		
Zynq(PL), Average					1.5		
DDR3		0.425					
DDR Term.		0.36					
PHY(DP83865)			0.43	0.19	0.01		
LMS6002D(Typ with WCDMA carrier)			0.1		0.6		
SGA6489						0.075	
RFPA3800							0.8
Total Current(A)	1.8	1.035	0.83	0.19	2.46	0.075	0.8
Total Power(Watt)	1.8	1.5525	1.494	0.475	8.118	0.45	5.6

TOTAL CONSUMPTION(Watt)	19.5
Power Available from PoE+(Watt)	25

Fig. 7.7 Overall power consumption

7.5.3 DDR3 Subsystem

The dual 256 MBs of DDR3 memory ICs are connected to the DDR bank of the Zynq device. Both the devices share 14 bit address lines, bank address and other control lines. The 16 bit data buses of each of the DDR3 SDRAMs together form the 32 bit data bus of the Zynq DDR3 bank. The DDR3 interface is a the high-speed SSTL 1.5 V logic family interface and the DDR3 ICs can operate at the maximum clock frequency of 1 GHz. Hence, this requires impedance matching and termination resistors. However, the termination resistors are placed only on the address, bank address, control line and the differential clocks. There are no termination resistors for the data bus. The signal integrity of the data bus is taken care by the ODT (On Die Termination) present in the DDR3 devices. The data bus requires this special technology because they are bidirectional and hence, static termination does not always provide best signal integrity. And, this is because the driver and the receivers have different input or output impedances respectively. Hence, a configuration of read and write termination resistances are selected which provide optimal signal integrity.

7.5.4 Peripherals Subsystem

The peripherals of the White RHINO platform are the Quad SPI flash (which is used as the boot flash), the debug interface (with USB-UART/JTAG/I2C bridge), the GPS receiver and the I2C temperature sensors.

7.5.4.1 Quad SPI Flash

This is a 128 MBit (16 MB) flash device which connects to the Zynq PS bank 500. The device is powered by a 3.3 V power supply. The digital interface consists of a clock, chip select and four data lines. The programming data rate is 1.5 Mbps.

7.5.4.2 GPS Receiver

The White RHINO receives a GPS signal centered at 1575.42 MHz through an onboard SMA connector which shall be connected to an external antenna. This GPS signal is taken by the Skyworks SE4110L device which directly downconverts and demodulates the BPSK signal. After the signal processing, the GPS information is provided in the form of magnitude and sign. The signal harness to the the Zynq PL bank 33 contains the mag, sign, the clock and the data enable signals. The GPS receiver requires two 3.3 V supplies, one RF and the other, digital.

7.5.4.3 USB-UART/JTAG/I2C

The host PC can get access to the debug interface through the micro USB port. There are ESD diodes for the circuit protection. The USB signal is taken to the FT4232 through 90 Ω differential traces. This signal is then converted by the FT4232 into the corresponding protocols and made available at the channels A–D. The channels A and B are dedicated to JTAG and I2C respectively and the channels C and D are dedicated to the PS and the PL UARTs respectively. Both I2C and the JTAG go through signal buffers to increase the drive strength of the signals. Also, there is an EEPROM connected to the FT4232 device which stores configuration information for the device. The EEPROM has 2 K Bits memory and is connected to the FT4232 device through a 3-wire serial IO interface consisting of clock, chip select and data. Finally, the clock is supplied to the FT4232 through a 75 MHz crystal.

7.5.4.4 Temperature Sensors

The temperature sensors shall be placed at various critical locations on the board. These temperature sensors are powered by 3.3 V and connected to the I2C bus. There are three of them on the White RHINO board and couple more of them are present on the RF board.

7.5.5 Ethernet Subsystem

There are two onboard Ethernet PHY devices which connect to the Zynq through the RGMII and to the RJ-45 connector through the MDIO interface. Now, the RGMII interface between the Zynq and the DP83865 is a 12-bit interface as shown in Fig. 7.8. The figure also shows the management interface bus which are used to configure the ethernet PHY chip. Since, DP83865 supports multi-protocol media independent interfaces(MII, GMII, RGMII-3COM, RGMII-HP), there are two jumpers dedicated for the mode select of the media independent interface. The DP83865 receives a 25 MHz clock from a crystal. The transmit and the recieve clocks are obtained and then reconstructed from the Ethernet hosts. In case of the 1 gbps mode, this shall be a 125 MHz clock signal. Now, the data in the media dependent interface (MDIO) is analog data which is carried through four pairs of 100 Ω difeerential traces onto the RJ-45 connector. However, the Ethernet1 schematic also includes an isolation transformer in between which shall provides access to the 48 V PoE+ through its center taps. These power lines are then carried over to a separate header which shall interface with a PoE+ board. The device receives the reset from the same power-on-reset signal from the power distribution schematics through a schematic port.

7.5.6 Low Power RF Subsystem

The low power RF subsystem on the White RHINO consists of the LMS6002 (with its digital and RF interfaces) and the transmit and the receive baluns (which convert differential transmission lines to the single ended 50 Ω transmission lines).The various power supplies required for the LMS6002 are as follows:

- **Digital core supply for ADC and DACs**—1.8 V
- **Analog supply for DAC**—3.3 V
- **Analog supply for ADC**—1.8 V
- **TX LO supply**—1.8 V
- **RX LO supply**—1.8 V

The RF transmit section consists of a lummped 65–100 Ω lumped filter and then a discrete transformer which converts the signal from 100 Ω differential to 50 Ω single ended signal. This 50 Ω signal is then launched on to the SMA connector so that it can be taken to the RF board where this signal shall be amplified. In the RF receive section, there is a discrete transformer B0322J5050 from Anaren which converts the 50 Ω unbalanced input from the antenna port to 50 Ω balanced signal required by the LMS6002.

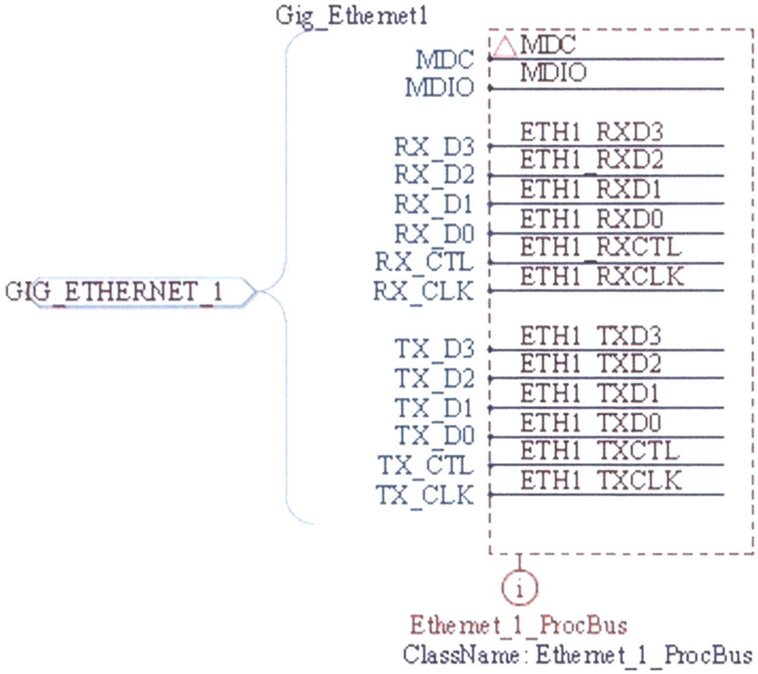

Fig. 7.8 Ethernet: RGMII harness and management bus

7.5.7 High Power RF Subsystem

The high power RF subsystem in contained in the White RHINO RF board which is a small board consisting of the transmit power amplifier and the TDD switch in the signal path. The White RHINO RF consists of a signal header which interfaces this board with the core board.The signal header provides the following signals from the White RHINO board (apart from 6 V supply and ground):

- ANTSW_CTL_1 and ANTSW_CTL_2 for controlling the Antenna switch Peregrine PE42440 [8].
- HMC_VCTL for controlling the Receive switch Hitite HMC849LP4CE [9].
- PA_EN for enabling and disabling the power supply to the final stage amplifier, RFPA3800 [4].
- I2C signals of SDA and SCL.

This board requires three power supplies. The main power amplifier RFPA3800 requires 7 V, the pre-amplifier SGA6489 requires 6 V and the switches require 3.3 V. Now, the signal header provides the 6 V supply from the White RHINO which inturn is used to generate the 3.3 V supply. The 7 V supply is generated from a 12 V supply which again is the main power supply for the White RHINO board. The 7 V regulator LT1376CS8PBF is enabled and disabled using the PA_EN signal.

7.5.7.1 RF Transmit Chain

The RF transmit chain has been shown in Fig. 7.9. As it can be observed, the transmit block provides about 35 dB gain to the signal from the White RHINO which is less than 6 dBm. Both the amplifiers operate at class A region to achieve maximum linearity. The amplifiers have 220 pF dc block capacitors.

7.5.7.2 RF Receive and Antenna Switches

There are two RF high power switches which together provide enough isolation between the transmit and the receive chains for the TDD application.These switches

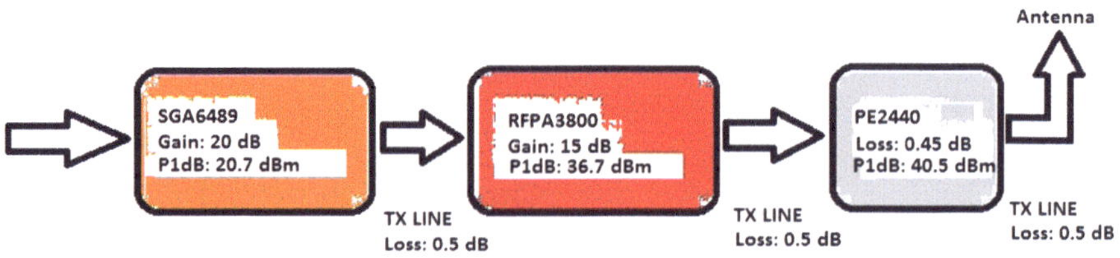

Fig. 7.9 White RHINO RF: transmit linup

are Hittite HMC849LP4CE and Peregrine PE42440. The Hittie switch has a one bit 3.3 V control and the Peregrine switch has a 2 bit 3.3 V control. The Peregrine switch is a 4-channel and hence, requires 2 bit control. However, we use only two of its channels [8, 9].

7.5.7.3 Temperature Sensors

Finally, there are two temperature sensors on the RF board. One of them shall be used to monitor the transmit chain temperature and the other one for the receive chain temperature. These temeperature sensors shall facilitate constant monitoring of the board and thus allow calibration routines to function effectively.

7.5.8 Overall RF Linup

All the RF parameters of the White RHINO design have been tabulated in Table 7.2. While the gain control range is a paramter of the LMS6002, the amplifier gain and the P1 dB emerge from the White RHINO RF transmit section as we have seen in the previous section. The balun amplitude and phase imbalance are internal parameters but they directly affect the error vector magnitude (EVM) and hence, the bit error of the transmitted ro the received data.

7.6 White RHINO Tests

The White RHINO testing is ongoing. However, operating system (U-Boot and linux) has been successfully ported and most of the digital interfaces (including the 1 Gbps Ethernet) have been successfully tested. The firmware development and testing of the

Table 7.2 Overall RF parameters

S.No	Parameter	Value
1	Transmit P1 dB	36.7 dBm
2	Amplifier gain	Over 30 dB
3	Transmit Gain Control Range	56 dB
4	Receive gain control range	61 dB
5	T/R isolation	Over 90 dB
6	Transmit Balun: Amplitude imbalance	Less than 1 dB
7	Transmit Balun: Phase imbalance	Less than 5°
8	Receive Balun: Amplitude imbalance	Less than 1 dB
9	Receive Balun: Phase imbalance	Less than 5°

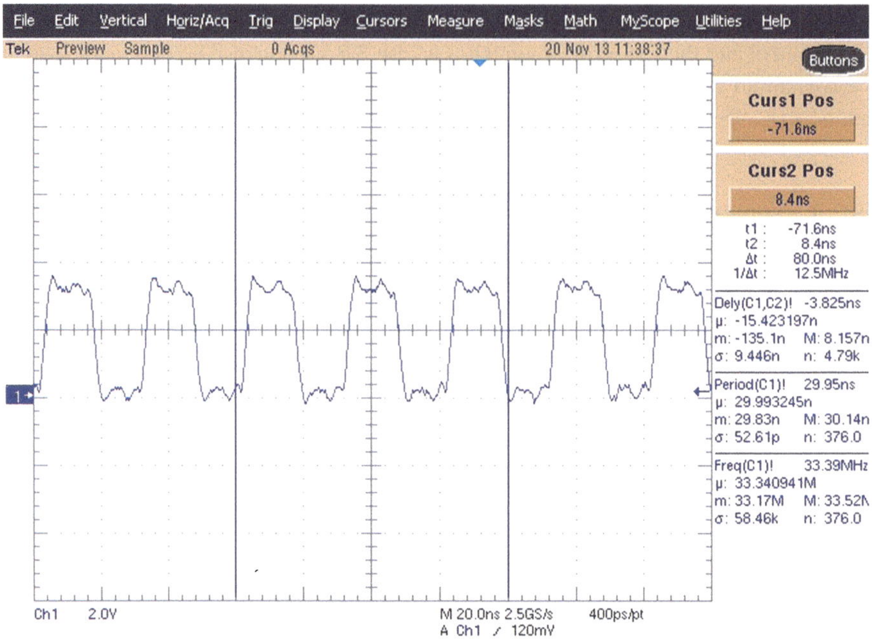

Fig. 7.10 PS clock: 33.3333 MHz

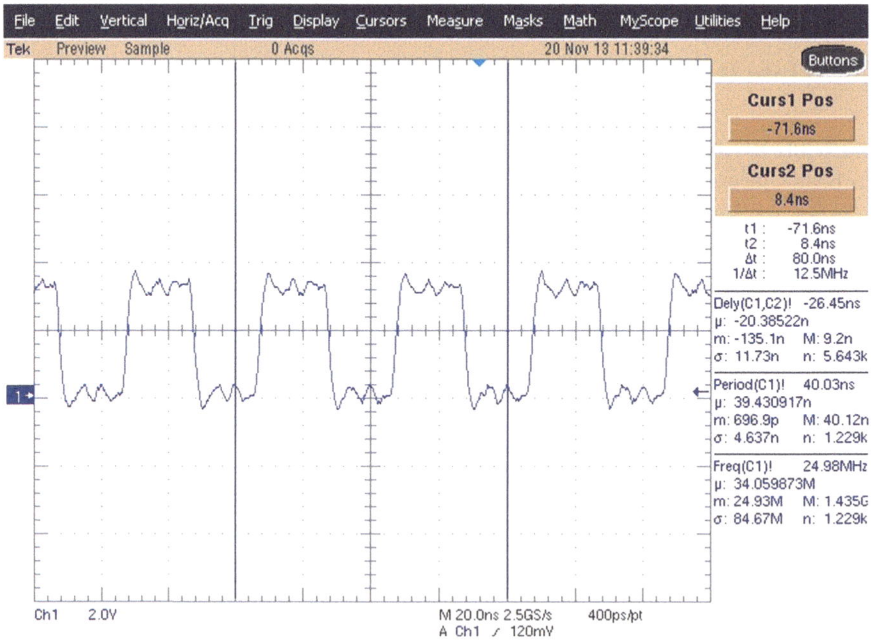

Fig. 7.11 Ethernet clock: 25 MHz

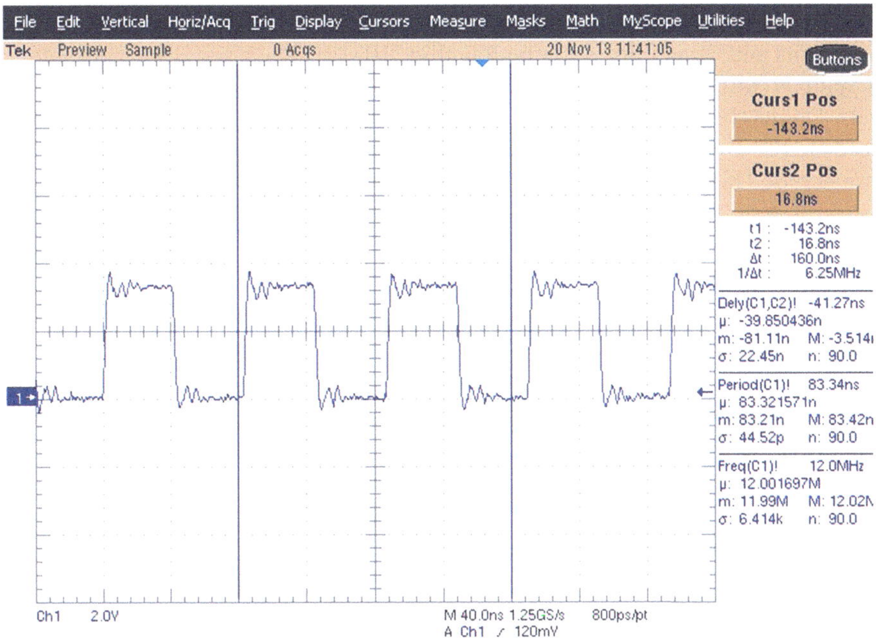

Fig. 7.12 FT4232 clock: 12 MHz

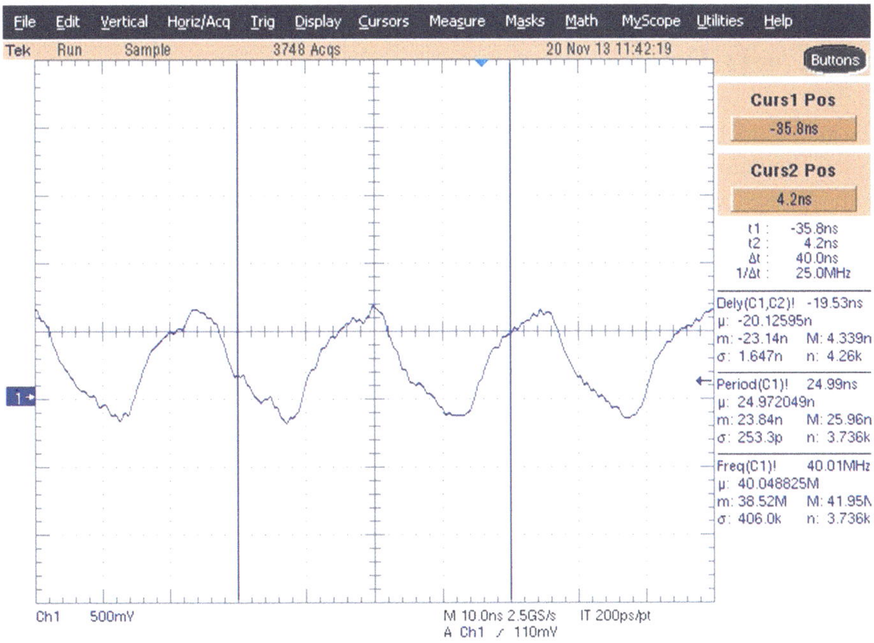

Fig. 7.13 LMS6002 TCXO: 40 MHz

Zynq7020 requires embedded software development tools. The information about these tools and design flows can be found in [10–12]. The Zynq device has multiplexed IOs (MIOs) which can be alternatively used to connect with different peripherals. This has been provided by the Zynq to facilitate flexible board design. However, since the IOs are multiplexed, our first task was to assign IO ports to various onbard peripherals. This process is called the PS7 initialization. Only after the PS7 initialization is done further software development can be done. The development tools required are as follows:

- Xilinx XPS—The Xilinx Platform Studio (XPS) is an embedded design tool which was primarily used to design of the Zynq MIO structure.
- Xilinx SDK—The Xilinx Software Development Kit (SDK) is an eclipse based Software development tool. Apart from stadard tools, it also contains the Xilinx Microprocessor Debugger(XMD) console which was extensively used to communicate with the processor. The debugging has been done through the JTAG port.
- Tera Term—This is an open source terminal emulator program. This software was also extensively used to observe serial console prints.

The running operating system on the White RHINO is Linux. So, after porting U-Boot on to the White RHINO, a Linux image was built using the Zedboard souce code. This image was ported into the Zynq memory along with the Root File System and the Device Tree.

7.6.1 Clocks Test

Clocks are the heartbeats of an electronic system. So, its necessary to check if the crystal oscillators are generating clocks to the respective devices because without them, the devices will not function. This test was done with a 5 GS/s Oscilloscope. The observed clocks are shown in Figs. 7.10, 7.11, 7.12 and 7.13.

- Zynq7020—33.33 MHz, 3.3 V
- Ethernet PHY—25 MHz, 3.3 V
- FT4232—12 MHz, 3.3 V
- LMS6002 TCXO—40 MHz, 2.5 V. This clock output is a clipped-sine waveform.

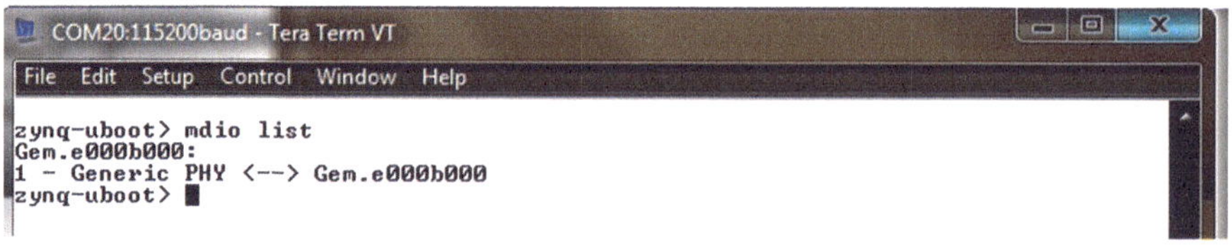

Fig. 7.14 MDIO info: U-boot

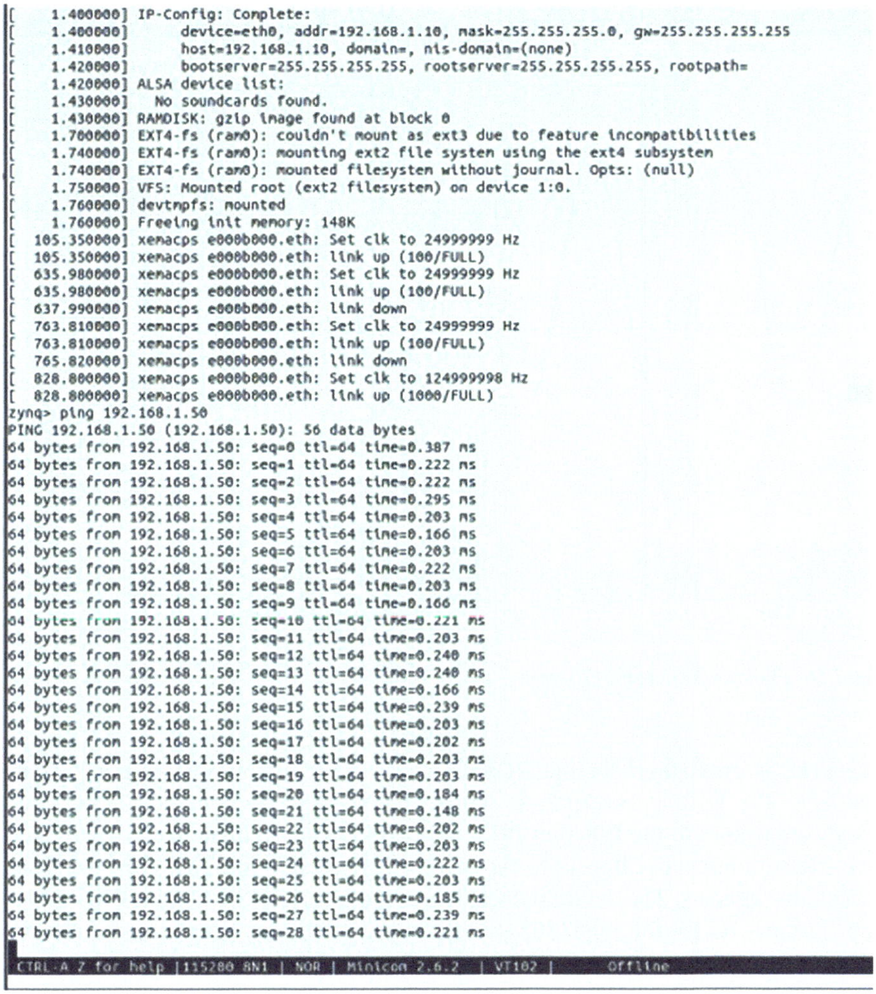

```
[    1.400000] IP-Config: Complete:
[    1.400000]      device=eth0, addr=192.168.1.10, mask=255.255.255.0, gw=255.255.255.255
[    1.410000]      host=192.168.1.10, domain=, nis-domain=(none)
[    1.420000]      bootserver=255.255.255.255, rootserver=255.255.255.255, rootpath=
[    1.420000] ALSA device list:
[    1.430000]   No soundcards found.
[    1.430000] RAMDISK: gzip image found at block 0
[    1.700000] EXT4-fs (ram0): couldn't mount as ext3 due to feature incompatibilities
[    1.740000] EXT4-fs (ram0): mounting ext2 file system using the ext4 subsystem
[    1.740000] EXT4-fs (ram0): mounted filesystem without journal. Opts: (null)
[    1.750000] VFS: Mounted root (ext2 filesystem) on device 1:0.
[    1.760000] devtmpfs: mounted
[    1.760000] Freeing init memory: 148K
[  105.350000] xemacps e000b000.eth: Set clk to 24999999 Hz
[  105.350000] xemacps e000b000.eth: link up (100/FULL)
[  635.980000] xemacps e000b000.eth: Set clk to 24999999 Hz
[  635.980000] xemacps e000b000.eth: link up (100/FULL)
[  637.990000] xemacps e000b000.eth: link down
[  763.810000] xemacps e000b000.eth: Set clk to 24999999 Hz
[  763.810000] xemacps e000b000.eth: link up (100/FULL)
[  765.820000] xemacps e000b000.eth: link down
[  828.800000] xemacps e000b000.eth: Set clk to 124999998 Hz
[  828.800000] xemacps e000b000.eth: link up (1000/FULL)
zynq> ping 192.168.1.50
PING 192.168.1.50 (192.168.1.50): 56 data bytes
64 bytes from 192.168.1.50: seq=0 ttl=64 time=0.387 ms
64 bytes from 192.168.1.50: seq=1 ttl=64 time=0.222 ms
64 bytes from 192.168.1.50: seq=2 ttl=64 time=0.222 ms
64 bytes from 192.168.1.50: seq=3 ttl=64 time=0.295 ms
64 bytes from 192.168.1.50: seq=4 ttl=64 time=0.203 ms
64 bytes from 192.168.1.50: seq=5 ttl=64 time=0.166 ms
64 bytes from 192.168.1.50: seq=6 ttl=64 time=0.203 ms
64 bytes from 192.168.1.50: seq=7 ttl=64 time=0.222 ms
64 bytes from 192.168.1.50: seq=8 ttl=64 time=0.203 ms
64 bytes from 192.168.1.50: seq=9 ttl=64 time=0.166 ms
64 bytes from 192.168.1.50: seq=10 ttl=64 time=0.221 ms
64 bytes from 192.168.1.50: seq=11 ttl=64 time=0.203 ms
64 bytes from 192.168.1.50: seq=12 ttl=64 time=0.240 ms
64 bytes from 192.168.1.50: seq=13 ttl=64 time=0.240 ms
64 bytes from 192.168.1.50: seq=14 ttl=64 time=0.166 ms
64 bytes from 192.168.1.50: seq=15 ttl=64 time=0.239 ms
64 bytes from 192.168.1.50: seq=16 ttl=64 time=0.203 ms
64 bytes from 192.168.1.50: seq=17 ttl=64 time=0.202 ms
64 bytes from 192.168.1.50: seq=18 ttl=64 time=0.203 ms
64 bytes from 192.168.1.50: seq=19 ttl=64 time=0.203 ms
64 bytes from 192.168.1.50: seq=20 ttl=64 time=0.184 ms
64 bytes from 192.168.1.50: seq=21 ttl=64 time=0.148 ms
64 bytes from 192.168.1.50: seq=22 ttl=64 time=0.202 ms
64 bytes from 192.168.1.50: seq=23 ttl=64 time=0.203 ms
64 bytes from 192.168.1.50: seq=24 ttl=64 time=0.222 ms
64 bytes from 192.168.1.50: seq=25 ttl=64 time=0.203 ms
64 bytes from 192.168.1.50: seq=26 ttl=64 time=0.185 ms
64 bytes from 192.168.1.50: seq=27 ttl=64 time=0.239 ms
64 bytes from 192.168.1.50: seq=28 ttl=64 time=0.221 ms
CTRL-A Z for help |115200 8N1 | NOR | Minicom 2.6.2 | VT102 |     Offline
```

Fig. 7.15 Host PC pinging the white RHINO at 1 Gbps

7.6.2 1Gig Ethernet Test

The 1Gig Ethernet subsystem is a datapath block of the White RHINO communications. This test includes the following subtests:

- Verifying the MII interface
- Verifying the MDIO interface
- Performing an MII Loopback
- Performing a ping from the White RHINO to the host PC and vice-versa.

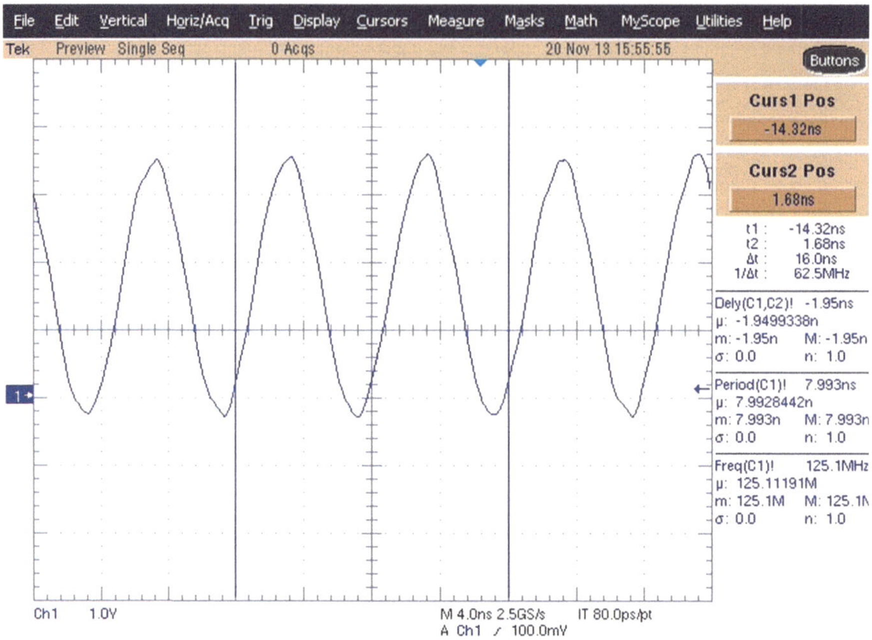

Fig. 7.16 Ethernet data clock: 125 MHz

The PHY is configured for the RGMII-3COM mode for the media-independant interface. The U-Boot incorporates some Ethernet PHY commands. These commands are passed to the Ethernet PHY through the management and control interface. These commands allow us to check if the PHY device is available and read and write to its registers. The device detected has a manufacturer ID of 0×80017 which is the correct ID for the DP83865 device. In the same figure, we can also see the standard PHY registers being read back. We can also see the MDIO info in Fig. 7.14. Now, after we ascertained that the PHY device is alive, we ported Linux into the Zynq and performed an MII loopback test using the local loopback IP address. Finally, after the MII loopback, we connected the White RHINO to the host PC using a CAT-5 Ethernet cable. After changing the IP address of the host PC to 192.168.1.50, PING was run on both White RHINO and the host PC. The PC ping replies from the host observed on the White RHINO linux prompt is shown in Fig. 7.15 and the 125 MHz data clock measured on the oscilloscope is shown in Fig. 7.16.

7.7 Summary

In this chapter, discussed the various features and design aspects of the new White RHINO SDR platform. We also discussed some of the tests that have been done. The optimizations are still going on. These optimizations shall be implemented in the next prototypes. However, it is an open source platform and the project can be tracked through the github website at https://github.com/ojonav/White-RHINO. Inspite of the various SDR platforms being currently available in the market, it is believed that the feature set offered by the compact and powerful White RHINO at a very competitive price of about 1,000 USD is unique.

References

1. IEEE standard for information technology: Telecommunications and information exchange between systems wireless regional area networks (WRAN): Specific requirements (2011)
2. Zynq-7000 EPP (XC7Z010 and XC7Z020) Data Sheet:DC and Switching Characteristics. http://www.xilinx.com/support/documentation/data_sheets/ds187-XC7Z010-XC7Z020-Data-Sheet.pdf
3. LMS6002 Multi-band multi-standard transceiver: Product Brief. http://www.limemicro.com/download/Lime_ProductBrief.pdf
4. RFPA3800. http://www.rfmd.com/CS/Documents/RFPA3800DS.pdf
5. MT41K128M16JT-125 Micron: Data Sheet. http://www.micron.com/parts/dram/ddr3-sdram/mt41k128m16jt-125&filter=DataSheet
6. FT4232H datasheet—FTDI. http://www.ftdichip.com/Documents/DataSheets/ICs/DS_FT4232H.pdf.
7. IEEE 802.3 ETHERNET WORKING GROUP. http://www.ieee802.org/3/
8. PE42440 DataSheet—Peregrine Semiconductor. http://www.psemi.com/pdf/datasheets/pe42440ds.pdf.
9. HMC849LP4CE—Hittite Microwave
10. Zynq-7000 All Programmable SoC: Concepts, Tools, and Techniques (CTT)
11. Zynq-7000 All Programmable SoC Software Developers Guide
12. Embedded System Tools Reference Manual

Chapter 8
TV White Space Channel Estimation and Equalisation: Challenges and Solutions

Vimal Bhatia

Abstract In recent years, IEEE 802.22 has emerged as a viable solution to bring broadband access to hard-to-reach, low population density areas, typical of rural environments. There has been successful trials around the world to test the viability of this standard with support from both Governmental and private companies. Since the communication is allowed in the TV White Space band, there has been concerns regarding the effect of interference in these communication channels. For reliable and fast data rates, it is desirable to have low noise and interference. This chapter presents an overview of the standard, from channel interference perspective, followed by a generalised scheme for mitigating the effects of noise and interference in TVWS communication.

8.1 Introduction

TV White Space is designed to work in Wireless Radio Access Network (WRAN). This standard is intended to enable deployment of interoperable IEEE 802 multivendor wireless regional area network products, to facilitate competition in broadband access by providing alternatives to wireline broadband access and extending the deployability of such systems into diverse geographic areas, including sparsely populated rural areas, while preventing harmful interference to incumbent licensed services in the TV broadcast bands. The Wireless Regional Area Networks (WRANs) for which this standard has been developed are expected to operate primarily in low population density areas in order to provide broadband access to data networks. The WRAN systems will use vacant channels in the very high frequency (VHF) and ultra high frequency (UHF) bands allocated to the Television Broadcasting Service in the frequency range between 54 and 862 MHz while avoiding interference to the broadcast incumbents in these bands. A typical application can be the coverage of the rural area around a village, as illustrated in previous chapters, within a radius of 10 to 30 km from the base station depending on its effective

V. Bhatia (✉)
Indian Institute of Technology Indore, Indore, India
e-mail: vbhatia@iiti.ac.in

A.K. Mishra and D.L. Johnson (eds.), *White Space Communication*,
Signals and Communication Technology, DOI 10.1007/978-3-319-08747-4_8

isotropic radiated power (EIRP) and antenna height. The medium access control layer (MAC) can also accommodate user terminals located as far as 100 km with proper scheduling of the traffic in the frame when exceptional radio frequency (RF) signal propagation conditions are present. With the physical layer (PHY) implesmented in this standard, WRAN systems can cover up to a radius of 30 km without special scheduling The IEEE standard 802.22 standard includes cognitive radio techniques to mitigate interference to incumbents, including geolocation capability, provision to access a database of incumbent services, and spectrum-sensing technology to detect the presence of incumbent services, other WRAN systems, and IEEE 802.22.1 wireless beacons. Since interference from Licensed user is a major concern, hence this is discussed in further details in IEEE standard 802.22.1—2010, IEEE Standard for local and metropolitan area networks—Part 22.1: "Methods to Enhance Protection of Low-Power, Licensed Device Operation" in the TV Broadcast Bands from Harmful Interference from License-Exempt Devices Operating in those frequency bands. According to the standard the definition of candidate channel (IEEE 802.22 Sect.–3.7): A channel that, once cleared from potential harmful interference to incumbents, can become a backup channel. Any device using TVWS does so on the basis that it must not cause interference to licensed users and that users of TVWS will not receive any interference protection from either licensed users or each other. Since interference effects the performance of the TVWS communication systems, the IEEE standard covers this with various definitions and designs. The IEEE 802.22.1 wireless beacon is defined, an RF device that provides enhanced protection to licensed low power auxiliary devices such as those used in the production and transmission of broadcast programs [e.g., devices licensed as secondary under Title 47 of the Code of Federal Regulations (CFR) in the USA and equivalent devices in other regulatory domains] from harmful interference potentially caused by WRAN operation. This chapter discusses about the various definitions pertaining to interference in TVWS and proposes a viable solution to mitigate this effect.

Following this introduction, Sect. 8.2 discusses an overview of IEEE 802.22 standard, architecture and interference temperature. Section 8.3 discusses the PHY layer channel estimation and equalisation with a generalised algorithm for channels effected by high interference temperature. Section 8.4 summarizes and concludes the chapter.

8.2 Design and Challenges

In this section, an introduction to some of the design and challenges in the TVWS is discussed. Overview of the reference architecture, spectrum manager, physical layer, and interference temperature is provided in this section, with detailed discussion in previous chapters. The overview provided in this section is primarily focused on the performance of TVWS in the presence of interference.

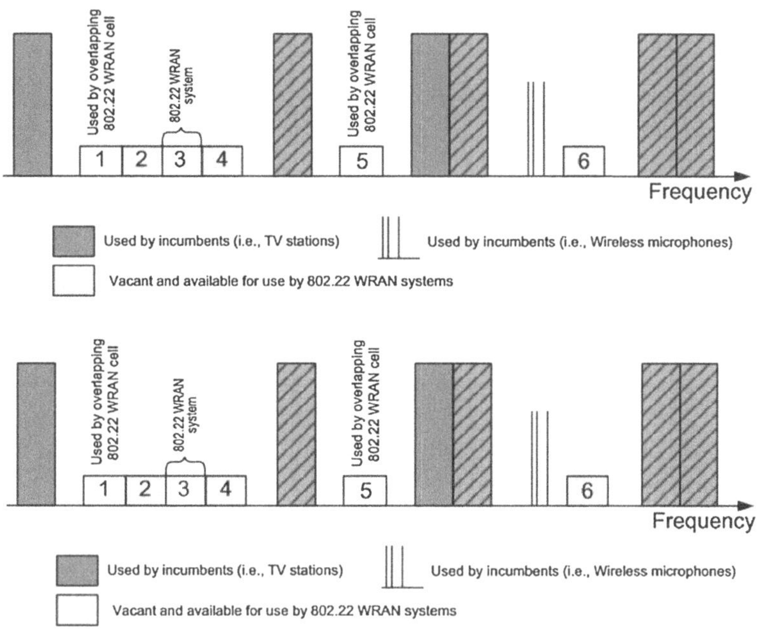

Fig. 8.1 Reference architecture [1]

8.2.1 Reference Architecture

The IEEE 802.22 reference architecture is described in details in previous chapters, where a unique characteristic of this architecture is its cognitive components. Spectrum availability in the TV bands can be fragmented, i.e., some channels can be occupied by incumbents in an area (see Fig. 8.1) while others can be available for WRAN transmission, and this availability can vary in time. As shown in Fig. 8.1, channel 3 is allocated to the IEEE 802.22 WRAN system, while channels 1 and 5 are in use by overlapping IEEE 802.22 cells. Channels 2, 4, and 6 would be available for more WRAN services in the area. Also, proper frequency separation is enforced in order to protect incumbent services using cognitive radio techniques—the air interface must be frequency agile, must adjust to the fragmented and time-varying spectrum availability, and must avoid interference to the TV band incumbent services. The required functional capabilities of these cognitive radio techniques are established in this standard (see IEEE 802.22 Clause 10). In particular, the cognitive components of the system architecture must keep track of multiple channels, know which of these channels are occupied by incumbents and which can be used for WRAN transmission, allow for dynamic frequency selection, and avoid interference to incumbents on a real-time basis [15].

8.2.2 Spectrum Manager

The spectrum manager (SM) shall reside in the cognitive plane of the base station (BS) at the same layer as the MAC common part sublayer (CPS) in the data plane as shown in previous chapter. The SM shall maintain spectrum availability information, manage channel lists, manage quiet periods scheduling, and implement coexistence mechanisms. The SM shall also take requests from the MAC/PHY. For example, the MAC must inform the SM if an interference situation has been detected (e.g., with incumbents or other IEEE 802.22 cells) during normal operation in the channel. The SM must then take appropriate actions to resolve the issue such as moving to another channel (as specified in IEEE 802.22 10.2.3 of the standard). In order to do this, the SM must first provide sufficient quiet period time for the CPEs to carry out in-band sensing (on channels N and $N \pm 1$) to clear these channels within the specified channel monitoring requirement (refer IEEE 802.22 Annex A), and then keep an updated list of backup and candidate channels in a prioritized order and make sure that the CPEs have enough idle time to clear a sufficient number of backup channels. If an incumbent is found from the in-band sensing, the MAC shall use the informed response from the SM to perform the switching operation.

8.2.3 IEEE 802.22 (PHY)

As discussed in previous chapters, the orthogonal frequency division multiplexing(OFDM) has been chosen as the modulation scheme. The system parameters for WRAN, as described in IEEE 802.22 Table-198, clearly specify the orthogonal frequency division multiplexing access (OFDMA) scheme. As the information about OFDMA is found in most of the literature, so we give relevant information regarding symbols and their packet structure [7, 9]. In the previous chapters, we have seen that the operation frequency range is from 56 to 867 MHz with single air interface of 2048 sub carriers. One symbol of OFDM is combination of inverse fast Fourier transform(IFFT) output with cyclic prefix. So one Symbol duration of OFDM is $T_{IFFT} + T_{CP}$. Cyclic prefix mode vary from $T_{IFFT}/4$ to $T_{IFFT}/32$, so based on that symbol duration also changes. This conveyed to all customer premise equipment (CPE) by BS in terms of T_{CP} to T_{IFFT} through super frame control header. In frequency domain an OFDMA symbols is defined in terms of its sub carrier. Based on functionality sub carries are grouped as (1) data sub carriers, (2) pilot sub carriers, (3) guard and null (including DC) sub-carriers [3, 21]. A sub-channel is the basic unit used for sub carrier allocation in both downstream and upstream. Among the 2,048 sub carriers in each OFDM symbol, 384 sub-carriers are null sub-carriers (left guard band, right guard band and DC subcarriers) with 0 amplitude and 0 phase. The remaining 1,680 sub carriers are partitioned into 60 sub channels. Each sub channel is of 28 sub carrier with 24 for data and 4 for pilot [7, 8, 10]. The modulation scheme, block diagram and pilot spacing is clearly defined in the standard. In the Sect. 9.6.1, the pilot pattern is shown for the OFDM frame.

8.2.4 Interference Temperature

In this section, we first discuss the definition of interference temperature and then look at the ways to measure it. Currently, the radio environment is *transmitter-centric*, in the sense that the transmitted power is designed to approach a prescribed noise floor at a certain distance from the transmitter [17]. However, it is possible for the RF noise floor to rise due to the unpredictable appearance of new sources of interference, thereby causing a progressive degradation of the signal coverage. To guard against such a possibility, the FCC Spectrum Policy Task Force [10] has recommended a paradigm shift in interference assessment, that is, a shift away from largely fixed operations in the transmitter and towards real-time interactions between the transmitter and receiver in an adaptive manner. The recommendation is based on a new metric called the interference temperature, which is intended to quantify and manage the sources of interference in a radio environment. Moreover, the specification of an interference-temperature limit provides a worst case? characterisation of the RF environment in a particular frequency band and at a particular geographic location, where the receiver could be expected to operate satisfactorily. The recommendation is made with two key benefits in mind:

(1) The interference temperature at a receiving antenna provides an accurate measure for the acceptable level of RF interference in the frequency band of interest; any transmission in that band is considered to be harmful? if it would increase the noise floor above the interference-temperature limit.

(2) Given a particular frequency band in which the interference temperature is not exceeded, that band could be made available to unserviced users; the interference-temperature limit would then serve as a cap? placed on potential RF energy that could be introduced into that band.

The interference temperature maybe measured in degrees Kelvin, following the well-known definition of equivalent noise temperature of a receiver [1, 4]. Moreover, the interference temperature limit $T(max)$ multiplied by Boltzmann constant $k = 1.3807 \times 10^{-23}$ joules per degree Kelvin yields the corresponding upper limit on permissible power spectral density in a frequency band of interest, and that density is measured in joules per second or, equivalently, watts per hertz. Interference temperature T_I is specified in Kelvin and is defined as $T_I(f_c, B) = P_I(f_c, B)/kB$, where $P_I(f_c, B)$ is the average interference power in Watts centres at f_c, covering bandwidth B measured in Hertz. For measuring interference temperature, the idea is that by taking a single measurement, a cognitive radio can completely characterise both interference and noise with a single number. Of course, it has been argued that interference and noise behave differently. Interference is typically more deterministic and independent of bandwidth, whereas noise is not. For a given geographical area, the FCC would establish an interference temperature limit, T_L. This value would

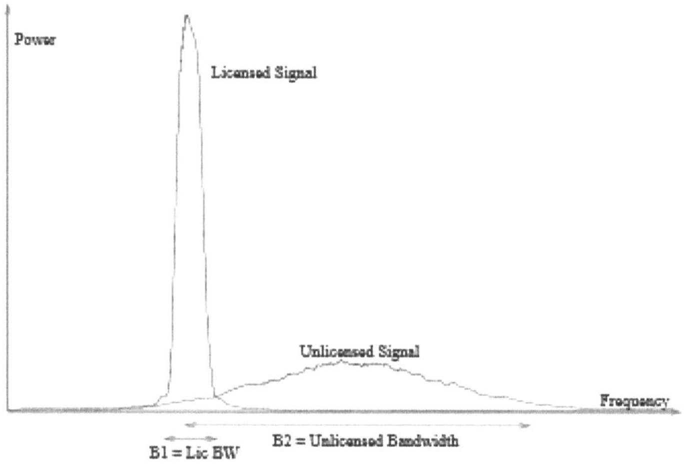

Fig. 8.2 Power spectral density [9]

be maximum amount of tolerable interference for a given frequency band in a particular location. Any unlicensed transmitter utilizing this band must guarantee that their transmissions added to the existing interference must not exceed the interference temperature limit at a licensed receiver. The Fig. 8.2 shows an example of the PSD for an unlicensed signal partially overlapping a licensed signal. While this may seem clear cut, there is ambiguity over which signals are considered interference, and which f_c and B to use [4, 5].

In [30], the authors suggest a technique to measure this cut-off in a controlled environment for IEEE 802.11 protocol. Similar studies have been taking place to study the effect of interference–plus–noise and to measure this in TVWS [8, 9]. The Cambridge White Space Consortium report on Cambridge TV White Space Trial—A summary of the technical findings, 2012 [11] highlights some of the challenges due to interference in the UK. The white space device manufacturers are looking at how some of the effects of interference might be mitigated and hence enhance the performance. Since OFDM communication system's performance is sensitive to the presence of interference, in the next section, a generalised solution for improving bit error rate (BER) performance is presented. An generalised OFDM channel estimation and equaliser for operation in interference limited channels is discussed next.

8.3 OFDM Channel Estimation and Equaliser

Orthogonal frequency division multiplexing is a promising multi-carrier digital communication technique for transmitting data at high bit-rates over wireless or wireline channels. The high-speed serial data are converted into many low bit rate streams that are transmitted in parallel, thereby increasing the symbol duration and reducing the ISI. These features have led to an increase in the use of OFDM or related techniques in many high bit rate communication systems. Discrete multi-tone modulation which is quite similar to OFDM is extensively used in DSL communication systems. OFDM has been chosen for digital audio broadcasting (DAB) and digital video broadcasting (DVB). It is also used for the 2.4 GHz wireless local area networks (WLAN) i.e. IEEE 802.11g, worldwide interoperability for microwave acccess (WiMAX) i.e. IEEE 802.16, and in TVWS communication system i.e. IEEE 802.22.

Coherent OFDM transmission invariable requires estimation of the channel frequency response (i.e. the gains of the OFDM tones). Currently there can be three possible solution: (1) blind, (2) semi-blind, and (3) pilot aided. In blind channel estimation techniques, the channel is estimated without the knowledge of the transmitted sequence. It is attractive as the throughput is higher as no bits are lost in training. However it requires large amount of data to be stored before channel estimation can begin, which invariably introduces delays. The pilot based technique estimates the channel by transmitting a known (at the receiver) training sequence along with the unknown data at the receiver. The receiver estimates the channel using some criterion based on comparing the change in these pilots due to channel. The semi-blind techniques try to reduce the size of the training sequence by exploiting both the known and the unknown (blind) portions of the data.

Channel estimation in OFDM is critical to the overall performance of the communication system. The insertion of pilots in OFDM symbols provides a base for reliable channel estimates. There has been considerable increase in channel estimation research over the years [5, 6, 26] to name a few. However most of the current work is based on channel estimation for Gaussian channels or assuming that the interference is very low. This assumption is however not always valid in scenarios where there are a small number of interferers (e.g. Bluetooth device [33] or microwave oven operating in the presence of a WLAN or in TVWS from Licensed or un-Licensed users). With the co-existence of various wireless equipment in home and office environments the interference from neighbouring devices has become a major concern [7]. In interference affected channels the algorithms designed for Gaussian assumption are not optimal [3, 24].

In IEEE 802.22 PHY Layer, information to down stream from BS or upstream from multiple CPEs is transmitted using OFDMA. Physical parameters used in WRAN system are tabulated in Table 8.1. As the information about OFDMA is found in most of literature,so we give relevant information regarding symbols and their timing information [1, 31]. Operation frequency range is from 56 to 867 MHz with single air interface of 2048 sub carriers. One symbol of OFDM is combination of IFFT output with cyclic prefix. So one Symbol duration of OFDM is $T_{IFFT} + T_{CP}$. Cyclic prefix mode vary from $T_{IFFT}/4$ to $T_{IFFT}/32$, so based on that symbol duration also changes. This conveyed to all CPE by BS in terms of T_{CP} to T_{IFFT} through super frame control header. In frequency domain, an OFDMA symbols is defined in terms of its sub carrier. Based on functionality sub carries are grouped as (1) data sub carriers, (2) pilot sub carriers, (3) guard and null (including DC) sub-carriers [1, 20]. A sub-channel is the basic unit used for sub carrier allocation in both downstream

Table 8.1 System parameters for WRAN

Parameters	Specification	Remarks
Frequency range	54 862 (MHz)	
Channel bandwidth	6, 7 or 8 (MHz)	According regulatory domain
Data rate	4.54–22.6 (Mbit/s)	
Spectral efficiency	0.76–3.78 [bit/(sHz)]	
Payload modulation	QPSK, 16-QAM, 64-QAM	BPSK used for preamble pilots and CDMA codes
Transmit EIRP	4W maximum for CPEs 4W maximum for BSs in the USA regulatory domainS	Maximum EIRP for BSs may vary in other regulatory domains
Multiple access	OFDMA	
FFT Size (NFFT)	2048	
Cyclic prefix modes	1/4, 1/8, 1/16, 1/32	
Duplex	TDD	

and upstream. Among the 2048 sub carriers in each OFDM symbol,384 sub carriers are null sub-carriers (left guard band, right guard band and DC subcarriers) with 0 amplitude and 0 phase. The remaining 1680 sub carriers are partitioned into 60 sub channels. Each sub channel is of 28 sub carrier with 24 for data and 4 for pilot as described in Table 8.1 [1, 20, 38].

The interferers are also assumed to be using OFDM modulation technique and can be either synchronous or asynchronous. The algorithm discussed in [22] specifically deals with synchronous interference and after modifications for asynchronous interference [21]. However in [19, 21, 22] it was noted that interference was modelled as Gaussian, which may not be the case if only a few or infact one major interferer as in [7, 33] is present.In this section,no such *a priori* assumption on the distribution of the interfering received signal is made. Moreover no individual parameter of the interferer is estimated specifically. The channel is assumed to be stationary for a given block. In this section, the fading channel is estimated in the presence of interference directly in time domain using an ML technique. In fact, the presence of interference along with Gaussian noise is jointly considered as a Gaussian mixture noise [3, 28]. In this section, it is noted that traditional zero forcing equalisation (usually used in OFDM receivers) technique falls short of the performance in the presence of interference. Thus a MAP symbol-by-symbol equaliser to improve the BER performance is proposed. Simulation results confirm the non-optimal estimates when least squares (LS) is used and improved BER performance by using the proposed channel estimator and equaliser algorithms.

This section is organised as follows: Sect. 8.3.1 the problem statement is formulated for a general OFDM communication system. A brief discussion on kernel density estimator for complex density estimation is described in Sect. 8.3.3. The iterative based on non-parametric maximum likelihood (NPML) channel estimator for OFDM is described in Sect. 8.3.4. Section 8.3.5 discusses the modified non-parametric symbol-by-symbol equaliser. To test the robustness of the algorithm, in Sect. 8.3.6 the simulations results are presented for both flat and multi-path fading.

8.3.1 A Generalised OFDM Communication System Model

The baseband equivalent representation of a general OFDM communication system as in Fig. 8.3 is considered here. The discussion on the estimation of one OFDM symbol instead of a sequence of symbols is justified below. At the transmitter side, the serial input data is converted into M parallel streams, and each data stream is modulated by a linear modulation scheme, such as QPSK, 16QAM or 64QAM. If QPSK is used, for instance, the binary input data of $2M$ bits will be converted into M QPSK symbols by the serial-to-parallel converter (S/P) and the modulator. The modulated data symbols, which are denoted by complex-valued variables $X(0), \ldots, X(m), \ldots, X(M-1)$, are then transformed by the inverse fast Fourier transform (IFFT), and the complex-valued outputs $x(0), \ldots, x(k), \ldots, x(M-1)$ are converted back to serial data for transmission. A guard interval is inserted between symbols to avoid ISI. If the guard interval is longer than the channel delay spread, and if the samples of the guard at the receiving end are discarded, the ISI will not affect the actual OFDM symbol. Therefore, the system can be analysed on a symbol-by-symbol basis. OFDM system is also exhibits high peak-to-average power ratio (PAPR). Namely, the peak values of some of the transmitted signals could be much larger than the typical values. This could lead to a necessity of using circuits with linear characteristics within a larger dynamic range, otherwise the signal clipping at high levels would yield a distortion of the transmitted signal and out-of-band radiations. In this chapter, it is assumed that the received OFDM signal is not affected by PAPR problem. At the receiver side, after converting the serial data to M parallel streams, the received samples $y(0), \ldots, y(k), \ldots, y(M-1)$ are transformed by the FFT into $Y(0), \ldots, Y(m), \ldots, Y(M-1)$ [6].

Using the notations for the OFDM symbols, the output of the channel can be written as

$$y(k) = \sum_{l=0}^{N_T-1} h^*(l)x(k-l) + \sum_{p=0}^{P_I-1}\sum_{l=0}^{N_T-1} g_p^*(l)u_p(k-l) + n(k), \quad 0 \leq k \leq M-1 \tag{8.1}$$

where h and x represents desired user's channel and data respectively. Without loss of generality, the complex conjugate h^* instead of h is chosen in the above equation [16]. N_T represents the channel length and $n(k)$ is the independent and identically distributed (i.i.d.) additive white Gaussian noise. P_I represents the number of interferers where g_p and u_p are the interfering channel and signal respectively. Note

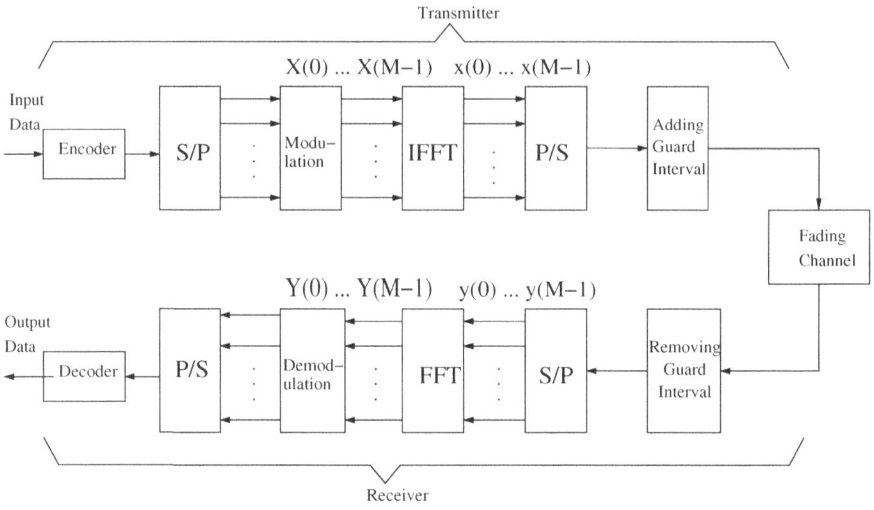

Fig. 8.3 A generalised OFDM communication system

that $y(k)$, $x(k)$, $n(k)$, $h(l)$, $u_p(k)$ and $g_p(l)$ are all complex valued. It is assumed that the channel and interference doesn't change during the block transfer (assuming quasi-stationary channel) and interference is synchronous which makes the above representation possible.

To maintain orthogonality between tones, it is necessary to ensure that the symbol time contains one or multiple cycles of each sinusoidal tone waveform. This is normally the case, because the system numerology is constructed such that tone frequencies are integer multiples of the symbol period. In absolute terms, to generate a pure sinusoidal tone requires the signal to start at time minus infinity. This is important, because tones are the only waveform than can ensure orthogonality. Fortunately, the channel response can be treated as finite, because multipath components decay over time and the channel is effectively band-limited. By adding a guard time, called a cyclic prefix, the channel can be made to behave as if the transmitted waveforms were from time minus infinite, and thus ensure orthogonality, which essentially prevents one subcarrier from interfering with another (called intercarrier interference, or ICI). Figure 8.4 shows three tones over a single symbol period, where each tone has an integer number of cycles during the symbol.

The cyclic prefix is actually a copy of the last portion of the data symbol appended to the front of the symbol during the guard interval. Multipath causes tones and delayed replicas of tones to arrive at the receiver with some delay spread. This leads to misalignment between sinusoids, which need to be aligned to be orthogonal. The cyclic prefix allows the tones to be realigned at the receiver, thus regaining orthogonality. As discussed above, if a cyclic prefix is used for the guard interval, intercarrier interference (ICI) in a multipath channel can also be avoided. Then it can be shown that the following simple relation between $Y(m)$ and $X(m)$ holds:

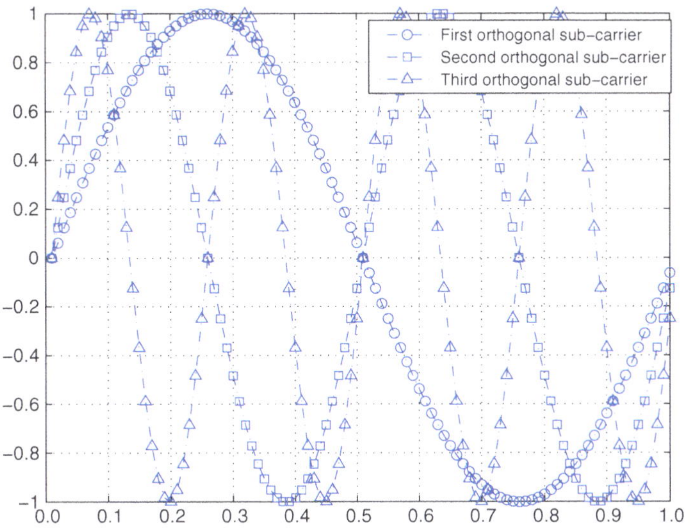

Fig. 8.4 Integer number of sinusoid periods

$$
\begin{aligned}
Y(m) &= \left(\sum_{l=0}^{N_T-1} h^*(l)\exp^{(-j2\pi\frac{ml}{M})}\right)X(m) \\
&\quad + \left(\sum_{p=0}^{P_I-1}\sum_{l=0}^{N_T-1} g_p^*(l)\exp^{(-j2\pi\frac{ml}{M})}\,U_p(m)\right) + N(m) \qquad (8.2)\\
&= H(m)X(m) + I(m) + N(m), 0 \le m \le M-1 \qquad (8.3)\\
&= H(m)X(m) + N'(m), 0 \le m \le M-1 \qquad (8.4)
\end{aligned}
$$

where $H(m)$ is the complex frequency response of the channel at the subchannel m, $I(m)$ be the complex interference at that subchannel m and $N(0), \ldots, N(M-1)$ are the discrete Fourier transform (DFT) of $n(0), \ldots, n(M-1)$. From [24]-example 15.3 it is observed that DFT of i.i.d. Gaussian $n(0), \ldots, n(M-1)$ are i.i.d Gaussian as well. It is assumed that the interfering signals $U_p(0), \ldots, U_p(M-1)$ are also OFDM signals, with same block and cyclic pre-fix lengths, and they are block synchronous with the desired signal. The system is depicted in Fig. 8.5. Equation (8.4) shows that the received signal is the transmitted signal attenuated and phase shifted by the frequency response of the channel at the subchannel frequencies due to fading in the presence of interference and noise [6]. It is assumed to be that noise is represented as complex i.i.d. with vector $\mathbf{n} = [n(0), n(1), \ldots, n(M-1)]^T$ with each component of $\mathbf{n}$ distributed as $\mathbb{C}\mathcal{N}(\mu_i, \sigma_{ni}^2)$ and are also independent. The multivariate complex Gaussian pdf is just the product of the marginal pdfs or

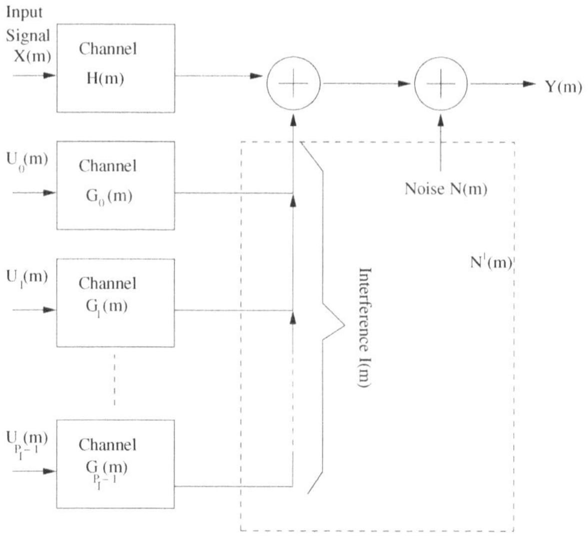

Fig. 8.5 OFDM communication system in presence of interference and noise

$$f(\mathbf{n}) = \prod_{i=0}^{M-1} f(n(i)) \tag{8.5}$$

which follows from the usual property of the pdf for real independent random variables, this can be written as

$$f(\mathbf{n}) = \frac{1}{\pi^M \prod_{i=0}^{M-1} \sigma_{ni}^2} \exp\left[-\sum_{i=0}^{M-1} \frac{1}{\sigma_{ni}^2} \mid n(i) \mid^2 \right] \tag{8.6}$$

Since the joint pdf depends on $\Re(n)$ and $\Im(n)$ (where $\Re$ and $\Im$ stands for real and imaginary components of a complex variable) only through $\mathbf{n}$, the joint pdf can be viewed to be that of the 'scalar random variable n'. This pdf Eq. (8.6) is called a 'complex Gaussian pdf' for a scalar complex random variable and is denoted by $\mathbb{C}$ $\mathcal{N}(0, \sigma_{ni}^2)$ [24] (without loss of generality noise is assumed to be zero mean).

8.3.2 Asynchronous Interferer

Asynchronous interference occurs when the cyclic prefix of the interferer does not align with the desired user's cyclic prefix. The receiver is usually synchronised with the desired user's transmitter. For an asynchronous interferer in a rich multipath environment, the received frequency-domain measurement is highly correlated. The frequency offset causes the loss of orthogonality and causes ICI in the interferer [37]. In many cases, the algorithms designed for synchronous interference fail when the asynchronous interference is encountered. The algorithm discussed later in this section does not require any change when asynchronous interference is encountered.

The interference model described in Eq. (8.4) is not valid for asynchronous interference. The interference in Eq. (8.6) is now modelled as:

$$I(m) = \sum_{p=0}^{P_I-1}\sum_{k=0}^{M-1} U_p(k)G_p(k)\frac{\sin(\pi(k-m+\triangle F))}{M\sin(\pi(k-m+\triangle F)/M)}\exp\left(\frac{j\pi(k-m+\triangle F)(M-1)}{M}\right) \tag{8.7}$$

where U_p, G_p represent the pth interferer's input signal and channel respectively. $\triangle F$ represents the (normalised) frequency offset [14, 23, 37], which is normalised by the subcarrier spacing which is roughly $\triangle F \approx f_{OFDM}/M$, where f_{OFDM} is the total occupied bandwidth by OFDM.

8.3.3 Kernel Density Estimation

Since both the noise and interference are complex, they can be modelled by a 'complex Gaussian mixture' pdf, where the real and imaginary components are assumed independent as discussed earlier. Parzen window or kernel density estimation assumes that the probability density is a smoothed version of the empirical sample. Its estimate $\hat{f}(y)$ of a complex random variable $y = \Re\{y\} + i\Im\{y\}$ is simply the average of radial kernel functions centered on the points in a sample M of the instance of y:

$$\hat{f}(y) = \frac{1}{M}\sum_{j=1}^{M} K(y - y(j)) \tag{8.8}$$

As in previous chapters, $K(.)$ is assumed to be Gaussian kernel (Parzen kernel) [28]:

$$K(y) = \mathcal{N}(0, \sigma^2) = \frac{1}{\sqrt{2\pi\sigma^2}}\exp\left(\frac{-\mid y\mid^2}{2\sigma^2}\right) \tag{8.9}$$

variance defined as σ^2. The joint pdf $\hat{f}(y)$ depends on the real and complex components through y, the pdf can be viewed as that of the scalar random variable y, as the notation suggest [24].

8.3.4 Non-parametric ML Channel Estimation

Performance of the OFDM system depends highly on the quality of the channel estimate, this makes the channel estimator one of the most useful part of a OFDM communication receiver. Due to increased bit-rate performance of the OFDM systems, there has been considerable interest in developing better channel estimators for various channel conditions. However, for Gaussian noise channels a simple technique based on LS (also known as zero-forcing estimator) [2] is usually used. The LS channel estimate is formed as,

$$H(m) = X(m)^{-1}Y(m), \quad 0 \leq m \leq M - 1 \tag{8.10}$$

for a training symbol, i.e. $X(m)$ is considered known at the receiver. There are various techniques for channel estimation in flat and Rayleigh fading channels ([12, 18, 26, 29] to name a few), however most of these techniques assume presence of Gaussian noise. The algorithm discussed in [22] can be used for channel estimation in the presence of interference, however it assumes interferers as Gaussian distributed. In this section, no assumption on the distribution of the interference is made, which makes this technique robust for both synchronous and asynchronous interference.

The typical channel impulse response components $\mathbf{h} = [h(0), \ldots, h(N_T - 1)]$ are independent complex-valued Gaussian random variables (which represents a frequency-selective Rayleigh fading channel). In a regular OFDM system, the channel delay spread N_T is much smaller than the number of subcarriers M, which leads to a high correlation among the channel frequency responses $H(m)$, even when $h(l)\forall l$, are independent [29]. Thus the channel impulse response $\mathbf{h} = [h(0), \ldots, h(N_T - 1)]$ is estimated directly as they are independently specified and the number of parameters in the time domain is smaller than that in the frequency domain [6].

The combined interference and AWGN $N'(m)$ in Eq. (8.4) is together considered as a noise that is non-Gaussian because of the presence of interference [3, 28], which can be considered equivalent to interference temperature [9, 35]. As also discussed in [3, 28] the LS estimator does not find the optimal solution in the case of Gaussian mixture noise. The mean square error criterion which minimises the energy between the desired and the system output does not converge to the optimal solution. Also, the LS estimator is not minimal sufficient statistics [39], and hence the ML estimator is used for this Gaussian mixture formulation.

As discussed in [24], if the noise is Gaussian then the LS estimate leads to the ML estimate. However, in communication systems where the interference temperature is high (which can be approximated as Gaussian mixture) then, no closed form ML solution exists for such Gaussian mixture distributions. Thus an iterative algorithm to find the ML estimate of the channel is proposed [4]. The proposed algorithm is initialised with the LS channel estimate based on the pilots. The first likelihood estimate is obtained by using the LS channel taps. After estimating this likelihood, the ML solution is sought iteratively exploiting the pilot symbols. The classical stochastic gradient algorithm is used with log-likelihood being the cost function i.e. the gradient here is the first derivative of the log-likelihood function with a constant multiplier (similar to the well known gradient ascent algorithm) [4]. The update equation is represented as:

$$\hat{\boldsymbol{h}}_k = \hat{\boldsymbol{h}}_{k-1} + \mu(k)\, \nabla_{\mathbf{h}}\, \mathcal{L}(\mathbf{h} \mid \mathbf{Y})|_{\mathbf{h}=\hat{\boldsymbol{h}}_{k-1}} \tag{8.11}$$

where $\mu(k)$ is the adaptation constant and $\nabla_{\mathbf{h}}$ represents the gradient of the cost function. Referring to eqs. (8.4) and (8.11) the likelihood function can be written as:

$$L(\mathbf{h} \mid \mathbf{Y})|_{\mathbf{h}=\hat{\boldsymbol{h}}_{k-1}} = f(\mathbf{Y} \mid \mathbf{h}) = \prod_{i=1}^{M} f_{N'}(E(i))$$

where $f_{N'}(.)$ is scalar pdf of 'complex Gaussian mixture' of data length from $i = 1, \ldots, M$ and the previous estimation error is defined as:

$$E(i) = Y(i) - \left(\sum_{l=0}^{N_T-1} h_k^*(l) \exp^{\left(-j2\pi \frac{il}{M}\right)} \right) X(i) \tag{8.12}$$

Kernel density estimators are known to be effective in estimating the pdf over short data record and also provide a differentiable smooth estimated pdf. From the definition of kernel density estimator [36]:

$$\hat{f}_{N'}(E) = \frac{1}{M} \sum_{j=1}^{M} K(E - E(j)) \tag{8.13}$$

where M is the number of subcarriers. The log-likelihood function can then be written as:

$$\begin{aligned} \hat{\mathcal{L}}(\mathbf{h} \mid \mathbf{Y})\Big|_{\mathbf{h}=\hat{\boldsymbol{h}}_{k-1}} &= \sum_{i=1}^{M} \log\left(f_{N'}(E(i))\right) && (8.14) \\ &= \sum_{i=1}^{M} \log \left(\frac{1}{M} \sum_{j=1}^{M} K(E(i) - E(j)) \right) \\ &= \sum_{i=1}^{M} \log \sum_{j=1}^{M} K(E(i) - E(j)) - \log M && (8.15) \end{aligned}$$

Maximising the log-likelihood function w.r.t the channel weight vector. By definition of complex vector differentiation [16],

$$\nabla_{\mathbf{h}} \hat{\mathcal{L}}(\mathbf{h} \mid \mathbf{Y})\Big|_{\mathbf{h}=\hat{h}_{k-1}} = \frac{\partial \hat{\mathcal{L}}(\mathbf{h} \mid \mathbf{Y})}{\partial \mathbf{h}}\Big|_{\mathbf{h}=\hat{h}_{k-1}} = \sum_{i=1}^{M} \frac{\sum_{j=1}^{M} \frac{\partial K(E(i)-E(j))}{\partial \mathbf{h}}}{\sum_{j=1}^{M} K(E(i)-E(j))}\Big|_{\mathbf{h}=\hat{h}_{k-1}} \tag{8.16}$$

Thereby substituting this gradient in Eq. (8.11) gives an iterative solution.

As with any stochastic gradient algorithm the choice of optimal $\mu(k)$ varies with the application and requirements. As discussed in [4], $\mu(k) = \frac{\sigma^2}{M}$ is chosen in Eq. (8.11) and convergence is witnessed in a few iterations. The σ is chosen as $\sigma = (1/M)^{(1/6)}\sigma_n$, as found analytically in [28] for estimating complex Gaussian distribution.

8.3.5 Non-parametric Symbol-by-Symbol MAP Equaliser

Once the frequency-domain channel response $\hat{H}$ is found, the estimate of the transmitted signal can be obtained by solving

$$\hat{X}(m) = \arg\min_{X \in C} \mid Y(m) - \hat{H}(m)X(m) \mid^2, \quad 0 \le m \le M-1 \tag{8.17}$$

which leads to the final estimate of the transmitted signals (assuming $X \in C$) as follows:

$$\hat{X}(m) = \text{Quantisation}\left\{\frac{Y(m)}{\hat{H}(m)}\right\}, \quad 0 \le m \le M-1 \tag{8.18}$$

The above formulation is equivalent to the zero-forcing equaliser and is optimal for Gaussian noise assumption. This forms a popular 'one-tap equaliser' used in many practical OFDM communication systems.

Similar to the channel estimator discussed before, the conventional detector (equaliser [6]) is not optimal for the interference affected channels. The performance of this zero-forcing equaliser [6] is highly sensitive to the quality of estimated channel and the ratio of interfering received signal with estimated channel, in addition to typical constraints of a zero-forcing equaliser.

$$\hat{X}(m) = \text{Quantisation}\left\{\frac{Y(m)}{\hat{H}(m)}\right\} = \text{Quantisation}\left\{\frac{X(m)H(m) + I(m) + N(m)}{\hat{H}(m)}\right\} \tag{8.19}$$

The impact of imperfect channel estimation in case of Gaussian noise is studied in details in [25]. The decision boundary for the said equaliser is clearly non-linear due to the presence of interference and the imperfect channel estimation. Thus the assumption that the decision boundary being linear (based on Gaussian assumption) is no longer valid. It can be observed from Fig. 8.6, that the noise in presence of

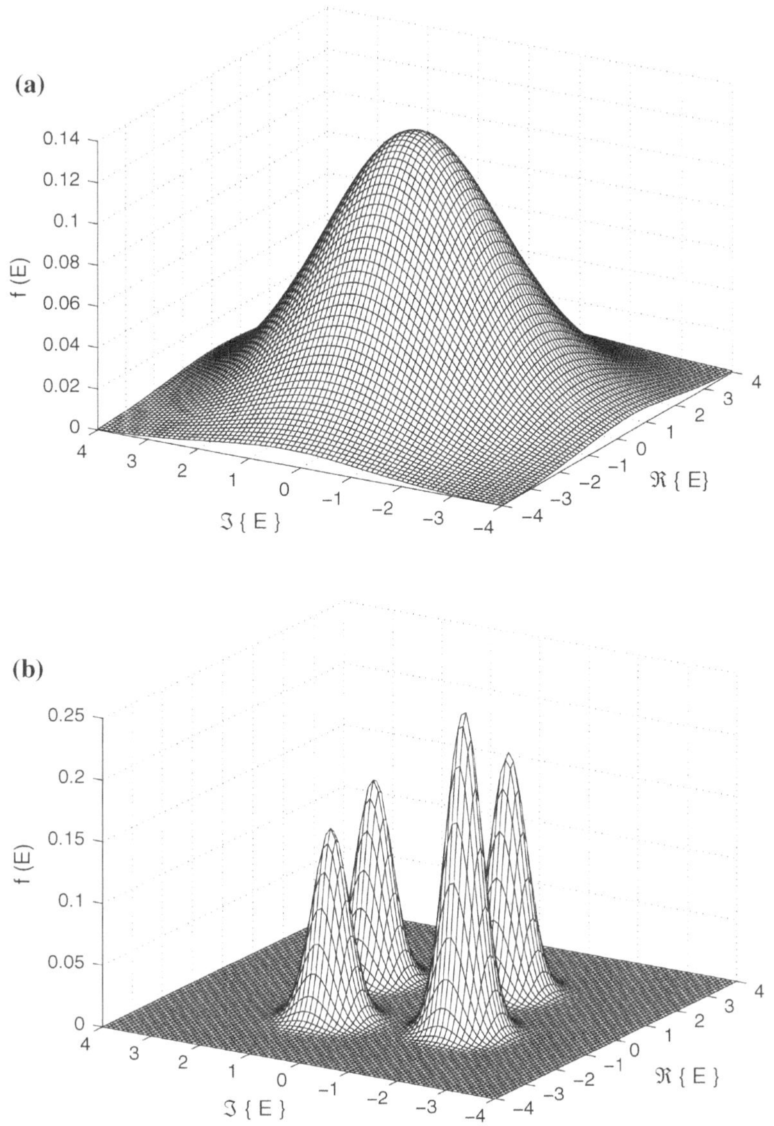

Fig. 8.6 Distribution plot

interference forms a Gaussian mixture. This makes the decision boundary non-linear and as can be observed from Eq. 8.19.

Here a probabilistic symbol-by-symbol MAP equaliser whose decision is based on the estimated likelihood is used. The endevour is to maximise the probability of correct decision for the received symbol i.e. maximise $F(X_c(m) \mid Y(m))$; $c = 1, \ldots, C$. The decision criterion is based on selecting the signal corresponding to the

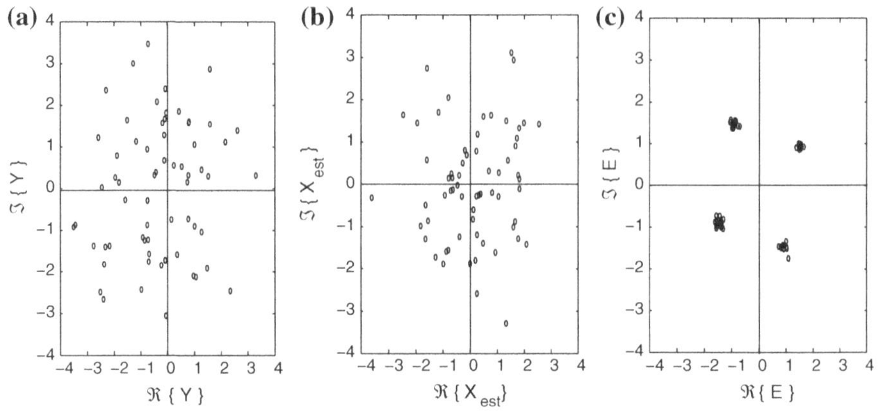

Fig. 8.7 Signals at various stages of receiver for synchronous multi-path fading channel in presence of synchronous interference **a** Received signal (Y) for training symbol **b** Estimated signal (X) for training symbol **c** Difference signal (E) for training symbol

maximum of the set of posteriori probabilities $\{F(X_c(m) \mid Y(m))\}$. This decision criterion is called the maximum *a posteriori* probability criterion. Using Bayes' rule, the posterior probability is expressed as:

$$F(X_c(m) \mid Y(m)) = \frac{f(Y(m) \mid X(m))F(X(m))}{f(Y(m))} \tag{8.20}$$

where $f(Y(m) \mid X(m))$ is the conditional pdf of the observed vector given $X_c(m)$, and $F(X_c(m))$ is the *a priori* probability of the cth signal being transmitted. The denominator of Eq. (8.20) may be expressed as

$$f(Y(m)) = \sum_{c=1}^{C} f(Y(m) \mid X_c(m))F(X_c(m)) \tag{8.21}$$

From Eqs. (8.20) and (8.21), it is observed that the computation of the posterior probabilities $F(X_c(m) \mid Y(m))$ requires knowledge of the *a priori* probabilities $F(X_c(m))$ and the conditional pdfs $f(Y(m) \mid X_c(m))$ for $c = 1, \ldots, C$.

Some simplification in the MAP criterion is possible when the C signals are equally probable *a priori*, i.e., $F(X_c(m)) = 1/C\ \forall c$. Furthermore, the denominator in Eq. (8.20) is independent of the transmitted symbol. Consequently the decision rule based on finding the signal that maximises $F(X_c(m) \mid Y(m))$ is equivalent to finding the signal that maximises $f(Y(m) \mid X_c(m))$ [34].

The conditional pdf $f(Y(m) \mid X_c(m))$ is called as the *likelihood function* as seen previously. Thus the decision criterion for the said problem, the MAP detection, reduces to detecting the ML symbol. For the estimated channel impulse response $\hat{\boldsymbol{h}}$ (after convergence) from Eq. (8.11) and taking its Fourier transform, the ML estimate of the transmitted signal can be obtained by

$$\hat{X}(m) = \arg \max_{X_c \in C} (\hat{f}_E(Y(m) \mid \hat{H}(m)))\Big|_{\mathbf{h}=\hat{\boldsymbol{h}}_k} \tag{8.22}$$

It should be noted that the estimated pdf $\hat{f}_E$ for detection is generated by using Eq. (8.13). Based on the higher probability of occurrence the hard-decision (or Quantisation) is taken on $\hat{X}(0), \ldots, \hat{X}(M-1)$ to generate the output data as shown in Fig. 8.3. From simulation results it is observed that significant BER improvement is achieved by using this probabilistic equaliser.

8.3.6 Simulation Results

A generalised packet based generalised OFDM communication system with pilots forming the first symbol known at the receiver is assumed. The channel estimation is done on this first OFDM symbol, while the remaining payload is the useful information. A generalised OFDM communication system is considered as shown in Fig. 8.3. The raw binary input data is fed to the rate 1/2 encoder. The encoder used is specified in [32], where a original 1/3 encoder is used and then puncturing is performed to form a 1/2 rate coder. The encoded serial bit stream is then converted to parallel by S/P. The parallel data are then mapped (modulated) to QPSK signal constellation, where it is then passed to IFFT processor. The output of IFFT $x(0), \ldots, x(M-1)$ is then converted into serial stream by P/S and after appending guard interval (to avoid ISI), the serial stream is then transmitted over a fading channel. It is assumed throughout the chapter that the guard interval is long enough to avoid any ISI and a cyclic prefix is introduced in the guard interval to avoid any ICI. At the receiver, the reverse process as that at the transmitter takes place. First the guard bits are removed followed by serial to parallel converter. Then fast Fourier transform (FFT) is performed on $y(0), \ldots, y(M-1)$ to obtain $Y(0), \ldots, Y(M-1)$, which is then demodulated by QPSK demodulator. The demodulated bits are then converted to serial stream and then decoded using the Viterbi decoder [13]. Input and output serial data streams are then compared to calculate the bit error rate.

The robustness of the above proposed adaptive algorithm is verified for two special case of fading channels; flat-fading and multi-path fading. In multi-path fading two special cases of synchronous and asynchronous interference are considered. The channel is considered to be slowly fading, i.e. the channel remains static (generally defined as quasi-static) for each OFDM packet (of size 64-subcarriers and 8-symbols). Using the formula discussed in [6], a bit rate of 1-Mbps at 1-GHz with vehicle speed of 6-mph (which is sufficient for indoor environment) can be achieved. To verify the robustness of the algorithm, simulations were carried out on Matlab for ensemble of 1000-runs. The performance measure is average BER for fixed SNR and for various values of SIR. The SIR is defined as (Eq. 8.3):

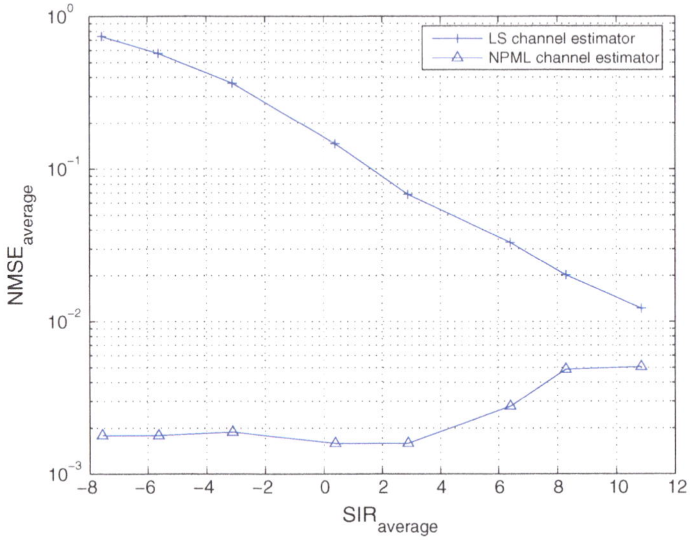

Fig. 8.8 Average NMSE performance in flat fading channel

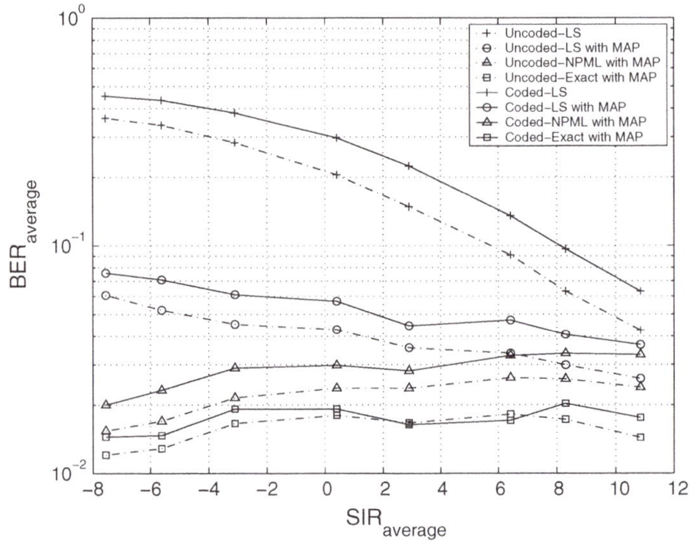

Fig. 8.9 Average BER performance in flat fading channel

$$SIR = \frac{\mathbb{E}\{(HX)(HX)^*\}}{\mathbb{E}\{II^*\}} \tag{8.23}$$

8.3.6.1 Flat Fading

The performance evaluation for the proposed estimator and equaliser begins by using a flat-fading channel. Figures 8.8 and 8.9 shows the performance for fixed SNR = 16.5dBs and SIR varied for range of values. The channel model is chosen as:

$$H(z) = \alpha_0 \exp(j\theta_0) \tag{8.24}$$

whereas the interfering channel is chosen as:

$$G_0(z) = \alpha_1 \exp(j\theta_1) \tag{8.25}$$

where α_0, α_1 are the i.i.d random variables with Rayleigh distribution, and θ_0, θ_1 are i.i.d. random variables with uniform distribution.

The legend, in Fig. 8.8, 'LS channel estimator' and 'NPML channel estimator' represent when the flat fading channel is estimated at the receiver by Eq. (8.10) and Eq. (8.11) respectively. It is seen that considerable improvement in terms of NMSE is achieved by using the proposed NMPL channel estimation algorithm. The legend, in Fig. 8.9, 'Uncoded-LS' refers to the LS estimator for channel estimation and Gaussian assumption MAP equaliser i.e. Eq. (8.18) is used for detection of symbols following the training (pilot) sequence. 'Uncoded-LS with MAP' refers to the scenario when the channel is estimated by LS estimator followed by a non-parametric equaliser (using this estimated channel and residual error) from Eq. (8.22). The 'Uncoded-NPML with MAP' refers to the NPML algorithm for channel estimation followed by non-parametric MAP equaliser. The 'Uncoded-Exact with MAP' represents when the exact channel is known at the receiver followed by a non-parametric MAP equaliser. The prefix 'Uncoded' is used when raw binary input is transmitted, whereas prefix 'Coded' is used when the raw binary input data stream is coded by 1/2-rate encoder. It is observed that the algorithm with NPML as channel estimator and non-parameteric symbol-by-symbol MAP equaliser gives the best performance, except in hypothetical case when receiver has *a priori* channel information. Comparing 'LS with MAP' and 'NPML with MAP' for both un-coded and coded input data highlights the performance gain achieved by using NPML channel estimator as compared to the LS estimator (as used in [27]). It is interesting to observe that gains achieved by using NPML decreases when the interference power decreases i.e. at high SIR. It is also observed that the coding for flat-fading doesn't improve the BER performance for the said simulation set-up, which can be attributed to the fact that the channel decoder is based on Gaussian noise assumption. Thus using NPML in conjunction with non-parametric MAP equaliser forms a robust communication receiver for interference limited channels.

8.3.6.2 Multipath Fading

The channel is defined as a two-path Rayleigh fading channel with transfer function [6]:

$$H(z) = 0.8\alpha_0 \exp(j\theta_0) + 0.6\alpha_1 \exp(j\theta_1)z^{-1} \tag{8.26}$$

The interfering channel is defined as:

$$G_0(z) = 0.5\alpha_2 \exp(j\theta_2) + 0.1\alpha_3 \exp(j\theta_3)z^{-1} \tag{8.27}$$

where $\alpha_0, \alpha_1, \alpha_2, \alpha_3$ are the i.i.d random variables with Rayleigh distribution, and $\theta_0, \theta_1, \theta_2, \theta_3$ are i.i.d. random variables with uniform distribution.

8.3.6.3 Synchronous Interference

The synchronous interference model as represented by Eq. (8.2) is considered for these set of simulations. The average NMSE plot is shown in Fig. 8.10 and the average BER plot is shown in Fig. 8.11. The legends define the similar techniques as discussed for flat-fading channel. The SNR is kept fixed at 17.63 dBs while SIR is varied over a large range. The significant BER improvement is obtained by using the non-parametric symbol-by-symbol MAP equaliser as compared to other techniques. By using the NPML channel estimator the performance gain of about 8dBs can be obtained by using NPML channel estimator as compared to LS estimator with non-parametric MAP equaliser estimating density on residual error from both the estimator respectively. It is also worth noting that there is a little difference when this equaliser is used with NPML estimator and exact channel knowledge, this also confirms that the NPML estimates are closer to the exact channel. As observed from the simulation results, the BER gain for NPML based channel estimator and equaliser amplifies for the coded bit stream as compared to the uncoded one. The received, estimated and interfering signal are illustrated in Fig. 8.7 for the said simulation environment. It is also interesting to observe that the performance plots follow the pattern as noted in [28].

8.3.6.4 Asynchronous Interference

Figures 8.12 and 8.13 shows the results for the asynchronous interference model as represented by Eq. (8.7). The frequency offset is set as $\triangle F = 0.1$ for the results. The SNR is kept fixed at 17.63 dBs while SIR is varied over a large range. There is a huge performance gain by using proposed techniques as compared to the traditional approach of doing channel estimation and equalisation based on Gaussian assumptions. As seen earlier, the NPML estimator with non-parametric symbol-by-symbol MAP equaliser enhances performance instead of using LS estimator and

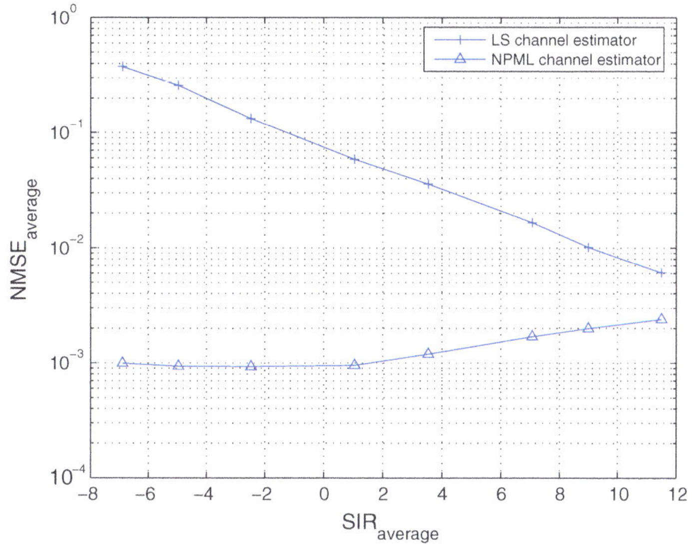

Fig. 8.10 Average NMSE performance in multipath frequency-selective fading channel with synchronous interference

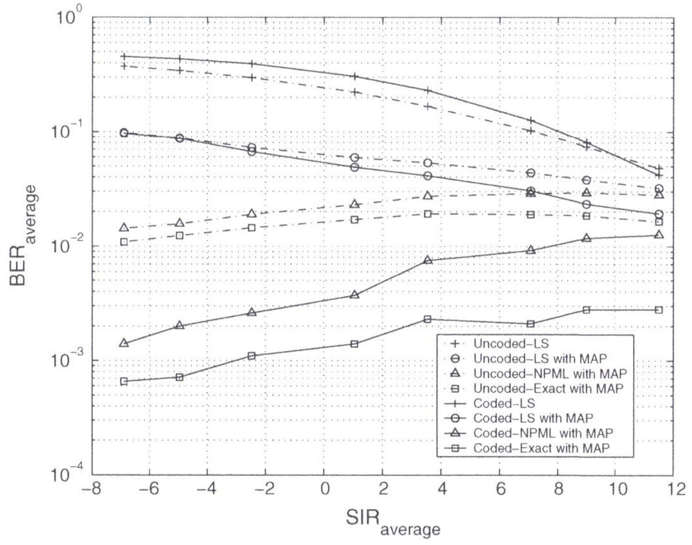

Fig. 8.11 Average BER performance in multipath frequency-selective fading channel with synchronous interference

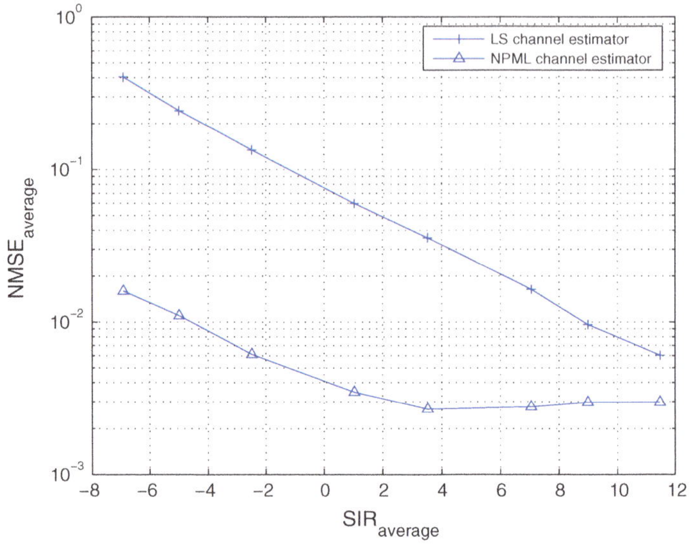

Fig. 8.12 Average NMSE performance in multipath frequency-selective fading channel with asynchronous interference, where $\triangle F = 0.1$

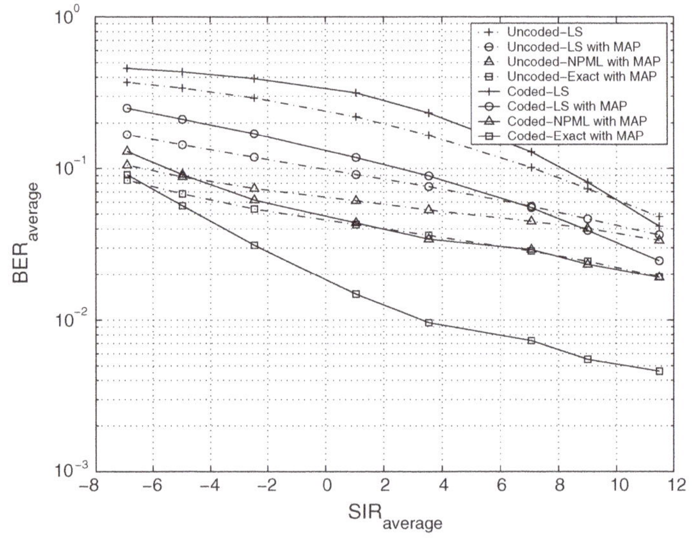

Fig. 8.13 Average BER performance in multipath frequency-selective fading channel with asynchronous interference, where $\triangle F = 0.1$

its residual error pdf for MAP equalisation. The receiver's *a priori* knowledge of the transmission channel forms a lower bound on the performance. It is observed that when SIR increases the Gaussian assumption detector performs better than non-parametric MAP equaliser (except when exact channel is known at the receiver), this is because at high SIR the N' in Eq. (8.4) can be approximated as Gaussian distributed. The asynchronous interference being more correlated than synchronous interference, at low SIR, a considerable difference in performance of 'Coded-NPML with MAP' is observed in both Figs. 8.11 and 8.13. However, from simulation results it is safe to conclude that the proposed algorithm and technique could be used for both synchronous and asynchronous interferer without any modification.

8.4 Conclusion

In this chapter, an overview of the challenges in the TVWS communication systems was provided. It was observed, that the effect of interference in the TVWS has been considered while formulating the standard. An overview and results for a generalised solution to improve performance of a TVWS communication systems, in the presence of high interference, was shown. It can be concluded that the it is important to develop algorithms and techniques for mitigating the effects of interference temperature in the TVWS communication systems.

Acknowledgments I wish to thank Prof. Bernard Mulgrew (The University of Edinburgh), Prof. David D. Falconer (Carleton University), IET Hudswell Bequest Fellowship and Dr. S. Nagappa for completion of this work. I would also like to thank The University of Edinburgh and Indian Institute of Technology Indore to facilitate this research and, my parents and family for support during writing this chapter.

References

1. IEEE standard for information technology-telecommunications and information exchange between systems wireless regional area networks(wran)-specific requirements part 22: Cognitive wireless ran medium access control (mac) and physical layer (phy) specifications: Policies and procedures for operation in the tv bands. IEEE Std 802.22-2011 pp. 1–680
2. van de Beek, J.J., Edfors, O., Sandell, M.: On channel estimation in OFDM systems. In: VTC 1995. vol. 2, pp. 815–819 (1995)
3. Bhatia, V., Mulgrew, B.: A EM-based channel estimator in non-Gaussian noise. In: VTC-2004 (Fall), LA, US (2004)
4. Bhatia, V., Mulgrew, B.: Non-parametric maximum likelihood channel estimator in the presence of non-Gaussian noise. Sig. Process **87**(11), 2569–2586 (2007)
5. Bradaric, I., Petropulu, A.P.: Performance of training-based OFDM systems in the presence of time varying frequency-selective channels. In: Proc. IEEE ICASSP-2004 (2004)
6. Chen, P., Kobayashi, H.: Maximum likelihood channel estimation and signal detection for OFDM systems. IEEE Int. Conf. Commun (ICC). **3**, 1640–1645 (2002)
7. Chiasserini, C.F., Rao, R.R.: Performance of IEEE 802.11 WLANs in a bluetooth environment. In: Proc. IEEE WCNC-2000. vol. 1, pp. 94–99 (2000)

8. Clancy, T.: Achievable capacity under the interference temperature model (2007)
9. Clancy, T., Arbaugh, W.: Measuring interference temperature. Virginia Tech,Symposium on Wireless Personal Communications(2006)
10. Commission, F.C.: Spectrum policy task force. Rep. ET Docket (02–135) (2002)
11. Consortium, C.W.S.: Cambridge TV white space trial—a summary of the technical findings. Cambridge White Space Consortium (2012)
12. Edfors, O., Sandell, M., van de Beek, J.J., Wilson, S., Borgesson, P.: OFDM channel estimation by singular value decomposition. IEEE Trans. Commun. **46**(7), 931–939 (1998)
13. Fleming, C.: A tutorial on convolutional coding with viterbi decoding (2003)[Online] Available:http://home.netcom.com/chip.f/viterbi/tutorial.html
14. Harada, H., Prasad, R. (eds.): Simulation and Software Radio for Mobile Communications. Artech House, London (2002)
15. Haykin, S.: Communication Systems, 4th edn. Wiley, New York (2001)
16. Haykin, S.: Adaptive Filter Theory, 4th edn. Pearson Education Asia, New York (2002)
17. Haykin, S., Thomson, D., Reed, J.: Spectrum sensing for cognitive radio. Proc. IEEE **97**(5), 849–877 (2009)
18. He, P., Schmidt, K.U., Ho, C.K., Sun, S.: Iterative channel estimator and equalizer for OFDM modulation systems. In: Proceedings IEEE VTC-2003 (Spring). vol. 2, pp. 1313–1317 (2003)
19. Hunziker, T., Hashiguchi, M., Ohira, T.: Reception of coded OFDM signals in broad-band fading environments with strong co-channel interference. In: Proceeding GLOBECOM'03. vol. 4, pp. 2310–2314 (2003)
20. Hwang, S.H., Um, J.S., Song, M.S., Kim, C.J., Park, H.R., Kim, Y.H.: Design and verification of ieee 802.22 wran physical layer. In: Cognitive Radio Oriented Wireless Networks and Communications, 2008. CrownCom 2008. 3rd International Conference on. pp. 1–6. IEEE (2008)
21. Jeremic, A., Thomas, T.A., Nehorai, A.: OFDM channel estimation in presence of asynchronous interference. Proc. ICASSP. **4**, 684–687 (2003)
22. Jeremic, A., Thomas, T.A., Nehorai, A.: OFDM channel estimation in presence of interference. IEEE Trans. Signal Process **52**(12), 3429–3439 (2004)
23. Jie, Z., Yun, Z., Ling, S., Ping, Z.: Effects of frequency-offset on the performance of OFDM systems. Proc. ICCT. **2**, 1029–1032 (2003)
24. kay, S.M.: Fundamentals of Statistical Signal Processing:Estimation Theory. Prentice Hall International, Lebanon (1993)
25. Leke, A., Cioffi, J.M.: Impact of imperfect channel knowledge on the performance of multicarrier systems. In: Proc. GLOBECOM 98. vol. 2, pp. 951–955 (1998)
26. Li, Y.: Pilot-symbol-aided channel estimation for OFDM in wireless systems. IEEE Trans. Veh. Technol. **49**(4), 1207–1215 (2000)
27. Luschi, C. (ed.): Probabilistic Techniques for Equalization of the Mobile Radio Channel in the Presence of Co-Channel Interference. PhD Thesis, Department of Electronics and Electrical Engineering, The University of Edinburgh (2002)
28. Luschi, C., Mulgrew, B.: Non-parametric trellis equalization in the presence of non-Gaussian interference. IEEE Trans. Commun **51**(2), 229–239 (2003)
29. Ma, X., Kobayashi, H., Schwartz, S.C.: EM-based channel estimation algorithms for OFDM. Eurasip J. Appl. Signal Process **2004**(10), 1460–1477 (2004)
30. MacDonald, J.T., Ucci, D.R.: Interference temperature limits of ieee 802.11 protocol radio channels (2007)
31. Mahmoud, H., Yucek, T., Arslan, H.: Ofdm for cognitive radio: merits and challenges. Wireless Commun., IEEE **16**(2), 6–15 (2009)
32. Multiband OFDM alliance SIG: Multiband OFDM physical layer proposal for IEEE 802.15 task group 3a ((2004, Sep) [Online] Available:http://www.multibandofdm.org/papers/MultiBand_OFDM_Physical_Layer_Proposal_for_IEEE_802153a_Sept_04.pdf)
33. Park, J., Kim, D., Kang, C., Hong, D.: Effect of Bluetooth interference on OFDM-based WLAN. Proceedings VTC (Fall) **2**, 786–789 (Oct 2003)
34. Proakis, J.G.: Digital Communications. McGraw-Hill, New York (1989)

35. S. J. Shellhammer, A.K.S., Zhang, W.: Technical challenges for cognitive radio in the white space spectrum. UCSD ITA Workshop (2009)
36. Schraudolph, N.N.: Gradient-based manipulation of non-parametric entropy estimates. IEEE Trans. Neural Networks 15(4) (Jul 2004)
37. Speth, M., Fechtel, S.A., Fock, G., Meyr, H.: Optimum receiver design for wireless broad-band systems using OFDM-part I. IEEE Trans. Commun **47**(11), 1668–1677 (1999)
38. Stevenson, C., Chouinard, G., Lei, Z., Hu, W., Shellhammer, S., Caldwell, W.: Ieee 802.22: The first cognitive radio wireless regional area network standard. Commun. Mag., IEEE 47(1), 130–138 (2009)
39. Bhatia, V.:B.M., Falconer, D.: Non-parametric maximum likelihood channel estimator and equalizer for OFDM in high interference. Proc. of. IET Commun. **1**(4), 647–654 (Aug 2007)

Chapter 9
VillageLink: A Channel Allocation Technique for Wide-Area White Space Networks

Veljko Pejovic, David Lloyd Johnson, Mariya Zheleva, Elizabeth M. Belding and Albert Lysko

This chapter is an extension of V. Pejovic, D. L. Johnson, M. Zheleva, E. M. Belding and A. Lysko, VillageLink: Wide-Area Wireless Coverage, COMSNETS'14, Bangalore, India, January 2014.

Abstract White spaces promise to revolutionize the way wireless connectivity is delivered over wide areas. However, large-scale white space networks face the problem of allocating channels to multiple contending users in the wide white space band. To tackle the issue, we first examine wireless propagation in a long-distance outdoor white space testbed and find that a complex combination of free-space loss and antenna effects impacts transmission in white spaces. Thus, a need arises for a strategy that goes beyond simple channel utilization balancing, and uses frequency probing to profile channels according to their propagation properties. We devise VillageLink, a Gibbs sampling-based method that optimizes channel allocation in a distributed manner with a minimum number of channel switching events. Through extensive simulations we demonstrate that VillageLink results in a significant capacity improvement over alternative solutions.

V. Pejovic
School of Computer Science, University of Birmingham, Birmingham, UK

D.L. Johnson (✉) · A. Lysko
Meraka Institute, CSIR, Pretoria, South Africa
e-mail: djohnson@csir.co.za

M. Zheleva · E.M. Belding
Department of Computer Science, University of California, Santa Barbara, CA, USA

A.K. Mishra and D.L. Johnson (eds.), *White Space Communication*,
Signals and Communication Technology, DOI 10.1007/978-3-319-08747-4_9

9.1 Introduction

Internet connectivity is available to merely 39 % of the world's population [7]. Lack of internet access is primarily restricted to developing regions, however, many remote communities in the developed world are currently disconnected as well [22]. The main cause of limited Internet penetration stems from the fact that more than three billion people live in rural areas. These areas are hard to connect via copper cables, fiber optic or cell phone base stations due to high deployment cost and low population density which renders these techniques economically infeasible. Rural areas are also hard to reach via cheap license-free solutions such as WiFi, as these technologies, operating in 2.4 or 5 GHz bands, have a very limited connectivity range.

In the 50–800 MHz band, a large block of frequencies has recently been freed due to the analog to digital TV transition. This spectrum, called white spaces, promises to deliver an affordable means of providing wide area coverage. It is extremely attractive for rural areas as the propagation range is an order of magnitude higher than in the bands used by competing technologies. However, the complexity of signal propagation over the wide white space band, and the economic necessity of a resource-efficient, unlicensed, distributed solution for rural areas renders wide area white space networks uniquely challenging to realize.

In a white space network a number of frequency selective effects will be present due to the topology, vegetation and antenna design. Because of the highly complex nature of electromagnetic propagation and the difficulty in accurately modeling fading, the analytical solution that provides a clear picture of frequency quality in white spaces is not practical. Frequency profile, however, is extremely important as it can be used as a basis for channel assignment in a white space network. However, even when frequency propagation information is available, channel allocation in a white space network is very hard. In a network where the span of available channels is not large, such as in WiFi networks, channel allocation can be cast to a graph coloring problem. In white spaces, the wide range of available channels leads to drastic differences in propagation among channels. These differences stem from the variation of free space propagation over frequencies, but also from antenna properties, as in practice antennas do not perform uniformly over a very wide span of white space frequencies. Finally, frequency assignment in such a wide band network has to satisfy conflicting goals: maximize useful transmission by allocating frequencies with superior propagation properties and minimize interference by allocating frequencies that propagate over a shorter radius.

In this chapter we successfully address the above challenges by designing a lightweight frequency profiling methodology to evaluate channel quality and a novel channel allocation method that assigns operating frequencies to base stations with the goal of minimizing the impact of interference over the useful signal levels in a network. We compile these contributions into a practical channel profiling and allocation scheme for wide area white space networks called VillageLink. We test VillageLink's frequency probing mechanism on a long-distance software-defined radio white space link we deployed and confirm that antenna effects and the environment are a significant reason for high propagation diversity among white space channels. Through simulations we evaluate VillageLink's channel allocation. We show that our frequency-aware channel allocation leads to up to twice as much network capacity than an alternative heuristic based on interference avoidance, and that with its high performance, efficient resource usage and distributed nature, VillageLink represents a practical solution for wide area white space coverage in rural areas.

9.2 Wide-Area White Space Networks

White spaces represent a historic opportunity to revolutionize wide area wireless networking. White spaces not only deliver much greater communication range than Gigahertz frequencies, they also support non-line of sight communication, including transmission through vegetation and small obstacles, which makes them highly suitable for various terrain configurations. However, white space networks have to deal with unique peculiarities of transmission over a wide band of relatively low frequencies, and should enable license-free unplanned deployments in rural developing regions.

9.2.1 TV Spectrum Availability

White spaces spectrum can also be used by primary users, typically television and wireless microphones. In order to understand the amount of available white space spectrum, we conducted a number of spectrum scans in rural and urban regions in South Africa, Zambia and the USA.

South Africa has five TV channels utilizing VHF and UHF TV bands; all channels are available in urban centres and a portion of these channels available in rural areas. Zambia only has two national TV programs with very limited coverage. The USA, in contrast, has anywhere between 10 and 25 terrestrial TV channels available depending on location. We carried out spectrum scans in rural and urban areas in the USA and Southern Africa using an Ettus USRP2,[1] 8dBi log-period antenna, and a laptop. More details on the measurement setup can be found in [12]. Figure 9.1 shows the spectrum scan of the lower portion of the UHF band used for television broadcasting. Urban areas of the USA have very limited white space available. Rural areas of the USA are comparable to urban areas of South Africa. Rural areas of Zambia and South Africa have an abundance of white space and are thus very well suited for rural connectivity solutions.

[1] www.ettus.com.

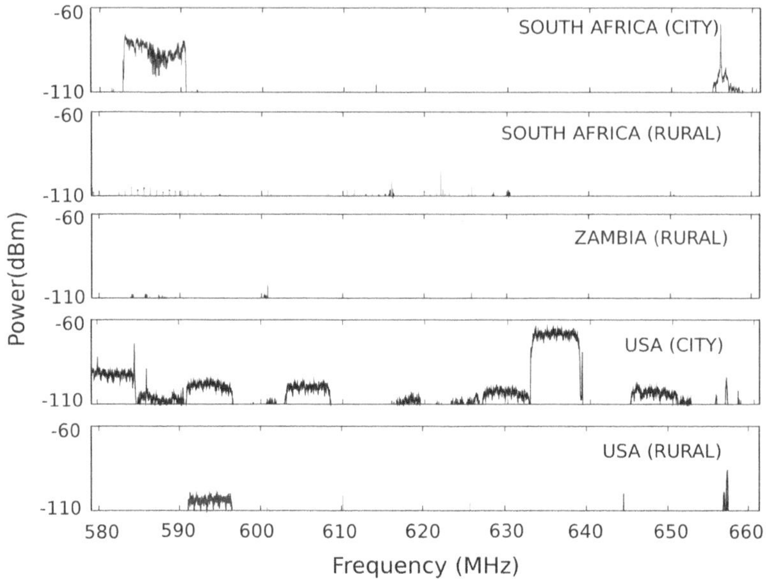

Fig. 9.1 Analysis of available spectrum in the lower UHF band for urban and rural areas in South Africa, Zambia and the USA

9.2.2 Wide Band Frequency Selectivity

One of the unique challenges of establishing wireless communication in the TV bands is the large amount of frequency selectivity due to variation in gain and free space loss across the operating band. This frequency selectivity complicates the choice of the optimal communication channel.

The large variation in free-space loss across the band also known as "dynamic range". Dynamic range is calculated as follows:

$$D_{dB} = 20 \log (f_U / f_L) \tag{9.1}$$

where f_L and f_U are the lowest and the highest frequency in the band, respectively. In Table 9.1 we summarize the dynamic range of a number of traditional wireless systems. Free-space loss in a traditional wireless network, such as WiFi or GSM, is relatively uniform over the range of frequencies these networks operate on. The reason for low dynamic range in these networks lies in the fact that they either operate over a relatively narrow band of frequencies, such as 50 MHz for GSM and 80 MHz for 2.4 GHz WiFi, or they operate on high central frequencies where the difference between the lowest and the highest frequency diminishes, as is the case with 5 GHz WiFi. White spaces, however, operate on a wide band of low frequencies, and the difference in propagation between white space frequencies can be large. Creating an antenna which has a flat frequency response across the entire white space band is,

Table 9.1 Dynamic range and fractional bandwidth of different wireless systems

Technology	f_L (MHz)	f_U (MHz)	D (dB)	FB (%)
802.11 (2.4 GHz)	2,412	2,484	0.26	2.9
802.11 (5 GHz)	5,170	5,700	0.85	9.8
GSM900	935	960	0.23	2.6
White spaces	43.25	797.25	25.31	179

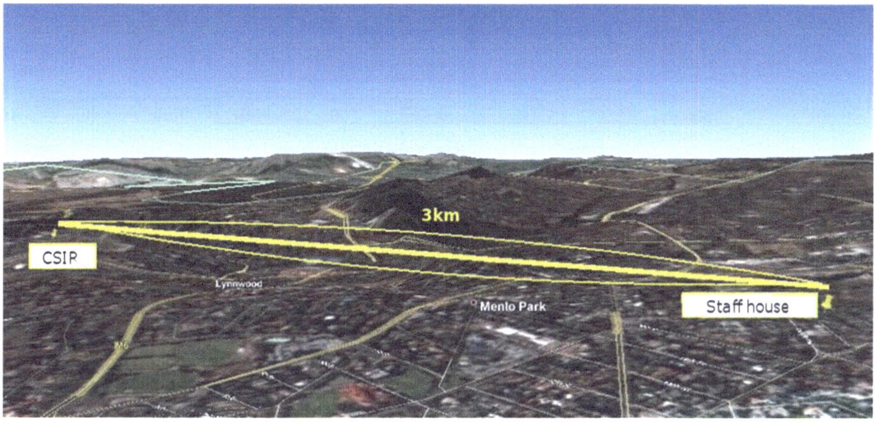

Fig. 9.2 White space 3 km experimental link in sub-urban area of Pretoria, South Africa between CSIR offices and a staff house

however, challenging. Note that the same issue does not arise in GSM (as well as 3G and 4G/LTE) networks, that can also operate on a wide range of frequencies (e.g. GSM850, GSM900, GSM1800). Unlike with white spaces, in these networks once the band selection is done, the operation is restricted to a single relatively narrow range of channels.

Besides the wide dynamic range, white space links experience uneven fading due to antenna patterns. The fractional bandwidth (FB) for a frequency band, calculated as a ratio of operating bandwidth and the central frequency, determines how wide-band an antenna should be in order to have the same gain over all frequencies with the band. From Table 9.1 we see that white spaces require significantly wider band antennas than GSM and WiFi. Such antennas are hard, if not impossible, to design. Consequently, white space links are highly prone to the effects of imperfect antennas.

To confirm this statement, we deployed a 3 km outdoor non-line-of-sight white space link in a sub-urban area in South Africa shown in Fig. 9.2. The terrain consists of undulating hills and dense foliage. Each of the link nodes consists of a USRP2[2] software defined radio and a dual core 2.4 GHz Pentium PC running GNUradio software. Each radio is equipped with a 8 dBi UHF yagi antenna. One node acts as

[2] http://gnuradio.org/.

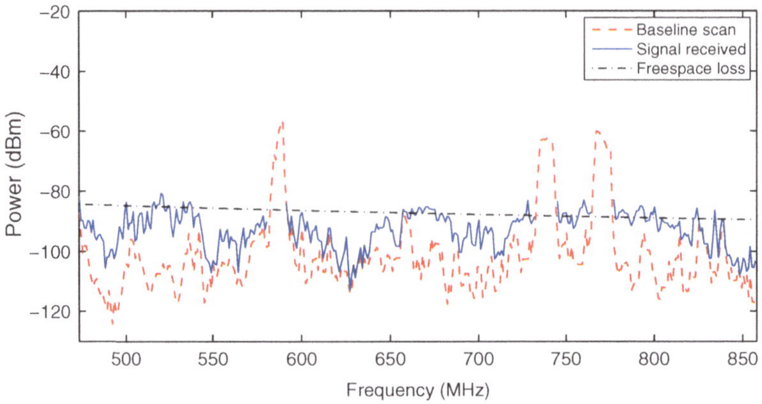

Fig. 9.3 Analysis of received signal strength over the UHF band using 8 dBi yagi antennas at transmitter and receiver. The plots demonstrate that received signal strength is far more dependent on the antenna gain pattern than on attenuation due to free-space loss. This is confirmed by antenna frequency profiles of a number of TV antenna models in Fig. 9.4

a transmitter and sends probes spaced 1 MHz apart over the white space spectrum. The transmitter radio signal is passed through a 1 W power amplifier. Another node, the receiver, scans the spectrum in 1 MHz steps both with the transmitter turned off (baseline scan) and with the transmitter sending probes (signal scan).

Figure 9.3 shows the received signal strength across the UHF TV band in South Africa. Three TV stations were detected and probes did not occur at these frequencies. We note that the received signal strength does not fall of monotonically with increasing frequency, which would be the case if only free-space loss determined the propagation loss. Instead, due to the antenna characteristics the propagation loss is non-uniform across the UHF band. In order to study the antenna characteristics, we use the WIPL-D antenna modelling package and create a model of the deployed antenna. The results are shown in Fig. 9.4. While an antenna with no surrounding structures has a more predictable gain patterns, when surrounding structures and antenna imperfections, such as bent elements, are introduced, the antenna gain pattern has far less predictability similar to what was seen in our received signal strength measurements. Predicting the type of TV antenna being used or the structures surrounding the antenna is not possible, which necessitates frequency probing in white spaces.

In addition to antenna effects, a part of frequency selectivity may stem from the environment and terrain effects. *Shadowing*, i.e. slow fading due to physical obstacles on the signal path would still be detected and accounted for with frequency probing to detect average channel gain described in Sect. 9.4. Unlike shadowing, detecting the level of *multipath*, which leads to rapid variation of propagation within a channel, requires a more sophisticated channel sounding process but can use the same frequency probes used to detect average channel gain. However, channel allocation

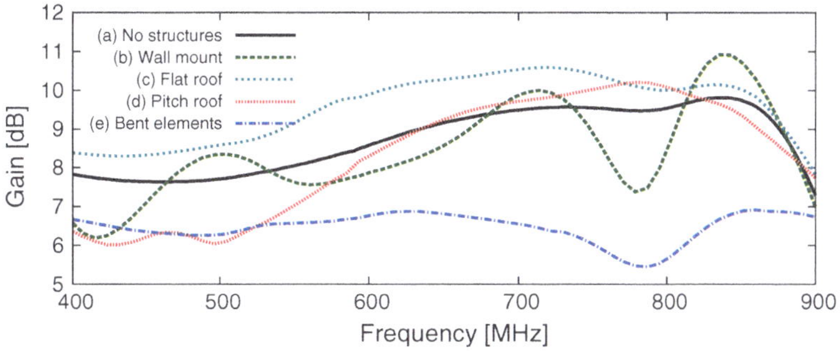

Fig. 9.4 Antenna gain for yagi antenna used in white space link in the outdoor testbed. The plot shows the following scenarios: (*a*) Antenna with no nearby structures, (*b*) Antenna mounted on the side of a wall, (*c*) Antenna mounted on a flat roof, (*d*) Antenna mounted on a pitch roof, and (*e*) Antenna with imperfections due to bent elements.

only requires knowledge of the average channel gain of the channel, and in Sect. 9.3 we devise a lightweight channel probing mechanism.

9.2.3 Channel Assignment in White Space Networks

The problem of channel assignment in wireless networks is often expressed with graph coloring, where each color represents a different channel. For a link, one of the available central frequencies is assigned so that a goal, such as maximum throughput, is achieved. In the channel allocation literature on traditional wireless networks all colors are considered equal in terms of their propagation properties [20]. In white spaces, due to the wide dynamic range and antenna effects, the transmission range varies significantly among frequencies in the band (see Table 9.1 and Fig. 9.3). Therefore, *selection of the operating frequency can impact the existence of a link itself*. This further complicates the problem of graph colouring, as now not all colours are equal. Figure 9.5 show one such example where a tradeoff between establishing links and avoiding interference is hard to achieve. In a white space network the color affects the graph structure, thus the existing approaches to frequency assignment are not directly applicable.

9.2.4 Network Architecture

The network scenario that describes the setting in which VillageLink will operate is given in Fig. 9.6. In this paper, we consider wide-area white space networks that consist of individual base stations (BSs), each with a set of associated customer-premises equipment (CPE) clients. We term one such BS with its CPEs *a cell*. A BS

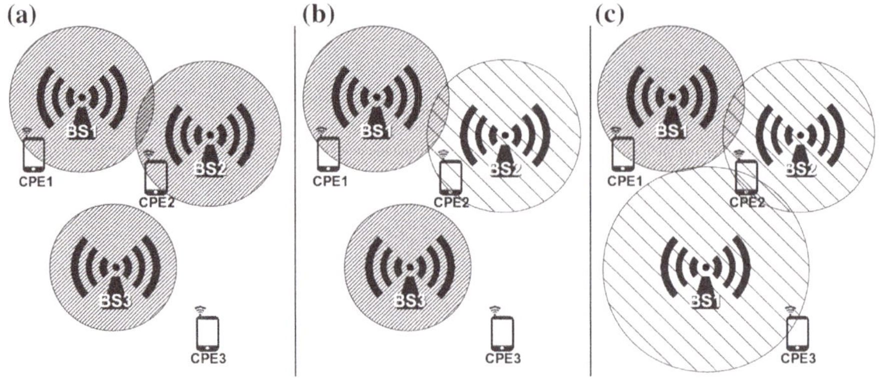

Fig. 9.5 A simple example of the challenges of frequency assignment in white spaces. We want to establish connections from base station (BS_1) to client premises equipment (CPE_1), BS_2 to CPE_2, and BS_3 to CPE_3. Two channels represented by different hatching patterns are given. In **a** all links operate on the one of the channels, however there is interference between BS_1 and BS_2, which may affect clients in the interference area; in addition, due to the channels poor propagation properties BS_3 signal is not reaching CPE_3. In **b** the interference is resolved by switching the frequency for BS_2. In **c** a link between CPE_3 and BS_3 is establishment through assignment of a different frequency to BS_3. However, this frequency propagates further, introducing interference at CPE_2 served by BS_2

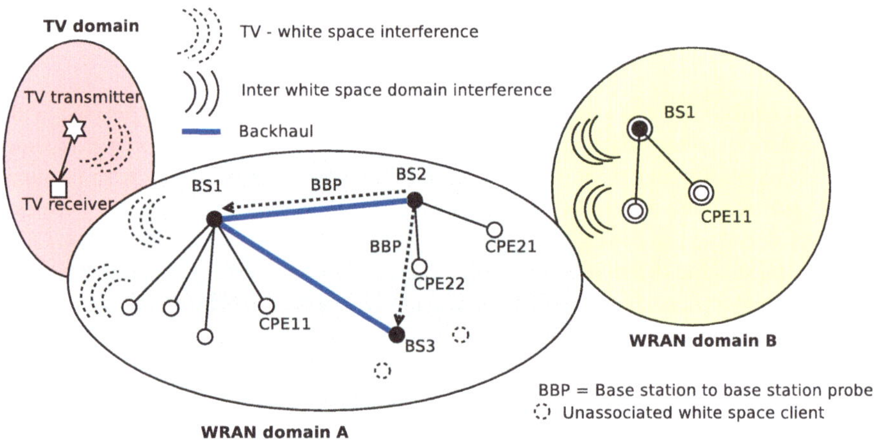

Fig. 9.6 Layout of a targeted white space network showing interference scenarios between television and white spaces, and between white space networks in different domains. White space base stations within the same domain send base station to base station probes (BBPs) to calculate the channel conditions among themselves

and all the CPEs within a cell operate on the same channel[3]; thus, when considering

[3] We envision OFDMA channel sharing among the CPEs of a BS. Such an approach is mandated by IEEE 802.16 and IEEE 802.22 standards. We leave the details of subcarrier allocation as the future work, and in this paper concentrate solely on channel allocation at the BS level.

channel allocation we use "BS" and "cell" interchangeably. All cells that are operated within the same administration are called a WRAN domain. The existence of TV transmission and other white space networks not in our service set reduces the number of channels available to the BSs within our domain. The aim of our work is to develop a channel allocation algorithm, as well as supporting structures such as a MAC layer and a frequency probing mechanism, so that the overall network performance within our white space WRAN domain is maximized.

We assume that base stations are connected with a back channel. This can be another white space radio tuned to a common channel that does not interfere with the set of channels available for base station to client connectivity. Moreover, because the amount of control data sent over the back channel is low, a VHF/UHF packet radio, or any other low bandwidth communication technology can be used.

9.3 Medium Access Control Modifications to Support Channel Probing

Channel probing is a necessary tool for propagation evaluation over a wide white space frequency range. Unfortunately, the existing MAC protocols proposed for wide area networks [2, 18, 19] do not explicitly support frequency profiling. The MAC protocol that most closely resembles our proposed system is IEEE 802.22. The protocol has built-in protection for primary users and mechanisms to move to new channels but has no built-in mechanism to choose from a set of available channels. It specifies that the channel may be chosen from the available list by an operator or by a "local routine". Thus, instead of rebuilding an entire MAC layer we propose to extend the 802.22 protocol to include a feature that performs frequency profiling on all available channels. The details of the technique to calculate the channel conditions using a probe are provided in Sect. 9.4. In this section we outline how we extend the 802.22 MAC protocol to support probing.

The MAC is organised into 160 ms superframes which consist of sixteen 10 ms frames. Each frame is divided into a downstream subframe and an upstream subframe; the size of these subframes depends on the amount of downstream and upstream data that needs to be sent between the base station and the clients. All base stations are equipped with a GPS and are able to synchronize their clocks to within 2ns, which is sufficiently accurate, given the timing requirements of IEEE 802.22. Base stations start their first superframe at the start of a GPS minute cycle; this greatly simplifies inter-base station communication and scheduling. Final synchronization is carried out using the superframe preamble.

The current 802.22 MAC specification already has sophisticated mechanisms to detect other 802.22 WRAN domains using co-existence beacons, move into a time division multiplex mode in cases where base stations are forced to share the same channel, allow clients to scan for and associate with base stations, and carry out ranging between clients and base stations to account for propagation delays. However

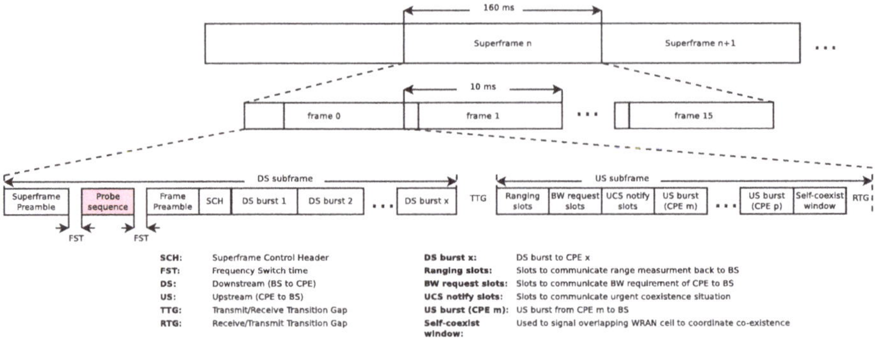

Fig. 9.7 General superframe structure of 802.22 showing additional probe sequence required for frequency profiling. The 1,024 bit probe sequence, in the worst case, increases overhead by 0.44 %; this includes the 10 μs Frequency Switch Time (FST)

none of these mechanisms allow a base station to discover channel conditions between WRAN cells at all available channels.

Figure 9.7 shows the superframe structure of 802.22 with our modification, an additional probe sequence after the superframe preamble to allow for frequency profiling at different wireless frequencies. The probe sequence uses a 1,024 bit PN sequence modulated using DBPSK with four samples per symbol. The probe sequence incurs low overhead, using an additional 0.44 % of the channel in the worst case. We place the probe sequence after the superframe preamble to ensure we do not break any timing synchronization. Clients make use of the superframe preamble to synchronize any clock offsets. The probe will be transmitted on a probe channel, p_i, where $\boldsymbol{p} = (p_1, p_2, \ldots, p_N)$, a set of N probe channels. The probe channel set is a subset of the complete set of available TV channels, $\boldsymbol{v} = (v_1, v_2, \ldots, v_M)$ with $N < M$ after eliminating non-vacant channels. Each base station in the WRAN domain would first have consulted its spectrum database and scanned all the channels for primary users and other white space domains to ensure availability.

In order to perform frequency profiling between base stations on all available channels, a mechanism is required to coordinate probing timing, channel probe senders and listeners. When a base station is in a probing state, it sends a probe at the beginning of each superframe. It sequentially steps through the full white space TV channel set and only sends a probe if the channel is contained in the probe channel set for that base station. The entire scan takes $160 * M$ ms, where M is the total number of TV channels. If the probe channel is not contained in the probe set, the base station does not send any transmission in the probe sequence slot. Base stations maintain their own probe sets as they may each generate interference to primary users on different sets of frequencies. We chose a mechanism where we step through the entire set of white space channels with interspersed quiet periods on non-vacant channels to avoid needing to maintain full consistency of all probe channels amongst all base stations and associated clients.

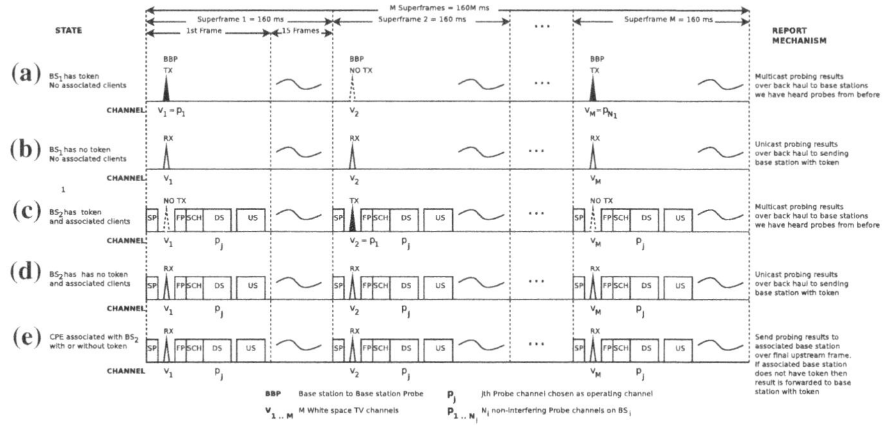

Fig. 9.8 Timing diagram showing the base station probes (BBPs) sent by the base station and received by other base stations and clients. **a** BS_1 has token with no associated clients. **b** BS_1 has no token and no associated clients. **c** BS_2 has token with associated clients. **d** BS_2 has no token and associated clients. **e** CPE associated with BS_2 with or without token

In order to coordinate probe transmission and reception, we propose a token approach in which a base station only transmits probes when it has a probe token and listens for probes when it does not have a token. A breath-first traversal of a spanning tree of the graph is used to ensure that a token traverses the graph of base stations when the back channel forms the edges. Once the base station has finished probing across the TV channel set, it sends the token to the next base station using a traversal algorithm; the probing process is completed when all base stations have used the token. Probes can take place when base stations are in the initialization phase and have not chosen an operating channel, or when they are in an operating phase and are communicating with associated clients and a new base station is added, for example. A request to probe is broadcast by the base station wishing to initiate probing in the domain.

The 802.22 specification makes use of clients to sense for primary users and extend the sensing coverage area. We propose to use a similar notion when listening for probes. Clients of one base station experience interference from all other base stations. To account for this interference clients can be instructed to listen for probes from base stations with whom they are not associated. Frequency profiling results for clients are sent back to the associated base station on the final upstream frame once the client has listened on the full set of white space channels. The maximum SNR value of a received probe heard at a base station and its associated clients is used to incorporate the worst case effect on the system. These SNR values from each of the cells that received the probes are unicast on the back channel to the sending base station. Results received at sending base stations are distributed to neighbouring base stations, where two base stations are defined as neighbouring if a probe can be exchanged between them on at least one frequency.

Figure 9.8 shows a number of timing diagrams for different base station and client states. Figure 9.8a shows a base station in the initialization phase in which it has a token. In this case only probes are sent by the base station. Figure 9.8c also shows a base station with a probe token but in this case the probes are interspersed within standard 802.22 frames as the base station is actively communicating with clients. Quiet periods, where no probe is sent, are different in Figs. 9.8a and c as they effect primary users on different channels. Figures 9.8b and d show a base station receiving probes and measuring the channel condition between the sending base station and itself. Figure 9.8e shows a client receiving probes from a base station with or without a token. In the case of a client hearing a probe from a base station it is not associated with, the frequency profile matrix is sent back to its associated base station and then forwarded on the back channel to the sending base station with the token.

Once the probing process is completed each BS i has information on signal propagation at different frequencies: (1) within its own cell, obtained through aggregation of probing results from the cell's CPEs, (2) between i and each of the BSs j that heard probes from i, and (3) within cells that are served by each of the neighbouring BSs j. We note that power failures are expected in rural areas [21], and should they be experienced, the base station can power up to its previous known state stored in non-volatile memory as the frequency profile matrix, which contains the frequency profile between all base stations at all available channels.

9.4 Channel Profiling

In this section we provide details of the probe that is sent at the beginning of each superframe. This probe allows us to determine the channel gain at each of the available white space channels between base stations and clients. The probe also allows us to determine interference levels between multiple base stations. For determining the optimal channel to use at a base station, we only require a calculation of the overall channel gain at each available frequency. However, once an optimal channel is chosen, the same probe can be used to calculate the detailed channel impulse response for the chosen channel—a process called channel sounding. The channel impulse response can be used for optimal allocation of OFDM sub-carriers to each client connecting to the base station.

9.4.1 Channel Gain Calculation

Each master probe is simply a pseudo-noise (PN) sequence modulated with DBPSK. The client calculates the average power measured, P_{avg}, over the probe listen window.

$$P_{avg}(dB) = 10\log\left(\frac{1}{N}\sum s(n)^2\right) + CF \qquad (9.2)$$

where $s(n)$ is the signal received, n is the sample number, N is the total number of samples and CF is the correction factor, which is calculated by calibrating the receiver. Average power is a low complexity ($O(N)$) calculation and the cognitive radio can carry this out in real time using a cumulative average.

The SNR of the probe can then be calculated by using a previously known noise level N_{avg} from an initial scan when no probes are present.

$$SNR(dB) = P_{avg}(dB) - N_{avg}(dB) \tag{9.3}$$

Figure 9.3 shows the result of this channel gain calculation across the entire UHF TV band. The average noise shown in the Figure is measured when no probes are sent and the average power is measured in 1 MHz steps when probes are sent from the base station.

From the measured SNR we can extract the channel gain:

$$H = \frac{SNR \cdot N_0 \cdot W}{P} \tag{9.4}$$

where N_0, W and P denote the noise constant, channel width and the transmission power, respectively.

The set of channel gains at each probe frequency are used for the channel allocation algorithm described in Sect. 9.5.

9.4.2 Channel Sounding

Once a channel is selected using the channel allocation algorithm, the system can carry out more detailed analysis of the channel. This is done by analyzing the channel impulse response at a specific frequency using the stored received probe sequence to calculate the channel gain at each available white space frequency.

The discrete channel impulse response is calculated using a correlation-based channel sounder. The received signal, $r[n]$, is given by the convolution of the channel, $h[n]$, and the transmitted signal, $s[n]$, where n is the sample number:

$$r[n] = s[n] * h[n] \tag{9.5}$$

In our case, we need to estimate $h[n]$ from the received signal $r[n]$ and a known signal $s[n]$ using deconvolution:

$$h[n] \approx \frac{1}{P_s} r[n] * s[-n]. \tag{9.6}$$

where P_s is the power spectral density of the input signal $s[n]$. As we are using digital signals, it is reasonable to assume that the input signal is approximately flat or

$P_s = |S(n))|^2$ is a known constant. Once $h(n)$ is known we can calculate the Fourier transform of $h(n)$ to find the frequency domain response, $H(n)$.

In order to achieve maximum discrimination in the correlator, a DBPSK modulated PN sequence is chosen such that the Index of Discrimination (ID) is high. ID is defined as the ratio of the first autocorrelation peak to the nearest neighboring autocorrelation peak. The receiver has prior knowledge of the PN sequence that was sent by the transmitter and can carry out the deconvolution with the received signal in order to calculate the channel impulse response. The frequency response of the channel is determined by Fourier transforming the channel impulse response.

The minimum multipath delay resolution that is possible is given by $\delta\tau = \frac{2}{\alpha}$, where α is rate of the PN sequence also known as the chipping rate. The minimum path difference, D_{pd}, that can be resolved between multipath components is given by $D_{pd} = c\alpha$. It is important to determine the level of multipath in a channel as increased levels of multipath cause less coherence in the frequency response of the channel. Although channel equalization can be achieved by adjusting the magnitude of the OFDM sub-carriers at the transmitter, selecting sub-carriers for each client is a complex problem due to different channel conditions between a base station and each of it's associated clients.

We store ten PN sequence lengths worth of received data in memory and carry out the deconvolution over this data set. This allows us to store ten different channel impulse responses for each repeated PN sequence and carry out transient analysis or averaging.

9.4.2.1 Practical Considerations

We use a chipping rate of 1 and 2 MHz for our experiments. We found that higher chipping rates do not perform well on the USRP radios. This translates to a minimum multipath delay resolution of 2 and 1 μs or a minimum path difference of 600 and 300 m respectively. From previous studies [4] rural open areas typically have delay spread of up to 1.3 μs or up to 11.6 μs if there are deep canyons or mountains. Suburban areas usually do not have delay spread over 2.4 μs and urban areas have delay spread up to 4.6 μs with many reflections. From this study our chipping rate should be sufficient to discriminate multipath in most environments.

To further improve the signal to noise ratio of our channel sounder, we sample at 4 MHz at the receiver. This receiver rate, between 2 and 4 times the PN sequence chip rate, creates an oversampled signal that improves the SNR. We also improve the SNR by using a long PN sequence (1,024 bits) taking multiple snapshots of the received PN sequence and calculating the average.

9.4.2.2 Results

Figure 9.9 shows the results for channel sounding at 530 MHz and a lower chip rate of 1 MHz. This is shown for a single deconvolution snapshot over one PN sequence.

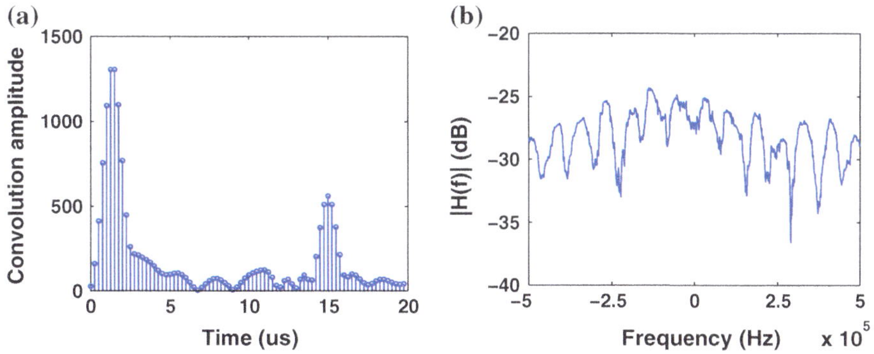

Fig. 9.9 Channel sounding at 530 MHz using 1 MHZ chipping rate and oversampling rate of 4. **a** Is the channel impulse response of the channel and **b** is the frequency response of the channel. This plot shows the result for deconvolution of a single received PN sequence to determine the channel impulse response

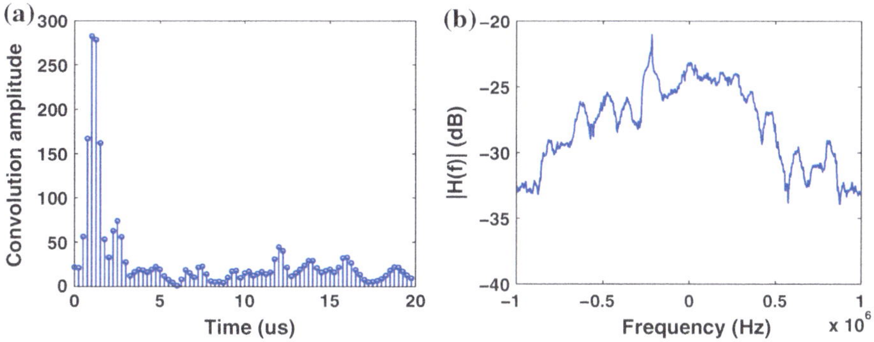

Fig. 9.10 Channel sounding at 530 MHz using 2 MHz chipping rate and oversampling rate of 2. **a** Is the channel impulse response of the channel and **b** is the frequency response of the channel. This plot shows the result for averaging the deconvolution of 10 received PN sequences to determine the channel impulse response

There is a very clear strong echo at 15 μs; there is also an echo at approximately 2 μs but it is hard to distinguish from the main direct path as the chipping rate is too low. Ideally you want the chipping rate resolution to be at least half that of the time to echo. The frequency response shown in Fig. 9.9b shows some deep fades due to the large echo present in the channel.

In Fig. 9.10, the chipping rate is increased to 2 MHz and we generate an average of 10 channel impulse responses. The first echo at 2 μs is now very visible, however the 15 μs echo is no longer visible. It is possible that this echo was temporary due to a change in the environment, such as a moving truck. Fewer deep fades are visible in the frequency response.

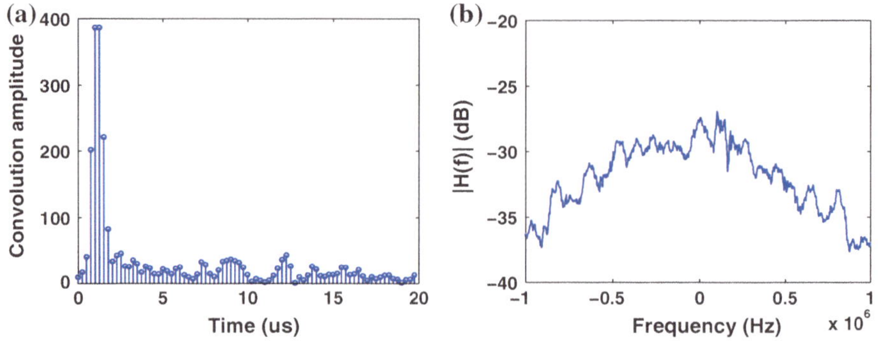

Fig. 9.11 Channel sounding at 834 MHz using 2MHZ chipping rate and oversampling rate of 2. **a** Is the channel impulse response of the channel and **b** is the frequency response of the channel. This plot shows the result for averaging the deconvolution of 10 received PN sequences to determine the channel impulse response

Figure 9.11 shows the channel impulse response for 834 MHz at the 2 MHz chipping rate. There are fewer echoes than in the lower UHF frequency plots, and as a result this frequency response is the most coherent. Environmental terrain has different absorption levels at different frequencies and this specific sub-urban environment had higher absorption levels for multipath signals at higher frequencies.

In the rest of the chapter, we only make use of the channel gain calculations for optimal channel selection and we leave optimal allocation of sub-carriers using the channel impulse response for future work.

9.5 Channel Allocation

In this section we devise a distributed channel allocation algorithm that uses information obtained through frequency profiling (Sect. 9.3) and does not incur channel switching overhead typical for other allocation schemes. Our approach is based on the annealed Gibbs sampler, a technique that can help us minimize a target function in a distributed way. In the next subsection we present the basics of Gibbs sampling. An interested reader can find more details about Gibbs sampling in [1]. We then cast our problem to the Gibbsian framework and sketch the channel allocation algorithm.

9.5.1 Gibbs Sampling

The Gibbs sampler is a Markov chain Monte Carlo (MCMC) technique for obtaining random samples from a multivariate probability distribution. The sampler is useful in situations where the joint distribution is unknown or difficult to sample, but the conditional distributions of variables are known and easy to sample. The Gibbs sampler draws samples from a multivariate probability distribution $p(x_1, \dots, x_N)$ as follows:

- Initialize all variables $x_1, \dots, x_N$ to (random) starting values $x_1^0, \dots, x_N^0$.
- In every iteration $j = 1, \dots, k$, sample each variable x_i from the conditional distribution $p(x_i | x_1^j \dots, x_{i-1}^j, x_{i+1}^{j-1}, \dots, x_N^{j-1})$ to obtain x_i^j.

After the above process is finished, we are left with $x_1^j, \dots, x_N^j; j \in [1, \dots, k]$ samples from the joint distribution p.

We can solve the channel allocation problem through Gibbs sampling, if we obtain the samples from a multivariate probability distribution that:

1. Is related to overall network performance.
2. Depends on the selected operating channel of each of the base stations.
3. Isolates the impact of each of the base stations on the total optimization function.
4. Can be calculated in a distributed way and sampled independently at each of the base stations.
5. Favors states that lead to maximum network performance.

In the following section we develop a network performance metric that can be used as a basis for a probability distribution that satisfies the above demands.

9.5.2 Network Performance Metric

Traditionally, the goal of a channel allocation protocol is to assign available channels to BSs so that the total network capacity is maximized. The capacity $C_i(c_p)$ of a single cell operating on the channel c_i is:

$$C_i(c_i) = \sum_{k \in \mathbb{K}_i} W_k \log(1 + SINR_{ik}(c_i))$$

where $\mathbb{K}_i$ is the set of CPEs within the cell, W_k is the width of a part of the channel c_i used by CPE k, and $SINR_{ik}(c_i)$ is the signal to interference plus noise ratio at the CPE k. We approximate the presence of all clients within the cell with a single *virtual CPE* with an SINR value $SINR_i(c_i) = \sum_k SINR_{ik}(c_i)/|\mathbb{K}|$. The cell capacity is now:

$$C_i(c_i) = W \log(1 + SINR_i(c_i))$$

where W is the full channel width, essentially a sum of all W_k as a cell operates in an OFDMA mode. This approximation hides channel distribution within the cell and helps us concentrate on inter-cell interaction.

If we consider a network with N cells, with a given channel assignment $\mathbf{c} = (c_1, c_2, \dots, c_N)$, $c_i \in \mathbb{C}$, where $\mathbb{C}$ is the set of available channels, the total network throughput is a sum of all individual capacities at their respective allocated channels:

$$C(\mathbf{c}) = \sum_i C_i(c_i) = \sum_i W \log\left(1 + SINR_i(c_i)\right) \tag{9.7}$$

A single BS's decision on the operating channel changes the interference level at all its neighboring BSs. In the above equation the interference is accounted for in the SINR, which is embedded within the logarithmic function. Thus, the impact of a single BS on the total sum is hard to isolate, and the total capacity is not a suitable metric for distributed computation using Gibbs sampling. Centralized optimization using known polynomial complexity techniques, such as linear programming, is not directly applicable either, since the target sum involves non-linear factors and discrete variables.

One of the ways to circumvent this is to revert to a tighter problem formulation that prevents interfering base stations from concurrent transmission [11]. While this can be enforced in a network that employs carrier sensing and collision avoidance, in our setting long distances between base stations render such coordination inefficient [18]. In addition, allowing some interference often yields more capacity than restricting concurrent transmissions [15]. Another approach is to modify the optimization function and instead of maximizing capacity concentrate on minimizing total network interference [9, 17]. This approach is attractive for networks, such as WiFi, where these two goals are essentially interchangeable. In a white space setting, where available channels can differ drastically in terms of their propagation properties, *a channel allocation that leads to minimal interference may not necessarily lead to maximum capacity*.

We propose a novel network performance metric—*Cumulative interference plus noise to signal ratio (CINSR)*—a sum of inverse of *SINR* experienced at each of the cells. *CINSR* can be seen as the overall ratio of the impact of harmful factors, noise and interference, to the beneficial one, received signal strength. Thus, our goal is to minimize it. Compared to metrics such as the total capacity or the overall level of interference, *CINSR* takes into account the frequency diversity that exists in white space networks, and allows distributed performance optimization with Gibbs sampling:

$$CINSR(\mathbf{c}) = \sum_{i=1}^{N} \frac{1}{SINR_i(c_i)} \tag{9.8}$$

$$= \sum_{i=1}^{N} \frac{N_0 W + \sum_{j=1..N}^{j \neq i} ch(i,j) PH_{ji}(c_i)}{PH_i(c_i)} \tag{9.9}$$

The first term in the numerator within the above sum is the thermal noise (a product of the channel width W and the noise constant N_0), whereas the second term is the sum of interference experienced at cell i, and originating from all other base stations that transmit at the same channel. Interference from a single source is a product of P—the transmission power and $H_{ji}(c_i)$—the propagation gain from base station j to cell i on channel c_i. The function $ch(i, j)$ is equal to 1 if i and j operate on the same channel and, otherwise it is equal to 0. The denominator in the above equation is the average signal strength received by the clients of the BS i that transmits with power P. The average channel gain towards the clients is denoted by $H_i(c_i)$.

We now isolate the impact of a single BS i on $CINSR(\mathbf{c})$ and term it *local CINSR*:

$$CINSR_i(\mathbf{c}) = \frac{N_0 W}{PH_i(c_i)} \tag{9.10}$$

$$+ \sum_{j \neq i} ch(i, j) \left(\frac{PH_{ij}(c_i)}{PH_i(c_i)} + \frac{PH_{ji}(c_i)}{PH_j(c_i)} \right) \tag{9.11}$$

Information needed for $CINSR_i(\mathbf{c})$ calculation, namely $PH_i(c_i)$, $PH_{ji}(c_i)$, $PH_{ij}(c_i)$ and $PH_j(c_i)$, is available locally at BSi, through channel probing described in Sect. 9.3.

9.5.3 The Gibbs Distribution

The Gibbs distribution associated with the function *CINSR* and a positive temperature parameter T is the probability distribution on c^N (the combined channel state space of all BSs) defined as:

$$\pi(\mathbf{c}) = \frac{e^{-\mathcal{CINSR}(\mathbf{c})/T}}{\sum_{\mathbf{c}' \in c^N} e^{-\mathcal{CINSR}(\mathbf{c}')/T}} \tag{9.12}$$

The above distribution is of special interest as it favors states in which *CINSR* is low. In addition, the channel selected by BS i is independent of all non-neighboring BSs and the distribution fulfils all the conditions listed Sect. 9.5.1.

The Gibbs sampler draws a sequence of samples from the above distribution by having each of the BSs i independently sample its local Gibbs distribution $\pi_i(\mathbf{c})$:

$$\pi_i(\mathbf{c}) = \frac{e^{-\mathcal{CINSR}_i(c_i,(c_j)_{j \neq i})/T}}{\sum_{\mathbf{c}' \in c^N} e^{-\mathcal{CINSR}_i(c_i',(c_j)_{j \neq i})/T}} \tag{9.13}$$

and transitions to the sampled local state, converging to the stationary distribution $\pi(\mathbf{c})$ (see Sect. 9.5.5).

Distribution $\pi(\mathbf{c})$ highly favors low *CINSR* states when the temperature T is low. While our goal is to minimize *CINSR*, by keeping the temperature low we risk getting stuck in a local minimum early in the process. *The annealed Gibbs sampler* introduces a slow decrease of temperature T to zero according to *a cooling schedule*. Therefore, in the beginning the probability of exploring a wide range of states is high, and as the time goes to infinity, the procedure converges to the minimum *CINSR* state.

9.5.4 Channel Allocation Algorithm

Algorithm 1 VillageLink channel allocation – distributed

1: {Executed at the base station i}
2: **while** $t < t_{end}$ **do**
3: $T = f(T_0, t)$ {f - schedule, T_0 - starting temperature}
4: **for all** channel $c_i' \in \mathbb{C}$ **do**
5: $\mathbf{c}' = (c_1, c_2, \ldots, c_i', \ldots, c_N)$
6: Calculate $CINSR_i(\mathbf{c}')$
7: **end for**
8: **for all** channel $c_i' \in \mathbb{C}$ **do**
9: $\mathbf{c}' = (c_1, c_2, \ldots, c_i', \ldots, c_N)$
10: Calculate $\pi_i(\mathbf{c}')$
11: **end for**
12: Sample a random variable according to the law π_i and choose the next channel of the BS i accordingly.
13: Send information about the newly selected channel to i's neighbors.
14: **end while**
15: Switch the network interface to the last selected channel.

Algorithm 1 is executed at each of the BSs. The temperature falls off with time, ensuring that the Gibbs sampler converges towards the global minimum of *CINSR*. The starting time for all the BSs has to be loosely aligned, and can be achieved with a standard synchronization scheme such as NTP.

Compared to some other distributed channel allocation schemes [9, 14], Algorithm 1 has an attractive property that no channel switching is needed until the convergence. To see why note that the calculation of the local *CINSR* is done after the probing process, and during the algorithm run the only variable parameter is $ch(i, j)$. At BS i this parameter can be updated irrespective of the actual operating channel of BS j. In every step a BS decides on its current channel and sends the decision to its neighbors, who then update their $ch(i, j)$ tables. Once the cooling schedule is completed base stations switch to their channel of choice (line 15 in Algorithm 1). This greatly speeds up the convergence, as the channel allocation process is not limited by the channel switching time.

9.5.5 Algorithm Convergence

Convergence of a Gibbs sampler, and its annealed version, is a well researched topic [1]. Here we prove the convergence of our method, indicating that it is a natural heuristic for solving the channel allocation problem.

Proposition 1 *The Gibbs distribution* π *(Eq.* 9.12*) represents a Markov random field.*

Proof *A Gibbs potential* V associates a real number $V_\Gamma(\mathbf{s})$ with each subset Γ of a set S. The potential is determined by the state $\mathbf{s}$ of the nodes in Γ and is defined as zero if Γ is not a clique. *An energy function* $\mathcal{E}(\mathbf{s})$ maps each of the graph states to a real number. We say that the energy function *derives from the potential* V if:

$$\mathcal{E}(\mathbf{s}) = \sum_{\Gamma} V_\Gamma(\mathbf{s}) \tag{9.14}$$

where the summation goes over all subsets of the set S. The Gibbs distribution where the energy derives from a Gibbs potential is a Markov random field (p. 260 in [1]), and we proceed with showing that the function that we use to construct the Gibbs distribution in Eq. 9.12—*CINSR(c)* derives from the Gibbs potential.

We can represent *CINSR* as a sum of local impact of cliques of the graph of BSs $\mathcal{A}$. *CINSR* then takes the form described by Eq. 9.14 and *CINSR* can be used as the energy function for Gibbs sampling:

$$\begin{aligned} CINSR(\mathbf{c}) &= \sum_{i \in \mathcal{A}} \frac{N_0 W}{PH_i(c_i)} + \\ &+ \sum_{\{i,j\} \in \mathcal{A}} ch(i,j) \left(\frac{PH_{ij}(c_i)}{PH_i(c_i)} + \frac{PH_{ji}(c_i)}{PH_j(c_i)} \right) \\ &= \sum_{\mathcal{B} \subset \mathcal{A}} V_{\mathcal{B}}(\mathbf{c}) \end{aligned}$$

Here V denotes the Gibbs potential. The potential is defined for all subsets $\mathcal{B}$ of the set of BSs $\mathcal{A}$ as:

$$V_{\mathcal{B}}(\mathbf{c}) = \begin{cases} N_0 W / PH_i(c_p) & \text{if } \mathcal{B} = \{i\} \\ ch(i,j) \left(\frac{PH_{ij}(c_i)}{PH_i(c_i)} + \frac{PH_{ji}(c_i)}{PH_j(c_i)} \right) & \text{if } \mathcal{B} = \{i,j\} \\ 0 & \text{if } |\mathcal{B}| \geq 3 \end{cases}$$

Note that the potential is non zero only for cliques of size one and two. Thus, energy $CINSR(\mathbf{c})$ derives from the Gibbs potential and, consequently π is a Markov random field.

For a network of N BSs, each running a Gibbs sampler over its local Gibbs distribution $\pi_i(\mathbf{c})$, channel allocation converges in variation[4] towards the Gibbs distribution π, since the process can be described as a Gibbs sampler on a finite state homogeneous Markov chain represented by the selected channel allocation, for which the Gibbs distribution (Eq. 9.12) is the invariant probability measure (Example 6.5, p. 288 in [1]). Note that direct sampling of the capacity (Eq. 9.7) does not provide any guarantees on the performance as the capacity equation cannot be transformed to an energy function that derives from the Gibbs potential. Thus, we develop *CINSR*. Finally, for a fixed network of N BSs implementing Algorithm 1, channel allocation converges in variation towards a limit distribution that only puts positive probability mass on the states of minimum global energy, as we rely on the annealed Gibbs sampler (example 8.8, p. 311 in [1]). Conditions that the cooling schedule has to satisfy in order for convergence to happen can be found in [5].

9.6 Evaluation

The VillageLink system consists of our frequency profiling method built on top of the 802.22 MAC protocol, and the channel allocation algorithm based on Gibbs sampling. Experimental evaluation of such a system is challenging due to the need for a wide area outdoor deployment. In addition, off-the-shelf 802.22 equipment is not yet commercially available, and software defined radio platforms cannot support the synchronization that the MAC protocol requires [16]. Therefore, we evaluate our protocol in a simulated setting. However, the initial experimental investigation of channel probing and frequency selectivity in white spaces, presented in Sect. 9.2.2, was performed on a 3 km outdoor link.

9.6.1 Simulation Setup

For a comprehensive evaluation of the channel allocation algorithm, we rely on a Matlab-based custom simulator. The simulator allows us to scale our experiments over a number of cells, and to model different network layouts. We explicitly take into account high variability of signal propagation in the white space band by modeling propagation with the Friis transmission equation:

$$P_r = P_t + G_t + G_r + 20\log\left(\frac{\lambda}{4\pi R}\right)$$

where $\lambda, R, P_r, P_t, G_t$ and G_r are the wavelength, distance between antennas, received power, transmitted power, transmitter antenna gain, and receiver antenna

[4] Convergence in variation describes convergence of an array of samples to a probability distribution and is defined in [1], p. 128.

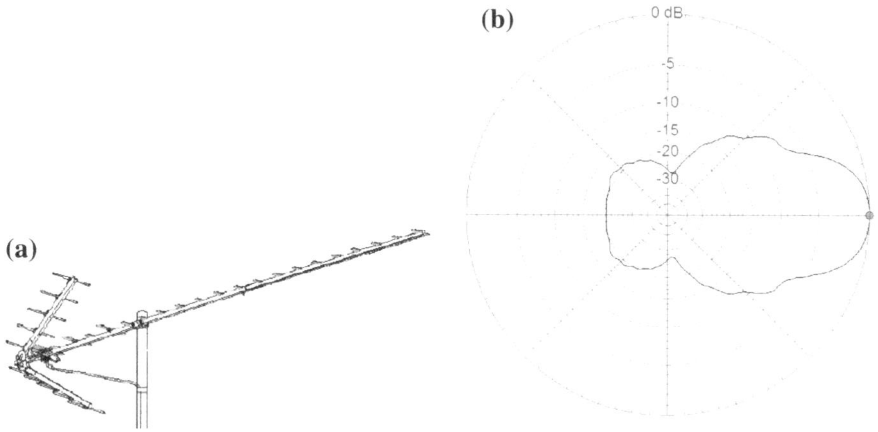

Fig. 9.12 Wineguard PR9032 UHF Yagi/corner reflector antenna used as a base station antenna in our evaluation. Showing **a** the antenna design and **b** its radiation pattern seen from the top of the antenna

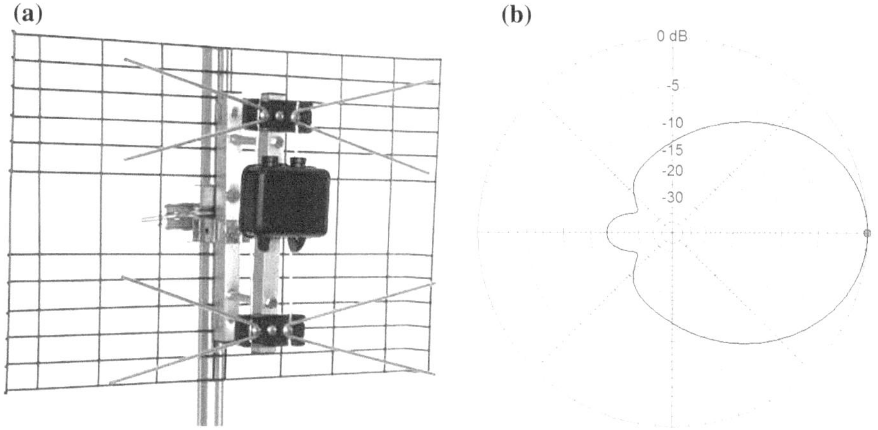

Fig. 9.13 AntennasDirect DB-2 2-Bay UHF antenna; one of the client antennas used for the evaluation. Showing the antenna in (**a**) and its radiation pattern seen from the top of the antenna in (**b**)

gain, respectively. Antenna gains depend on specific devices used and their orientations. Earlier, in our outdoor testbed, we confirmed that frequency dependence of antenna gain is the most dominant factor that leads to the frequency diversity in white spaces (Fig. 9.3, Sect. 9.2), thus we model antenna effects in detail.

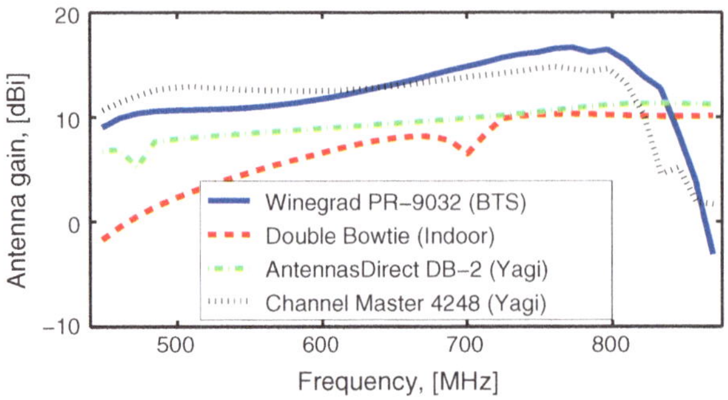

Fig. 9.14 Antenna profiles of four of the antennas used in our evaluation. One of the profiles, Wineguard PR-9032, corresponds to the BS antenna and, the other three, to client antennas

We use publicly available antenna models[5] and the Numerical Electromagnetics Code (NEC)[6] antenna modeling software to examine propagation over different frequencies with different antennas. Figures 9.12b and 9.13b show the radiation patterns seen from the the center frequency (598 MHz) of the white space band for two different antennas. In Fig. 9.14 we plot frequency dependence of antenna gain. We found that the shape of the antenna pattern does not change significantly for different frequencies. The gain, on the other hand, changes significantly and unpredictably, as seen in Fig. 9.14. Thus, in the simulations we use the antenna pattern shape of the center frequency to account for antenna orientation, and we use the full gain over frequency diversity.

All base stations in our simulations use the Yagi antenna from Fig. 9.12, as this antenna exhibits the best performance of all the antennas that were modeled. In our simulation we assume clients make use of existing TV antennas used to receive terrestrial TV broadcast signals. Operators have no control over the variety of antennas used by clients and we randomly select antennas from a set of 17 possible client TV antennas ranging from outdoor Yagi antennas with a gain of 15 dBi to simple indoor loop antennas with a gain of 3 dBi.

We run our experiments over a white space band from 443 to 875 MHz as the antenna models we use perform reasonably well within this range. The band is divided into 36 TV channels, each 6 MHz wide, with a 6 MHz guard band between adjacent channels. In all the experiments we simulate a 100 km $\times$ 100 km field with random BS placement and random antenna orientation. Each of the BSs has a single associated virtual client at a distance uniformly picked from 0.2 to 20 km and with its antenna pointed directly towards the BS. We also simulate a TV station that covers a part of the field with its signal and occupies two adjacent channels.

[5] http://www.hdtvprimer.com/ANTENNAS/comparing.html.

[6] www.nec2.org.

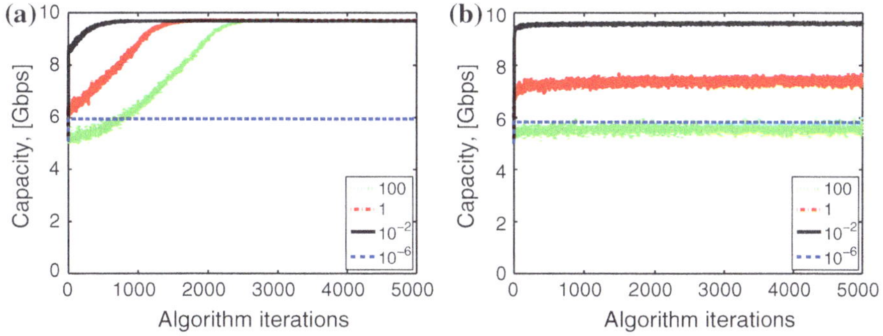

Fig. 9.15 Algorithm convergence with the (**a**) exponential, and (**b**) logarithmic cooling schedule. Different lines correspond to different starting temperatures

9.6.2 Channel Allocation Convergence

We simulate Algorithm 1 behavior in a network of ten base stations and five white space channels that are available for communication. We are interested in the algorithm convergence under different Gibbs sampling parameters. We experiment with two common cooling schedules:

- Logarithmic: $T = T_0/\log(t+2)$, proposed in [3].
- Exponential: $T = T_0\alpha^t$, proposed in [10].

where T denotes the temperature at time t, T_0 is the starting temperature, and α is a real number between zero and one; we empirically find value 0.995 to work well in our experiments.

The selection of the starting temperature is important for proper annealing. In Fig. 9.15 we plot total network capacity achieved with the two schedules and four different starting temperatures for each. Each point in the graphs is an average over 100 runs. The impact of the starting temperature is clearly visible: the higher T_0 is, the more time it takes for the algorithm to converge. At the same time, higher temperatures ensure exploration of a large part of the solution space, and generally lead to a better solution. We can also see that $T_0 = 10^{-6}$ does not result in any variation of capacity as the algorithm progresses—the sampler is "frozen" and BSs will stick to the initial channel allocation without exploring the full solution space. There is a trade-off, dictated by the starting temperature, between the convergence time and the assurance that the optimal value will be found. In the rest of the evaluation section we fix the starting temperature to 1, a value that allows full exploration of the solution space and converges in a reasonable amount of time.

We observe much faster convergence with the exponential schedule, that converged in all but one case ($T_0 = 10^{-6}$). The logarithmic schedule did not converge in 5,000 iterations for $T_0 = 100$ and $T_0 = 1$. In the rest of this section we rely exclusively on the exponential schedule.

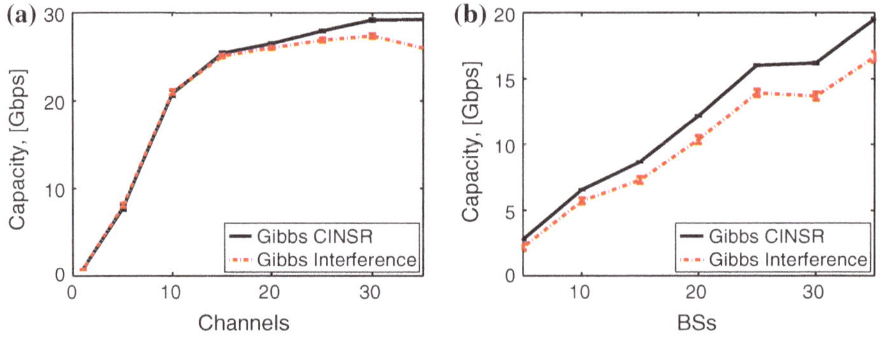

Fig. 9.16 Comparison of the total network capacity achieved with CINSR and Interference metrics. We simulate under-provisioned and over-provisioned number of channels with respect to the number of base stations in the network. **a** Channel under-povisioning. **b** Channel over-povisioning

9.6.3 CINSR as a Performance Metric

To confirm that *CINSR* is a good choice for the network performance metric, we compare it with an alternative—overall interference and noise in the network—which is often used as a metric in channel allocation algorithms [9, 17].

The total network interference and noise is defined as:

$$I(\mathbf{c}) = \sum_{i=1}^{N} \left(N_0 W + \sum_{j=1..N}^{j \neq i} ch(i,j) PH_{ij}(c_i) \right) \tag{9.15}$$

The impact of a single BS on the sum is defined as *the local interference*:

$$I_i(\mathbf{c}) = N_0 W + \sum_{j=1..N}^{j \neq i} ch(i,j) \left(PH_{ij}(c_i) + PH_{ji}(c_i) \right) \tag{9.16}$$

We modify the Gibbs distribution (Eq. 9.12) to include $I(\mathbf{c})$ instead of $CINSR(\mathbf{c})$, and the local Gibbs distribution (Eq. 9.13) to include $I_i(\mathbf{c})$ instead of $CINSR_i(\mathbf{c})$. The necessary conditions for the Gibbs sampler convergence still hold, and we apply an algorithm analogous to Algorithm 1.

Note that, defined this way, the interference function still uses the results of channel probing, yet it does not account for the balance between well propagating channels that are preferred by the CPEs and inferior channels that minimize inter-cell interference.

9.6.3.1 Channel Under-Provisioning

In the first scenario we simulate a network with a number of contending BSs higher than the number of available channels. This can be the case in the urban developed world, for example. We put 50 cells in the same 100 km $\times$ 100 km region. We experiment with a varying number of available channels. The total network capacity is plotted in Fig. 9.16a. Each point represents an average value of 20 runs of the algorithm with a different metric, *Gibbs CINSR* or *Gibbs Interference*, over the same topology.

When multiple cells operate on the same frequency the network is in a low SINR mode, and capacity can be increased by interference minimization. From Fig. 9.16a we see that the two versions of the Gibbs sampler perform equally well with a small number of available channels. As we increase the amount of available spectrum, BSs have more freedom to operate at different channels with minimal interference. Therefore, frequency-dependent performance of CPEs associated with the BSs becomes an important factor that impacts total capacity. Since this factor is not accounted for in Eq. 9.15, this version of the Gibbs sampler results in a channel allocation that delivers less capacity than the version that uses *CINSR*.

9.6.3.2 Channel Over-Provisioning

We now fix the number of available channels to 36 and compare the performance of the two versions of the algorithm with the number of BSs varying from 5 to 35. The total network capacity is plotted in Fig. 9.16b. Each point represents an average value of 20 runs of the algorithm (*Gibbs CINSR* or *Gibbs Interference*) over the same topology.

When the number of channels is greater than the number of BSs there is more than one allocation that leads to minimal interference. However, not all of the allocations are favored by the CPEs. Through the factor $H_i(c_i)$ CINSR accounts for the frequency dependent intra-cell preferences, and assigns channels that maximize capacity within each of the cells. The results presented here point out that *channel allocation in white spaces remains important even in rural areas where the channel availability is high* [6].

9.6.4 Comparison to Alternative Channel Allocation Methods

Channel allocation is a difficult problem to solve in a distributed setting. Heuristics are often used instead of a rigorous solution and we compare our approach with:

- *Least congested channel search (LCCS)*—a heuristic where each of the BSs individually scans for a channel with the least number of other BSs assigned to it [14].

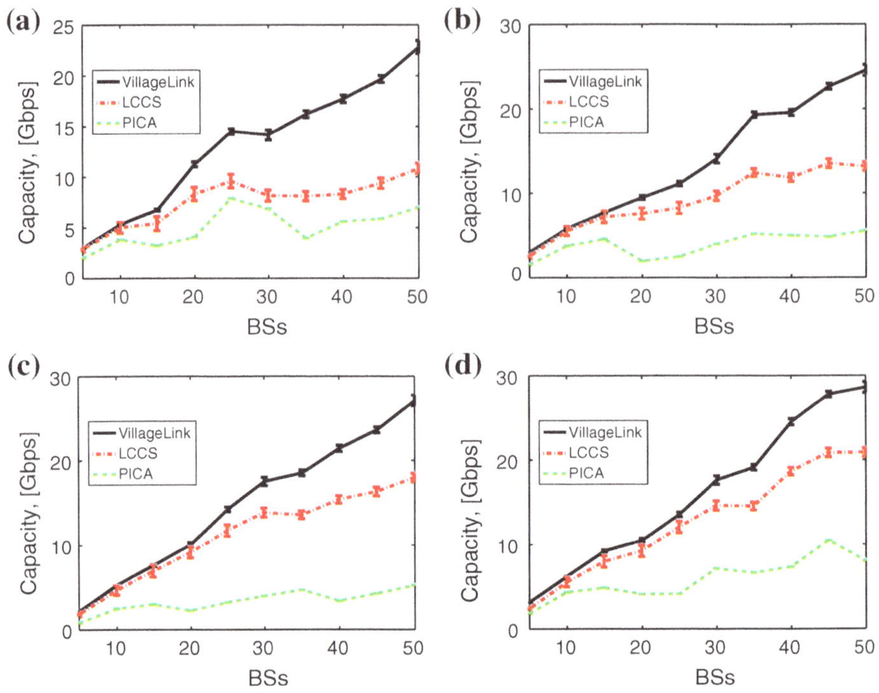

Fig. 9.17 Total network capacity with varying number of channels and base stations. **a** 10 available channels. **b** 15 available channels. **c** 20 available channels. **d** 25 available channels

- *Preferred intra-cell channel allocation (PICA)*—in this greedy method each of the BSs selects the channel for which it observes the highest channel gain towards its own CPEs ($\arg\max_{c_i} H_i(c_i)$).

These heuristics optimize a non-submodular capacity function in a greedy manner, therefore may settle for a solution that is arbitrarily far from the optimal. VillageLink's convergence to the states of minimum *CINSR* is proven in Sect. 9.5.5. We compare the experimental behavior of different solutions in a number of scenarios encompassing various numbers of BSs and available white space channels. We run each of the algorithms 100 times in each of the scenarios.

9.6.4.1 Total Network Capacity

In Fig. 9.17 we plot the total network capacity as we increase the number of cells in the system from 5 to 50. To ensure consistency among points in the graph, we do not generate a new topology every time we increase the number of cells, but add randomly placed cells to the existing topology. Each of the topology sequences are evaluated in environments with 10, 15, 20 and 25 available channels. We plot average values and two standard deviations (represented by error bars) for each data point.

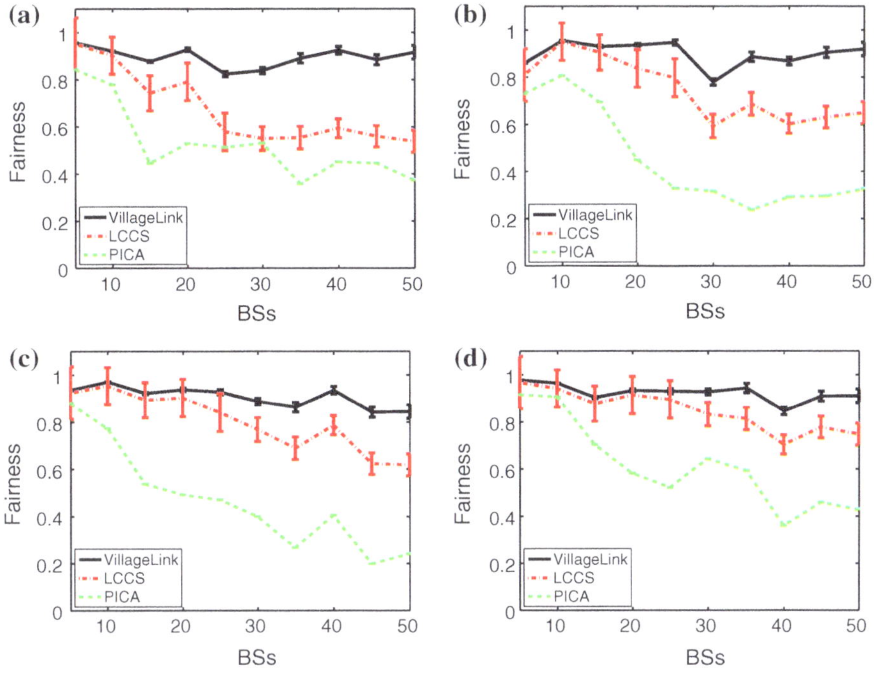

Fig. 9.18 Fairness with varying number of channels and base stations (the closer the fairness index value is to one—the better). **a** 10 available channels. **b** 15 available channels. **c** 20 available channels. **d** 25 available channels

VillageLink performs better or equal to the alternatives in all scenarios. The benefits of frequency-probing based channel allocation grow with the number of cells. In some scenarios, such as 50 BSs—10 channels and 50 BSs—15 channels, VillageLink delivers twice as much capacity as the next best alternative, LCCS. A comprehensive comparison of LCCS and PICA could unravel the importance of two conflicting goals in channel allocation: minimizing interference and maximizing intra-cell capacity in isolation, and is left for future work.

9.6.4.2 Fairness

In Fig. 9.18 we plot the Jain fairness index [8] for cell capacity with channel allocations determined by VillageLink, LCCS and PICA. We plot average values and two standard deviations (represented by error bars) for each data point. Although we designed VillageLink as a method to optimize total network capacity, it also ensures a remarkably fair allocation of resources. As the number of cells grows, the fairness of VillageLink is more pronounced as it stays close to 1 while the fairness indices of PICA and LCCS drop.

9.7 Related Work

Efforts to provide broadband connectivity to remote rural regions with low-cost unlicensed options, such as modified WiFi, have been proposed in the last decade [18, 19]. While numerous rural WiFi deployments provide useful general guidelines for wide-area coverage, the propagation characteristics in white spaces are drastically different than in the WiFi bands, and networking protocols have to be reconsidered for the new spectrum. IEEE 802.22 [2] is a standardized protocol for wide area white space coverage. In VillageLink we embrace the 802.22 frame structure, and augment it with novel channel probing and operating frequency selection mechanisms.

Previous work related to channel assignment in wireless networks usually casts the problem of channel assignment as an NP-hard graph coloring problem [14]. Numerous heuristics have been proposed to provide an approximate solution ([20] and references therein). Ma and Tsang [11] recognize the channel heterogeneity in the case of wide bands and propose an integer linear programming solution for the frequency allocation problem. However, they restrict frequency reuse to well defined interference domains, thus no two BSs are allowed to transmit at the same time if they interfere. Motivated by [15], we rely on a more sophisticated representation of interference—we directly measure its impact through probing and account for any interference during the allocation process.

The Gibbs sampling, under this name, was first proposed in 1984 for image manipulation [3]. Its applicability to distributed channel allocation, client association and power control in wireless networks has been realized over the last twenty years [9, 13]. VillageLink differs from these by two important properties. First, we propose a novel network performance metric called *CINSR*, that takes into account frequency dependence of both useful signal transmissions and interference. Second, our algorithm does not require subsequent channel switching and environment sensing after each local decision is made. Rather, only control information has to be exchanged among neighboring nodes, and once the algorithm terminates only a single channel switch is made per node.

9.8 Conclusion

The heterogeneity of white space frequencies imposes unique challenges when it comes to channel allocation in a wireless network. Rather than simply minimizing interference, a channel allocation policy has to account for transmission quality over different channels as well. In this work we develop VillageLink, a channel allocation protocol that relies on the knowledge of signal propagation in the whole white space band before it performs distributed channel assignment that converges towards a network-wide optimum.

White space networks are largely unexplored, and their straightforward implementation might prove difficult due to unique characteristic they exhibit. For example, experiments we performed on VillageLink demonstrate the necessity of careful channel allocation in white spaces even when the number of devices operating in the same interference domain is low, which is a stark contrast to WiFi networks. Our work examines only one aspect of network adaptation. The complex nature of signal propagation over a wide frequency band opens up new possibilities for protocol design and further refinement of channel access in white spaces.

References

1. Bremaud, P.: Markov Chains, Gibbs Fields. Monte Carlo Simulation and Queues. Springer, New York (1999)
2. Cordeiro, C., Challapali, K., Birru, D., Manor, B., Diego, S.: IEEE 802.22: an introduction to the first wireless standard based on cognitive radios. J. commun. **1**(1), 38–47 (2006)
3. Geman, S., Geman, D.: Stochastic relaxation, Gibbs distributions, and the Bayesian restoration of images. IEEE Trans. Pattern Anal. Mach. Intell. **6**, 721–741 (1984)
4. Greenstein, L.J., Erceg, V., Yeh, Y.S., Clark, M.V.: A new path-gain/delay-spread propagation model for digital cellular channels. IEEE Trans. Veh. Technol. **46**(2), 477–485 (1997)
5. Hajek, B.: Cooling schedules for optimal annealing. Math. Oper. Res. **13**, 311–319 (1988)
6. Harrison, K., Mishra, S. M., Saha. A.: How much white-space capacity is there? In: DySpan'10, Singapore (2010)
7. ITU World Telecommunication/ICT Indicators Database. International Telecommunication Union (2013)
8. Jain, R., Chiu, D.M., Hawe, W.R.: A Quantitative Measure of Fairness and Discrimination for Resource Allocation in Shared Computer System. Digital Equipment Corporation, Eastern Research Laboratory (1984)
9. Kauffmann, B., Baccelli, F., Chaintreau, A., Mhatre, V., Papagiannaki, K., Diot, C.: Measurement-based self organization of interfering 802.11 wireless access networks. In: INFOCOM'07, Anchorage (2007)
10. Kirkpatrick, S., Gelatt, C.D., Vecchi, M.P.: Optimization by simulated annealing. Science **220**, 671–680 (1983)
11. Ma, M., Tsang, D.H.K.: Joint design of spectrum sharing and routing with channel heterogeneity in cognitive radio networks. Phys. Commun. **2**(1–2), 127–137 (2009)
12. Masonta, M., Johnson, D.L., Mzyece, M.: The white space opportunity in Southern Africa: measurements with meraka cognitive radio platform. In: e-Infrastructure and e-Services for Developing Countries, Volume 92 of Lecture Notes of the Institute for Computer Sciences, Social Informatics and Telecommunications Engineering, pp. 64–73. Springer, Heidelberg (2012)
13. Mhatre, V., Papagiannaki, K., Baccelli, F.: Interfernece mitigation through power control in high density 802.11 WLANs. In: INFOCOM'07, Anchorage (2007)
14. Mishra, S., Hwang, J., Filippini, D., Moazzami, R., Subramanian, L., Du, T.: Economic analysis of networking technologies for rural developing regions. Lect. Notes Comput. Sci. **3828**, 184–194 (2005)
15. Moscibroda, T., Wattenhofer, R., Weber, Y.: Protocol design beyond graph-based models. In: HotNets'06, Irvine (2006)
16. Nychis, G., Hottelier, T., Yang, Z., Seshan, S., Steenkiste, P.: Enabling MAC protocol implementations on software-defined radios. In: NSDI'09, Boston (2009)
17. Ramachandran, K.N., Belding, E.M., Almeroth, K.C., Buddhikot, M.M.: Interference-aware channel assignment in multi-radio wireless mesh networks. In: INFOCOM'06, Barcelona (2006)

18. Raman, B., Chebrolu, K.: Design and evaluation of a new MAC protocol for long-distance 802.11 mesh networks. In: MobiCom'05, Cologne (2005)
19. Sheth, A., Nedevschi, S., Patra, R., Surana, S., Subramanian, L., Brewer, E.: WiLDNet: design and implementation of high performance WiFi based long distance networks. In: NSDI, Anchorage (2007)
20. Stojmenovic, I.: Handbook of Wireless Networks and Mobile Computing. Wiley-Interscience, New York (2002)
21. Surana, S., Patra, R., Nedevschi, S., Ramos, M., Subramanian, L., Ben-David, Y., Brewer, E.: Beyond pilots: Keeping rural wireless networks alive. In: NSDI, San Francisco (2008)
22. Sydell, L.: FCC Eyes Broadband For Indian Reservations. http://www.npr.org/templates/story/story.php?storyId=128004928 (2010)

Chapter 10
A Non-cooperative TV White Space Broadband Market Model for Rural Entrepreneurs

Sindiso Mpenyu Nleya, Antoine Bagula, Marco Zennaro and Ermmano Pietrosemoli

Abstract Access to broadband Internet services for rural households and businesses in the world has generally been lacking. Broadband helps lift the world's poorest out of poverty, brings the benefits of education and health care closer to rural and remote populations, and delivers social and economic benefits to all. However, from the service providers (rural entrepreneurs) perspective, there have traditionally been few incentives to provide broadband access to low-income customers. One of the key obstacles has been spectrum scarcity and the high cost of spectrum licenses. White space technology provides a promising approach to provide spectrum, unused by primary users, to secondary users for broadband provisioning. This chapter discusses the challenges of producing an optimal pricing scheme for leasing spectrum from primary users. It proposes a game-theory based solution to the problem of several primary users competing with each other to sell spectrum to secondary users that use QoS measures of delay and throughout to regulate price.

S.M. Nleya (✉)
Computer Science, University of Capetown, 18 University Avenue Rhodes Gift
Cape Town 7707, South Africa
e-mail: snelya@gmail.com

A. Bagula
Computer Science Department, University of Western Cape, P. Bag X17 Cape Town
7535, Bellville, South Africa
e-mail: bbagula@uwc.ac.za

M. Zennaro · E. Pietrosemoli
Telecommunications/ICT Development Lab, Abdus Salam International Centre for Theoretical
Physics, Strada Costiera 11, 34151 Trieste, Italy
e-mail: mzennaro@ictp.it

E. Pietrosemoli
e-mail: epietros@ictp.it

A.K. Mishra and D.L. Johnson (eds.), *White Space Communication*,
Signals and Communication Technology, DOI 10.1007/978-3-319-08747-4_10

10.1 Introduction

Access to broadband Internet services for rural households and businesses in the world over has generally been truly lacking. Broadband internet according to the United Nations [17] delivers access to the sum of human knowledge as it opens doors to the future. It helps lift the world's poorest out of poverty, brings the benefits of education and health care closer to rural and remote populations, and delivers social and economic benefits to all. However, from the service providers (rural entrepreneurs) perspective [5], there have traditionally been few incentives to provide access to low-income customers, who are presumed to have limited demand for new services, and to rural and remote regions, where the cost of extending or upgrading facilities and services is assumed to be higher than expected revenues. To this end, wireless technologies have generally proved to be a far more cost effective option for serving remote and rural areas, the core challenges are (a) the scarcity and cost of spectrum licenses, and (b) base station infrastructure deployment and operational expenses. However, technological innovations, many of which were initially designed for other applications, are now creating opportunities to reduce costs and/or increase revenues in these populations. A notable innovation in this regard,is that of Wireless Mesh Networks (WMNs) which have been touted as a candidate technology set for the ubiquitous connectivity of the end user. The WMNs basically comprise of wireless mesh routers and mesh clients as well as an endowed capability to dynamically self organize as well as self configure to the degree that the network nodes are able to establish and maintain connectivity among themselves. The WMN are characterized by low-upfront costs, ease of maintenance, robustness as well as reliable service coverage. WMN have found suitable application solutions spanning in the range of broadband home networking, community and neighborhood networks, enterprise networking, community, building automation, public safety etc. In spite of the numerous applications, the growth, performance and spread of WMN has been hampered by several limitations such as limited spectrum availability [23]. Limited spectrum availability is a consequence of the adoption of the Industrial, Scientific and medical (ISM) band for backbone communications. The adoption in turn leads to a scenario of devices existent in this particular band affecting the WMN with the case in point being nearby WLANS and bluetooth devices. Consequently the limited spectrum in this particular band cannot cope with the limited network applications leading to artificially high spectrum prices. Pursuant to understanding the high spectrum prices, findings on the empirical spectrum measurements have revealed a gross underutilization of licensed spectrum, called White Space. Moreover with the transition from analog to digital Television has also led to the release of large chunks of spectrum referred to as TV White Space (TVWS). Certainly, overcoming spectrum scarcity of the WMNs and enhancing the performance of these networks requires the full harnessing and exploitation of TVWS. TVWS exploitation can however be made feasible by leveraging on the technological development in Smart Radio (SM). Smart radio (SM) has the ability to observe, learn, optimize, and adapt transmission parameters according to the ambient environment. Moreover, the flexibility of this device renders feasible spectrum sharing between licensed (Primary Users-PUs) and unlicensed (Secondary Users-SU) services. Consequently Dynamic spectrum Access (DSA) is made possible when secondary users are permitted to opportunistically access licensed spectrum. DSA is thus a promising approach for reusing the underutilized spectrum as the spectrum is shared among the PUs and SUs improving flexibility and efficiency in the process [9]. Moreover, studies on SM based networks have revealed spectrum marketing as an effective way to realize spectrum sharing with economic modeling being among the main fundamental issues [1]. In tackling economic modeling where a scarce resource such as spectrum is concerned, game theory has often been adopted to model the behaviors of rational and self interested entities [19].

10.2 Motivation, Contribution and Organization of Chapter

Game theory is widely regarded as a useful tool implorable in the analysis of resource allocation as well as mathematical models of conflict and cooperation among rationally intelligent decision makers from a microeconomic perspective. Specifically, this tool is applicable in scenario of dynamic spectrum sharing, particularly with regards to the planning and decision making in a smart radio based System. The system environment comprises multiple entities that objectively interact to achieve self interests. Game theory thus provides the conflict resolution mechanism so as to satisfy all concerned entities.

In this chapter, we investigate the problem of spectrum sharing and pricing within the context of a smart mesh wireless network using a game theoretic oligopoly framework from microeconomics. The oligopoly is contextual defined as a market scenario in which a small number of enterprise producers independently compete with each other in their quest to maximize profits either through controlling and varying the quantity or via price setting. The quantity/price offered by a particular enterprise producer will in general, likely impact on the profit of other enterprise producers. Guided by the law of supply and demand in economics, if a single enterprise producer offers increased quantities of its commodity, the market price drops and subsequently reduces the profits of other enterprise producers. Specifically a Bertrand game model is used to maximize the payoff of individual enterprise producers.

In applying this Bertrand model of competition to spectrum sharing and pricing in Smart mesh wireless network system, we analytically model several licensed telecommunication enterprises competing with each other to offer services to unlicensed enterprise systems so as to maximize profits under Quality of Service (QoS), constraints of licensed users. The QoS in question is a composite metric assembled from a combination of delay [13] and throughput QoS performance metrics. The spectrum demand of the secondary routers is derived from a quadratic utility function dependant on quality of transmission in the available spectrum as well as substitutability. The substitutability depicts the flexibility with which the secondary router is able to switch to different frequency spectra offered by various licensed enterprises. Ultimately a numerical analysis is carried out so as to evaluate our analytic model's performance in allocating TV White space. The rest of the chapter is organized as follows. Section 10.3 presents the related work in the literature. The general characteristics of the Bertrand oligopoly model. The Bertrand game model for spectrum sharing under competition is presented in Sect. 10.4. Numerical analysis of the model is carried out in Sects. 10.5 and 10.6 concludes the chapter.

10.3 Related Work

A number of studies regarding pricing in networking issues have been presented in the last few years. The efforts have concentrated largely on characterizing quality of service parameters (QoS), such as delay, jitter, packet loss and throughput on the operating point of a wireless network system. Moreover these efforts have attempted to demonstrate how the operating point can be set a priori by appropriately choosing the parameter. However, with the advent of smart radios and subsequent emergence of wireless mesh networks as an economical solution to support broadband services, efforts have intensified in the quest for efficient resource utilization. An important development in pursuit of the efficient spectrum utilization has been the introduction of Authorized Shared Access (ASA) as an enabler to unlock access to additional frequency bands for mobile broadband under individual licensed regimes. The ASA approach has been extended as Licensed Shared Access (LSA) to involve license holders authorizing secondary usage of spare spectrum within their licensed bands but under tight controls to prevent any disruption. The approach was conceived with a view to supporting business cases for the build-out of mobile broadband network infrastructure, where it is both economically and technically feasible [4, 7]. A major benefit of LSA is the guarantee of controlled as well as predictable Quality of Service (QoS) for both incumbent spectrum users and the LSA licensees by considering a couple of entities involved in the sharing agreement. Moreover the LSA notion provides an effective and harmonized way to utilize existing assets and achieve economies of scale by making International Mobile Communications (IMT) bands available worldwide with existing user equipment and minimal modifications to the infrastructure [11]. A successful trial with a live Long Term Evolution (LTE) network in the 2.3 GHz shared band has been achieved in Finland in April 2013. A major drawback of this technology is that it excludes concepts such as opportunistic spectrum access, secondary use or secondary service where the applicant has no protection from the primary users. Clearly the opportunistic spectrum access (OSA) approach is an alternative to the ASA/LSA approach which also contributes significantly to the efficient use of spectrum. According to [8] Opportunistic and dynamic spectrum access is a novel access model designed to extract unused spectrum from allocated but underutilized spectrum, supporting newcomer traffic without affecting existing owners. The approach implores the use of smart radios to identify unused portions of licensed spectrum, and utilize that spectrum without adverse impact on the primary user licensees. This approach renders feasible an abundantly higher level of spectrum utilization and near zero-deployment time [16] in cooperative and non-cooperative wireless network systems with coexistence of Primary and Secondary users.

To this end, both cooperative [15] and non-cooperative network systems have been scrutinized. Non cooperative networks systems characterize competition both among licensed (PUs) spectrum holders and between non-license holders (SUs). The authors in [22] explain network throughput as a more important concern for the emerging multi-hop wireless networks such as wireless mesh networks. These efforts are complimented in [21] in which throughput maximization using a negotiation based algorithm is extended to a non-cooperative scenario. Alternative QoS parameters such as delay, have also been considered [24] when the authors assert that, supporting delay sensitive real-time traffic such as video and voice over wireless mesh networks is a challenging and attractive task [18]. Considers both parameters and acknowledges the spectrum allocation problem as requiring maximization of throughput and also a minimization of the delay. To this end, dynamic mesh network systems need to consider the trade-off between immediate costly transmission and low cost but delayed transmission [6]. The idea of a trade-off between throughput and delay has been studied in [9] for an ad-hoc mesh network. Clearly the concept needs to be considered for smart wireless mesh network scenario which are a future generation network candidate solution.

10.4 TV White Space Market Pricing Model

We consider a system of non-cooperative Mesh routers within the context of spectrum management wherein licensed routers which are called primary users, compete to offer services to an unlicensed system of routers called secondary users. From a primary user's point of view, the cost of offering a service to a secondary user is modeled as a function of QoS degradation. This interaction between the primary and secondary users as well as the outcome can best be understood using game theory by generalizing it as a game. The solution of the game is obtained by imploring John Nash's concept of a 'Nash Equilibrium' (NE) [12] which is an organizing concept in game theory. This concept represents the optimal solution in such interactive games.

10.4.1 System Model

Our system model envisages the existence of N primary users (Fig. 10.1 (where N = 2)) operating on different frequency spectra and a multitude of secondary users aspiring to share the precious spectrum resource with the primary users in question. To this end, we introduce P_i as the tariff/pricing policy and QoS guaranteed by primary user i with all other symbols defined in Table 10.1. Then each of the secondary users is motivated to subscribe at this given tariff so as to attain a QoS

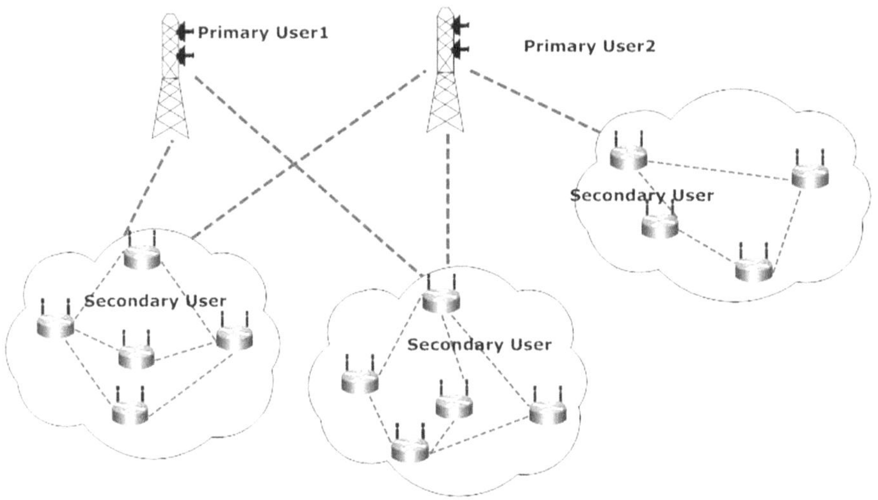

Fig. 10.1 Smart mesh network

sufficient to satisfy individual needs. In the process, the secondary users make use of adaptive modulation for transmissions by exploiting channel state information. With this type of modulation, transmission rate is influenced by channel quality while the bit error rate must be maintained at target levels. Consequently, the spectral efficiency of transmission for a secondary user i is according to [3] given by:

$$k_i = log2(1 + Ky_i) \tag{10.1}$$

where

$$K = \frac{1.5}{ln\left(\frac{0.2}{BER_i^{tar}}\right)} \tag{10.2}$$

On being allocated the spectrum, the secondary user i transmits with spectral efficiency k_i with its demand being a function of transmission rate in the allocated frequency spectrum as well as price charged by the allocating primary user.

10.4.2 QoS Measure and Cost

In the Next generation telecommunications industry, systems are required where cost is dependent on QoS. Subsequently in our system model, the QoS performance of a primary user is degraded when spectrum is shared with the secondary user. This translates to cost function being considerate of the QoS performance of the primary user. Ultimately, we consider the use of a composite QoS performance metric derived

Table 10.1 Notation summary

Symbols	Description
λ_i	Arrival rate
Q_i	Spectrum size (secondary user)
W_i	Spectrum size (primary user)
$P^{(i)}$	Price
P_j	Price
$k_i(p)$	Spectral efficiency (primary users)
$k_i(s)$	Spectral efficiency (secondary users)
C_i^D	Cost function (delay)
C_i^T	Cost function (throughput)
d_i	Constant (elasticity)
D_i	Delay
ζ	Composite metric
ε	Constant = 0.5
Λ	Constant = 0.5
ψ	Utility
W_i	Primary user spectrum
Δ	Substitutability
$\phi_i(T)$	Profit (throughput)
$\phi_i(D)$	Profit (delay)
$\phi_i(TD)$	Profit (composite)
y_i	Channel quality (player i)
y_j	Channel quality (player j)
T_i	Throughput
n	Number of users
β	Constant

from a combination of the delay and Throughput QoS metrics. Moreover we define and introduce the individual QoS Performance measures metrics and then finally combine them into a single composite QoS measure. We begin with the average delay QoS measure obtained from [14] and defined as:

$$D_i(Q_i) = \frac{1}{2}\frac{\lambda_i}{\left(k_i^{(p)}(W_i - Q_i)^2 - \lambda_i k_i^{(p)}(W_i - Q_i)\right)} \tag{10.3}$$

with the symbols meaning as given in the Table 10.1, it is worth to note that $k_i^{(p)}$ $(W_i - Q_i)$, denotes the service rate. Thus the average delay is a function of the total spectrum due to the primary user, the spectrum demand from the secondary user, the arrival rate and most importantly the spectral efficiency which subsequently leads to channel quality. The cost function is defined as:

$$C_i^D = dD_i(Q_i) \tag{10.4}$$

Intuitively, the cost function is a function of the average delay QoS measure and given its dependence on channel quality the cost can thus be subsequently varied through the channel quality parameter. Furthermore conscious and cognizant of the coexistence between the primary and secondary networks we focus on throughput as an alternative QoS measure. The achievable throughput in a secondary network system is given by [10] as:

$$T(Q_i) = \sum_{i=1}^{N} \frac{\beta Q_i}{\sqrt{nlogn}} \tag{10.5}$$

The throughput measure thus is a function of the spectrum demand by the secondary and the number of nodes. The spectrum demand is also dependent on spectral efficiency which in turn is a function of channel quality. Consequently the cost due to the throughput measure is expressed as:

$$C_i^T = dT_i(Q_i) \tag{10.6}$$

The implication of this cost function is that, the cost can again be made to vary with channel quality. We proceed to bring together the average delay and throughput to form a composite metric which we conveniently refer to as the composite QoS metric. The Composite QoS Measure is expressed as:

$$\zeta(Q_i) = \varepsilon(D(Q_i) + \Lambda(T(Q_i)) \tag{10.7}$$

The relation between Λ and ε is given by:

$$\Lambda = 1 - \varepsilon$$

Consequently, the cost due to this composite QoS metric is

$$C_i^\zeta = d\zeta(Q_i) \tag{10.8}$$

Since both metrics forming the composite metric are dependent on the channel quality, this Composite QoS parameter can also be varied by correspondingly varying the channel quality. To this end, we conveniently defer this exercise to Sect. 10.5.

10.4.3 Quadratic Utility Function

The utility gained by the secondary users makes it possible to ascertain the level of spectrum demand. A quadratic utility function defined as in [20]:

$$\Psi(\mathbb{Q}) = \sum_{i=1}^{M} Q_i k_i^{s} - \frac{1}{2}\Big(\sum_{i=1}^{M} Q_i^2 + 2\Delta \sum_{i=1}^{M} Q_i Q_j\Big) + J \tag{10.9}$$

where

$$\mathbb{Q} = Q_1, \ldots, Q_i, \ldots, Q_M \tag{10.10}$$

and J is given by:

$$J = -\sum_{i=1}^{M} P_i Q_i \tag{10.11}$$

The spectrum substitutability is included in the utility function by way of parameter ∇. This parameter permits the secondary users to switch between frequencies depending on the offered price. The demand function of the secondary user is obtainable from differentiating the utility function w.r.t Q_i as follows:

$$\frac{d\psi(\mathbb{Q})}{dQ_i} = 0 \tag{10.12}$$

The demand function is the size of shared spectrum that maximizes the utility of the secondary user given the prices offered by the primary service

$$Q_i = \frac{k_i^{(s)} \quad P_i - \Delta(k_j^{(s)} - P_j)}{1 - \Delta^2} \tag{10.13}$$

The revenue generated from the demand function is a function of the demand function and the relevant price.

$$R_i = Q_i P_i \tag{10.14}$$

10.4.4 Bertrand Game Model

The Bertrand oligopoly is formulated as in Table 10.2.

The profit due to the composite metric is obtainable from the revenue less the cost of the TV white space. Thus Profit

$$profit(\phi_i^{TD}) = Revenue(Eqn\ 10.14) - Cost(Eqn\ 10.18) \tag{10.15}$$

Table 10.2 Bertrand game formulation

Entity	Description
Players	Primary users
Strategies	Price per unit of spectrum (P_i)
Payoffs	The payoff for each player is the profit of primary user

$$\begin{aligned}\phi_i^{TD} &= Q_i P_i - d(\varepsilon(D(Q_i) + \Lambda(T(Q_i)) \\ &= \frac{P_i k_i^s - P_i^2 - P_i \delta(k_j^s - P_j)}{1-\Delta^2} - d(\varepsilon(D(Q_i) + \Lambda(T(Q_i)\end{aligned} \tag{10.16}$$

To obtain Nash Equilibrium (NE), the equation

$$\frac{d\phi_i^{TD}}{dP_i} = 0 \tag{10.17}$$

for all i. Therefore

$$\begin{aligned}\phi_i^{TD} &= \frac{P_i k_i^s - P_i^2 - P_i \delta(k_j^s - P_j)}{1-\Delta^2} - \Big(\frac{d\lambda_i}{(2W_i - Q_i)^2 - 4\lambda(W_i - Q_i)} \\ &+ \frac{dk_i^s - P_i - \Delta(k_j^s - P_j)}{(1-\Delta^2)(\sqrt{nlogn})}\Big)\end{aligned} \tag{10.18}$$

The derivative thus becomes

$$0 = \frac{k_i^s - 2P_i - \Delta(k_j^s - P_j)}{1-\Delta^2} + \frac{\frac{d\lambda_i(4Q_i - \lambda_i)}{1-\Delta^2}}{(2Q_i^2 - 2Q_i\lambda_i)^2} - \frac{1}{(1-\Delta^2)\sqrt{nlogn}} \tag{10.19}$$

where

$$Q_i = W_i - \frac{k_i^s - P_i - \Delta(k_j^s - P_j)}{1-\Delta^2} \tag{10.20}$$

10.5 Performance Evaluation

10.5.1 Parameter Setting

The parameters are set as in Table 10.3.

Table 10.3 System parameters

Parameter	Value
Primaryuserspectrum	5 (MHz)
BER	10^{-4}
Trafficarrivalrate	1 (Mbps)
d	1
Channelqualityspan	10–20 (dB)
λ_i	4
y_1	15
y_2	18
Δ	0.4
P_2	1
Primaryusers	2

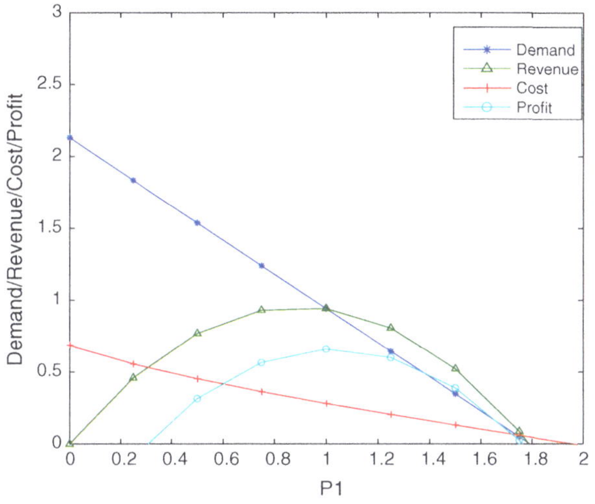

Fig. 10.2 Demand-revenue-cost and profit

10.5.2 Numerical Analysis

In this section, we present numerical results to validate the efficacy of our TVWS broadband market model developed within the context of low cost Smart wireless Mesh network.

Figure 10.2 depicts the demand function of the secondary user, the revenue, cost and profit of the primary user under various pricing options. As the first primary user increases its price, the secondary user responsively demands a small amount of spectrum owing to a decrease in the utility of the allocated spectrum. This is clearly shown by the negative gradient line which represents the demand function of the secondary user. However, this demand function behavior impacts on the cost for the primary user. This is to say, the cost for the primary user decreases with a decrease in the demand function this therefore translates to a larger amount of residual spectrum corresponding to smaller delay. The revenue and profit functions of the primary user are all traversing a parabolic path as depicted on the same graph. Clearly both functions in question initially increase with an increase in price up to the optimal point where both functions begin to show a decline in both the revenue and profit. To this end, the primary user is able to sell a larger amount of spectrum to the secondary user at a smaller price thereby giving an increase in revenue as well as profit. Conversely, when the price increases, a small amount of spectrum is sold due to a decrease in the level of demand by the secondary user ultimately resulting in dwindling profit. Certainly an optimal price exists upon which the profit is maximized and this denotes an apparent best response for the corresponding primary user. This best response is further investigated wherein the best responses of the two primary users are analyzed. In Fig. 10.3, we analyze the best response functions of the two primary users under variable channel quality for the secondary user. When the channel quality increases, the spectrum demand correspondingly increases. The individual primary user then consequently offers a higher price. However, the best response function graphs intersect at some point called Nash Equilibrium. We analyze this Nash equilibrium under varied channel quality. Figure 10.4 depicts a scenario in which the Nash equilibrium is higher for higher channel quality, this emanates from an increased demand of spectrum by the secondary user. Moreover, we realize that the channel quality offered by one individual primary user impacts on the other individual primary user. This is to say, channel quality offered by one primary user impacts the strategy adopted by the other primary user. Consequently when the demand for spectrum from an individual player is varied, the other player adapts the price so as to maximize profit.

10.6 Conclusion

In this chapter a non cooperative TVWS analytic business model for rural telecommunications entrepreneurs for broadband internet provision has been developed. The analytic model is based on a proposed composite QoS performance constraint which is assembled from the Delay and Throughput QoS constraints. The performance of the analytic model is evaluated within the context of a smart wireless mesh network wherein routers that have licensed spectrum lease out the spectrum to unlicensed routers/clients. The interaction of these licensed routers with unlicensed client routers is modeled as a bertrand oligopoly. In the oligopoly the licensed routers compete to sell their spectrum to clients using price as the strategy. The optimal responses of the routers (PUs) are analyzed as well the dependence on variable channel quality. When the channel quality increases spectrum demand from SUs also increases triggering an increase in price. Further planned efforts in this regard will involve extension of these

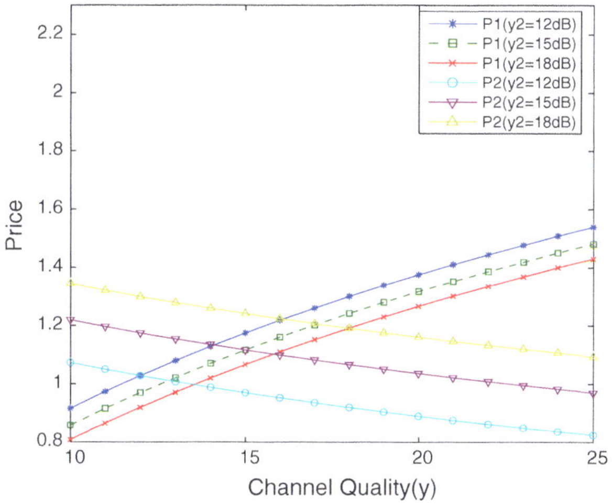

Fig. 10.3 Channel quality

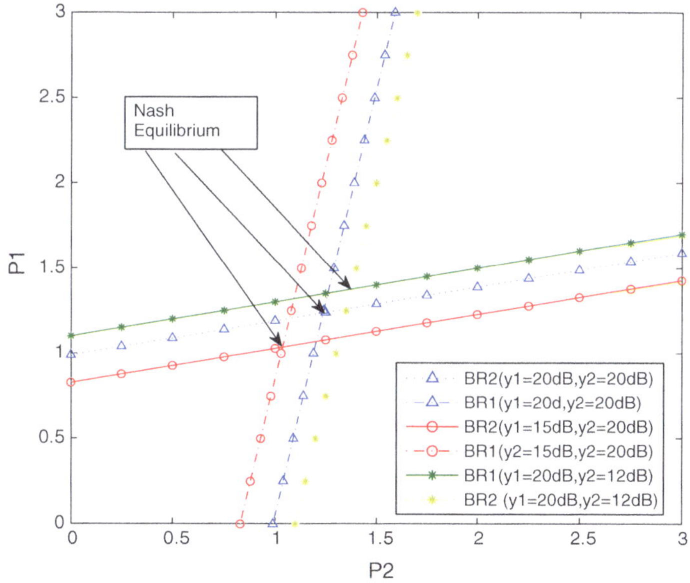

Fig. 10.4 Best response

efforts from networking engineering to traffic engineering as in [2]. This approach could possible help in simultaneous routing of delay sensitive and real-time traffic.

References

1. Akyildiz, I.F., Wang, X.: A survey on wireless mesh networks. IEEE Communications Magazine **43**(9), S23–S30 (2005)
2. Bagula, A.B.: Hybrid routing in next generation ip networks. Comput. commun. **29**(7), 879–892 (2006)
3. Goldsmith, A.J., Chua, S.-G.: Variable-rate variable-power mqam for fading channels. IEEE Trans. Commun. **45**(10), 1218–1230 (1997)
4. Holland, O., De Nardis, L., Nolan, K., Medeisis, A., Anker, P., Minervini, L. F., Velez, F., Matinmikko, M., Sydor, J.: Pluralistic licensing. In: Proceedings of the IEEE International Symposium on Dynamic Spectrum Access Networks (DYSPAN 2012), pp. 33–41 (2012)
5. Hudson, H.E.: Understanding the Digital Economy. Extending access to the digital economy to rural and developing regions. The MIT Press, Cambridge (2000)
6. Jagannathan, K., Menache, I., Modiano, E., Zussman, G.: Non-cooperative spectrum access the dedicated vs. free spectrum choice. IEEE J. Sel. Areas Commun. **30**(11), 2251–2261 (2012)
7. Khun-Jush, J., Bender, P., Deschamps, B., Gundlach, M.: Licensed shared access as complementary approach to meet spectrum demands: Benefits for next generation cellular systems. In: ETSI Workshop on reconfigurable radio systems (2012)
8. Kone, V., Yang, L., Yang, X., Zhao, B.Y., Zheng, H.: On the feasibility of effective opportunistic spectrum access. In: Proceedings of the 10th ACM SIGCOMM conference on Internet measurement, pp. 151–164 (2010)
9. Lin, P., Jia, J., Zhang, Q., Hamdi, M.: Dynamic spectrum sharing with multiple primary and secondary users. IEEE Trans. Veh. Technol. **60**(4), 1756–1765 (2011)
10. Lu, N., Shen, X.S.: Scaling laws for throughput capacity and delay in wireless networks—a survey. IEEE Communications Surveys and Tutorials (2014)
11. Matinmikko, M., Palola, M., Saarnisaari, H., Heikkila, M., Prokkola, J., Kippola, T., Hanninen, T., Jokinen, M., Yrjola, S.: Cognitive radio trial environment: First live authorized shared access-based spectrum-sharing demonstration. IEEE Vehicular Technology Magazine **8**(3), (2013)
12. Nash, J.F., et al.: Equilibrium points in n-person games. Proc. Natl. Acad. Sci. **36**(1), 48–49 (1950)
13. Niyato, D., Hossain, E.: Microeconomic models for dynamic spectrum management in cognitive radio networks. In: Cognitive Wireless Communication Networks, pp. 391–423. Springer, New York (2007)
14. Niyato, D., Hossain, E.: Competitive pricing for spectrum sharing in cognitive radio networks: dynamic game, inefficiency of nash equilibrium, and collusion. J. Sel. Area Commun. **26**(1), 192–202 (2008)
15. Nleya, S. M., Bagula, A., Zennaro, M., Pietrosemoli, E.: A tv white space broadband market model for rural entrepreneurs. In: IEEE Global Information Infrastructure Symposium, pp. 1–6 (2013)
16. Santivanez, C., Ramanathan, R., Partridge, C., Krishnan, R., Condell, M., Polit, S.: Opportunistic spectrum access: Challenges, architecture, protocols. In: Proceedings of the 2nd annual international workshop on Wireless internet, p. 13 (2006)
17. Sarah, P.: An update on the state of broadband. URL http://www.unfoundation.org/blog/an-update-on-broadband.html (2012)
18. Shin, M., Lee, S., Kim, Y.-A.: Distributed channel assignment for multi-radio wireless networks. In: IEEE International Conference on Mobile Adhoc and Sensor Systems (MASS), pp. 417–426 (2006)
19. Silverman, S. J.: Games theory and software defined radios. In: IEEE Military Communications Conference (MILCOM), pp. 1–7 (2006)
20. Singh, N., Vives, X.: Price and quantity competition in a differentiated duopoly. RAND J. Econ. **15**, 546–554 (1984)
21. Song, Y., Zhang, C., Fang, Y.: Joint channel and power allocation in wireless mesh networks: A game theoretical perspective. IEEE J. Sel. Area Commun. **26**(7), 1149–1159 (2008)

22. Tang, J., Xue, G., Chandler, C., Zhang, W.: Link scheduling with power control for throughput enhancement in multihop wireless networks. IEEE Trans. Veh. Technol. **55**(3), 733–742 (2006)
23. Zennaro, M., Pietrosemoli, E., Bagula, A., Nleya S.: On the relevance of using affordable tools for white spaces identification. In: IEEE 8th International Conference on Wireless and Mobile Computing, Networking and Communications (WiMob), pp. 606–611 (2012)
24. Zheng, S., Tian, H., Hu, Z., Chen, L., Zhu, J.: Qos-guaranteed radio resource allocation with distributed inter-cell interference coordination for multi-cell ofdma systems. In: IEEE 71st Vehicular Technology Conference (VTC 2010-Spring), pp. 1–5 (2010)

The manufacturer's authorised representative in the EU is Springer Nature Customer Service Centre GmbH, Europaplatz 3, 69115 Heidelberg, Germany. If you have any concerns regarding our products, please contact ProductSafety@springernature.com

Printed and bound by CPI Group (UK) Ltd, Croydon, CR0 4YY
30/09/2026
02225657-0004